Camels and One Health

Samantha Alex Gordine
Bernard Faye • Pamela Burger
Editors

Camels and One Health

Advancements in Animal Health, Human Well-Being and the Future Environment

NAHDI

Springer

Editors
Samantha Alex Gordine
Lead Scientist, One Health Camel Program, Arcondis
Allschwil, Switzerland

Pamela Burger
Research Institute of Wildlife Ecology
University of Veterinary Medicine
Vienna, Austria

Bernard Faye
CIRAD-SELMET (French Agricultural Research Centre for International Development)
Montpellier, France

ISBN 978-981-92-2245-2 ISBN 978-981-92-2246-9 (eBook)
https://doi.org/10.1007/978-981-92-2246-9

This Springer imprint is published by the registered company Springer Nature Singapore Pte Ltd.
The registered company address is: 152 Beach Road, #21-01/04 Gateway East, Singapore 189721, Singapore

This book is a gift. A gift from camel enthusiasts to fellow camel lovers, from researchers to the community and from loving sons to a cherished father. This book is also an invitation to discover, learn and appreciate the many ways in which the camel is truly a remarkable creature, too often underestimated in its significance depending on cultural context.

With a camel, there is always more than meets the eye.

Preface

It has been just over 2 years since I, Samantha Gordine, received the side question: "Could you help us write a book about camels?" The first idea for this book was to create an encyclopedia about camels, like a dictionary or reference book that would contain everything we know about camels with the latest scientific insights. As we developed the concept for this book, we started discovering many interesting nuggets of information about camels and their influence that were not captured comprehensively or elsewhere. For example, the Arabic and Kazakh languages have extensive camel-related vocabularies that could fill a book of its own, full of cultural and often of biological meaning. The concept for this book has therefore evolved significantly from its initial conception, building on input and ideas from various sources such as potential publishers and their reviewers. What we present in the following pages is the result of this evolutionary process.

For centuries, camels have sustained life in some of the world's harshest environments, yet their importance certainly remains underappreciated in large parts of this world. Anyone from the outside stepping into the world of camels will notice though that camel owners as well as researchers form a close-knit community that is truly passionate and loving when it comes to camels. Camels have long been essential to the cultural identity and economic survival of communities in arid landscapes across Africa, the Middle East, Asia and beyond. Yet, their importance is undergoing a scientific renaissance. With rising global attention on zoonoses, antimicrobial resistance, sustainable food systems, biodiversity loss, and climate change, camels emerge as both subjects of scientific interest and agents of adaptation. Their unique physiology, ecological role, cultural importance and emerging biomedical applications (such as nanobodies and genetic innovations) make camels a highly relevant subject for further study from the One Health perspective.

The request for this book came from loving camel owners who have spent years talking to other camel owners, handlers and breeders as well as nomadic families, veterinarians and biologists with various backgrounds ranging from breeding to infectious diseases. These conversations led to the realisation that there was no one, single compendium that reflected this existing knowledge whilst also painting future paths as to the direction of cutting-edge science and camel-relevant policy. This was coupled with the desire to help the camel community through disseminating this wealth of knowledge appropriately. This book should be seen alongside of and in addition to some of the latest scientific books about camels (see e.g. Padalino and Faye 2004; Faye et al. 2003; Khalafalla and Hussein 2021).

At its core, this book integrates perspectives from veterinary science, public health, genetics, food science and ecology to offer an evidence-based roadmap for camel-inclusive developments. We specially wanted to frame this scientific core by looking back through providing the historical and cultural setting, as well as looking forward through the lens of conservation. When viewed through the One Health lens, camels reveal interconnected stories about people, pathogens, ecosystems and resilience. It fills a major gap by uniting these elements into one accessible volume, guiding researchers, policymakers and practitioners towards a more camel-conscious future defined by a sustainable and symbiotic coexistence.

Camels and One Health provides a multidisciplinary, science driven exploration of how camel health intersects with human and environmental well-being. Unlike existing animal welfare or farming-oriented texts, this volume positions camels as a critical nexus in addressing global health, food security and climate change challenges across arid and semi-arid regions with a dedicated outlook on the future development in these areas. Across its chapters, readers will discover a unique interdisciplinary exploration of camels through the lens of the One Health framework linking human, animal and environmental health. It highlights cutting-edge topics such as camelid nanobodies, omics technologies for studying camels, zoonoses such as MERSCoV, and the role of genetic diversity in disease resilience. By connecting biomedical research with sustainability, conservation and climate adaptation, this volume offers a forward-looking perspective on the future of camel-reliant health systems and One Health solutions. It further illustrates how camels are central to sustainable development and the health of pastoralist communities.

This book is for anyone interested in the latest advancements in camel science and the future of camels in a changing world. Furthermore, it should serve as essential reading for practitioners and researchers in veterinary science, public health, ecology, One Health, food science and policy development. It will also serve as a valuable reference for postgraduate students and practitioners working in dryland development, livestock innovation, zoonotic disease management and climate adaptation strategies in emerging economies. Our extended editorial team paid particular attention to adopting a language suitable for the educated layperson to make this knowledge more widely accessible to audiences outside of the scientific community. We sincerely hope that we succeeded in achieving this goal.

The future of camels and the communities and ecosystems that rely on them ultimately depends on the decisions we make today. This book is intended to help guide these decisions scientifically, policy-wise and from an animal welfare perspective. We hope to make the knowledge contained herein as understandable and accessible as possible to reach not only an academic audience but also people working with camels on the ground. We appreciate that dissemination of knowledge hinges on several factors, including science communication and the inclusion of diverse voices as well as removing financial and language barriers for accessibility. We sincerely tried our best to address this by making this book open-access, focusing on accessible language as well as enlisting a diverse group of authors from different cultural and ethnic backgrounds. In the future, we would welcome translations to further facilitate improved understanding.

Allschwil, Switzerland — Samantha Alex Gordine
Montpellier, France — Bernard Faye
Vienna, Austria — Pamela Burger

References

Padalino B, Faye B (2024) Dromedary Camel Behavior and Welfare. Camel Friendly Management Practices. Springer, Cham. https://doi.org/10.1007/978-3-031-48600-5

Faye B, Konuspayeva G, Magnan C (2023) Large Camel Farming. A Care-Management Guide from Breeding to Camel Products. Springer, Dordrecht. https://doi.org/10.1007/978-94-024-2237-5

Khalafalla A, Hussein M (2021) Infectious Diseases of Dromedary Camels. A Concise Guide. Springer, Cham. https://doi.org/10.1007/978-3-030-79389-0

Acknowledgments

We express our deepest gratitude to the Al-Nahdi Family, whose generous sponsorship made the development and completion of this book possible. Their dedication to increasing the knowledge and research about camels has been instrumental in bringing this project to fruition.

We warmly thank our *reviewers* for their expert insights, constructive feedback and commitment to scientific rigour:

- **Omar Al-Ghamdi**, Al-Rwahil Al-Arabia
- **Dr. Pamela Burger**, University of Veterinary Medicine, Vienna
- **Dr. Bernard Faye**, UMR SELMET, CIRAD–France
- **Dr. Samantha Alex Gordine**, Arcondis Group
- **Dr. Leia Hazelwood**, Arcondis Group
- **Dr. Daniel Jimenez-Fernandez**, Arcondis Group
- **Dr. Floriane Leseur**, Arcondis Group
- **Prof. Serge Muyldermans**, Vrije Universiteit Brussels
- **Dr. Aiperi Otunchieva**, Arcondis Group

Their thoughtful reviews significantly strengthened the quality, accuracy and coherence of the chapters.

We also wish to acknowledge the exceptional dedication of our *extended editorial team*, whose collective hard work made this book possible:

- **Dr. Aiperi Otunchieva**, for her outstanding management of author correspondence, administrative coordination and overall editorial organisation. Her leadership guided the project from early concept development, through publisher selection, to final submission.
- **Dr. Leia Hazelwood**, for her remarkable ability to enhance clarity, readability, and consistency of the English text across chapters. This book would not exist in its present form without her invaluable contributions. Whatever you read in the following chapters has gone over Leia's desk at least once, but more often multiple times.
- **Dr. Mark Polinkovsky**, for helping humanise and refine the narrative voice throughout the volume.
- **Dr. Omar Hammouda**, for his Arabic translations and cross-checks, which enriched the multilingual accuracy and accessibility of the work.

- **Leonard Albert**, for ensuring logical coherence across chapters and compiling the glossary with precision and care.

To all those mentioned, and to the many others who supported this project in ways both visible and behind the scenes, we extend our sincere thanks.

Contents

Editors and Contributors

About the Editors

Samantha Alex Gordine, MRSB, AFHEA, is an ecologist by training, now working as a sustainability professional. She completed her PhD in Marine Biology at the University of St Andrews in 2017, using inter-disciplinary research to link the movement and behaviour of a large, migratory mammal with in situ and remote sensing data from extreme environments. Shortlisted for a Fulbright scholarship in 2018, Samantha instead continued her career in the non-profit sector and the life science industry. Currently, Samantha works with healthcare-related organisations to develop sustainable solutions and enhance environmental practices to improve One Health outcomes. She currently serves as the Programme Manager for a flagship One Health Programme focusing on camels in the Middle East and supports the organisation Legal Atlas to combat wildlife crime.

Dr. Samantha Alex Gordine's complete research portfolio and publication record can be accessed via his ORCID profile: [ORCID ID: https://orcid.org/0000-0003-3509-7044].
*Corresponding editor
Private email address: sagordine@gmail.com
Work email address: Samantha.gordine@alnahdi-office.com, Samantha.gordine@arcondis.com

Bernard Faye is a veterinarian and renowned camel expert with a career spanning over four decades. He earned his Doctorate in Veterinary Medicine in 1974, then a PhD in 1995, and later specialized in animal nutrition, epidemiology and endocrinology. Dr. Faye has worked in more than 50 countries, focusing on camel research, development and health. He was consultant for FAO in different camel countries (Saudi Arabiaé, Oman, Morocco, Mauritania, Chad) and emeritus expert at CIRAD (France). He is a founding member of ISOCARD (International Society of Camelid Research and Development) and has authored over 600 scientific papers, 56 books and chapters in book, and numerous technical reports. His expertise includes camel clinical biochemistry, mineral metabolism and disease epidemiology, contributing significantly to the global understanding of camelid sciences.

Dr. Bernard Faye's complete research portfolio and publication record can be accessed via his ORCID profile: [https://orcid.org/0000-0002-5762-5453].
Email address: bjfaye50@gmail.com

Pamela Burger is a veterinarian and population geneticist and interested in the evolutionary history and immune genetic response of domestic animals and their wild relatives to their environments. With next-generation sequencing becoming cheaper, Dr. Burger is moving towards a One Health genomic approach to understand species dispersal behaviour and associated zoonotic disease risk factors in relation to geographic and climatic variables, with effects on environmental, animal and human health.

Dr. Pamela Burger's complete research portfolio and publication record can be accessed via his ORCID profile: [https://orcid.org/0000-0002-6941-0257].
Email address: Pamela.Burger@vetmeduni.ac.at

Contributors

Sulaiman Aljasir Department of Veterinary Preventive Medicine, College of Veterinary Medicine, Qassim University, Buraydah, Saudi Arabia

Ibtisam Al-Wahaibi Business Communication Unit, College of Economics and Political Science, Sultan Qaboos University, Muscat, Sultanate of Oman

Mehdi Arbabi Ghahroudi Human Health Therapeutics Research Centre, Life Sciences Division, National Research Council Canada, Ottawa, Ontario, Canada

Department of Biochemistry, Microbiology and Immunology, Faculty of Medicine, University of Ottawa, Ottawa, Ontario, Canada

Department of Biology, Carleton University, Ottawa, Ontario, Canada

Bilal Aslam Department of Veterinary Preventive Medicine, College of Veterinary Medicine, Qassim University, Buraydah, Saudi Arabia

Silvia Bruno Department of Biosciences, Biotechnologies and Environment, University of Bari, Bari, Italy

Pamela Burger Research Institute of Wildlife Ecology, University of Veterinary Medicine, Vienna, Austria

Abdelmadjid Chehma Laboratoire des bioressources sahariennes, Préservation et valorisation, Ouargla University, Ouargla, Algeria

University of Ghardaïa, Bounoura, Algeria

Elena Ciani Department of Biosciences, Biotechnologies and Environment, University of Bari, Bari, Italy

Nieves Cisneros Rodríguez Department of Genetics, Faculty of Veterinary Sciences, University of Córdoba, Córdoba, Spain

Bernard Faye CIRAD-SELMET (French Agricultural Research Centre for International Development), Montpellier, France

Pauline Njoki Gitonga Centre for Epidemiological Modelling and Analysis (CEMA), The University of Nairobi, Nairobi, Kenya

Department of Biomedical Sciences, Colorado State University (CSU), Fort Collins, CO, USA

ASAL eXtension, Nairobi, Kenya

Jordan Hampton School of Veterinary Medicine, Harry Butler Institute, Murdoch University, Murdoch, WA, Australia

Faculty of Science, University of Melbourne, Parkville, VIC, Australia

Dana Hyláková Veterinary Clinic Vyvet Vyškov, Vyškov, Czech Republic

Czech Association of Zoo and Wildlife Veterinarians (CAZWV), Brno, Czech Republic

Carlos Iglesias Pastrana Andalusian Institute for Agricultural, Fishing, Food and Ecological Production Research and Training (IFAPA), Centro Alameda del Obispo, Alameda del Obispo, Córdoba, Spain

Anna Jemmett Wild Camel Protection Foundation, University of Kent, Kent, UK

Abdelmalik Ibrahim Khalafalla Department of Microbiology, Faculty of Veterinary Medicine, University of Khartoum, Khartoum, Sudan

Oscar Kipchirchir Koech Department of Land Resource Management and Agricultural Technology (LARMAT), The University of Nairobi, Nairobi, Kenya

Gaukhar Konuspayeva Biology and Biotechnology Faculty, Department of Biotechnology, Farabi University, Almaty, Kazakhstan

Rashid Mehmood Department of Life Sciences, College of Science and General Studies, Alfaisal University, Riyadh, Saudi Arabia

Noha Kamel Moussa Independent Researcher in Arabian and Islamic Archaeology, Cairo, Egypt

Tungalag Ulambayar Zoological Society Luujin, Ulaanbaatar, Mongolia

Abbreviations

16S rRNA	16S Ribosomal RNA
2D-GE	Two-dimensional Gel Electrophoresis
AES	Agroecosystem
AFCMP	Australian Feral Camel Management Project
AI	Artificial intelligence
AI (reproduction)	Artificial insemination
AMR	Antimicrobial resistance
ARDS	Acute respiratory distress syndrome
ARSO	African Organisation for Standardisation
ART	Assisted reproductive technique
ASAL	Arid and semi-arid lands
aTTP	Acquired Thrombotic Thrombocytopenic Purpura
BBB	Blood-brain barrier
BCE	Before the Common Era
BCMA	B-Cell maturation antigen
BLUP	Best linear unbiased prediction
BW	Body weight
C1/4	First/Fourth gastric compartment
CAGR	Compound annual growth rate
CAR-T	Chimeric antigen receptor T cell
CBM	Constraint-based modelling
CCHF	Crimean-Congo hemorrhagic fever
CDC	Centers for Disease Control and Prevention
cDNA	Complementary DNA
CDR	Complementarity-determining region
CH1/CH2/CH3	Constant heavy-chain domain 1/2/3
ChIP-seq	Chromatin immunoprecipitation sequencing
CHO	Chinese hamster ovary
CIDRAP	Center for Infectious Disease Research and Policy
CIRAD	Centre de Coopération Internationale en Recherche Agronomique pour le Développement
CL	Constant light-chain domain
CMC	Community Management Council
COI	Cytochrome c oxidase subunit I

COVID-19	Coronavirus disease 2019
CPD	Camel prion disease
CRISPR-Cas9	Clustered regularly interspaced short palindromic repeats-associated Protein 9
DIC	Disseminated intravascular coagulation
DM	Dry matter
DPP4	Dipeptidyl peptidase 4
EAEVE	European Association of Establishments for Veterinary Education
EBV	Estimated breeding value
ECDC	European Centre for Disease Prevention and Control
eDNA	Environmental DNA
ELISA	Enzyme-linked immunosorbent assay
EMA	European Medicines Agency
F(ab')$_2$	Divalent antigen-binding fragment
F1/2	First/Second filial generation
Fab	Fragment antigen-binding
FAO	Food and Agriculture Organization of the United Nations
FDA	Food and Drug Administration (United States)
GBS	Genotyping-by-sequencing
GCC	Gulf Cooperation Council
GGASPA	Great Gobi A Strictly Protected Area
GI	Gastrointestinal
GNLY	Granulysin
GnRH	Gonadotropin-releasing hormone
GPIIb/IIIa	Glycoprotein IIb/IIIa
GS	Genomic selection
GSO	GCC Standardization Organization
GWAS	Genome-wide association study
GZMA/B/H/K/M/O	Granzyme A/B/H/K/M/O
HBV	Hepatitis B virus
HCAb	Heavy-chain-only antibody
HCV	Hepatitis C virus
HGP	Human genome project
Hi-C	High-throughput chromosome conformation capture
HIV	Human immunodeficiency virus
HSA	Human serum albumin
ICU	Intensive Care Unit
IDF	International Dairy Federation
IGF1R	Insulin-like growth factor 1 receptor
IgG	Immunoglobulin G
IL6-R	Interleukin-6 receptor
INF	Interferon
IP	Intellectual Property
ISAH	International Society of Animal Hygiene

ISG	Interferon-stimulated gene
ISOCARD	International Society of Camelid Research and Development
ITS	Internal transcribed spacer
IUCN	International Union for Conservation of Nature
IUFoST	International Union of Food Science and Technology
IVF	In-vitro fertilization
KLR	Killer cell lectin-like receptor
KSA	Kingdom of Saudi Arabia
LC-MS	Liquid chromatography-mass spectrometry
lncRNA	Long noncoding RNA
LPAM-1	Lymphocyte Peyer's patch adhesion molecule 1
LRC	Leukocyte receptor complex
LW	Live weight
mAb	Monoclonal antibody
MAdCAM-1	Mucosal vascular addressin cell adhesion molecule 1
MAP	*Mycobacterium avium* subsp. paratuberculosis
MECC	Mongolian Environmental Citizen's Council
MeDIP-seq	Methylated DNA immunoprecipitation sequencing
MENA	Middle East and North Africa
MERS	Middle Eastern respiratory syndrome
MERS-CoV	Middle East respiratory syndrome coronavirus
MHC	Major histocompatibility complex
miRNA	MicroRNA
MOE	Ministry of environment (Mongolia)
MRL	Maximum residue limit
mRNA	Messenger RNA
MRSA	Methicillin-resistant *Staphylococcus aureus*
MS	Mass spectrometry
mtDNA	Mitochondrial DNA
MUFA	Monounsaturated fatty acid(s)
Mya	Million years ago
Nb	Nanobody
NBSAP	National Biodiversity Strategy and Action Plan
ncRNA	Noncoding RNA
NDVI	Normalized difference vegetation index
N_e	Effective population size
NGO	Nongovernmental Organization
NGS	Next-generation sequencing
NK	Natural killer
NKC	Natural killer complex
NRAP	Nebulin-related anchoring protein
NSO	National Statistics Office (Mongolia)
OHHLEP	One Health High-Level Expert Panel
OIE	Office International des Épizooties
ORF	Open reading frame

PAA	Protected Area Administration
PBUH	Peace Be Upon Him (honorific used after the name of Prophet Mohammed)
PCR	Polymerase chain reaction
PD-L1	Programmed death-ligand 1
PEG	Polyethylene glycol
PET/CT	Positron emission tomography/computed tomography
PGF2α	Prostaglandin F2-alpha
PGRP	Peptidoglycan recognition protein
PMDA	Pharmaceuticals and Medical Devices Agency (Japan)
PRF1	Perforin 1
PRNP	Prion protein gene
PTM	Posttranslational modification
RBD	Receptor-binding domain
RNA-seq	RNA sequencing
ROH	Runs of homozygosity
RSV	Respiratory syncytial virus
RVF	Rift valley fever
RVFV	Rift valley fever virus
SARS-CoV-1/2	Severe acute respiratory syndrome coronavirus 1/2
scFv	Single-chain variable fragment
scRNA-seq	Single-cell RNA sequencing
sdAb	Single-domain antibody
SDG	Sustainable Development Goal
SELMET	Systèmes d'Élevage Méditerranéens et Tropicaux
sgRNA	Single guide RNA
SMART	Spatial monitoring and reporting tool
SNP	Single nucleotide polymorphism
STEC	Shiga toxin-producing *Escherichia coli*
TAD	Topologically associating domain
TB	Tuberculosis
TLR	Toll-like receptor
TNF/TNFα	Tumor necrosis factor/Tumor necrosis factor alpha
TRD	T cell receptor delta
TRG	T cell receptor gamma
TSE	Transmissible spongiform encephalopathy
UAV	Unmanned aerial vehicle
UHT	Ultrahigh temperature
UNC-WFS	United Nations Committee on World Food Security
UNEP	United Nations Environment Programme
UNESCO	United Nations Educational, Scientific, and Cultural Organization
VCAM	Vascular cell adhesion molecule
VEGF	Vascular endothelial growth factor

VHH	Variable heavy domain of heavy-chain antibodies (nanobody)
VIB	Vlaams Instituut voor Biotechnologie
VL	Variable light-chain domain
vmNGS	Viral metagenomic next-generation sequencing
VSLA	Village Savings and Loan Association
vWF	von Willebrand factor
WAP	Whey acidic protein
WCPF	Wild Camel Protection Foundation
WCS	Wildlife Conservation Society
WES	Whole exome sequencing
WGS	Whole-genome sequencing
WHO	World Health Organization
WOAH	World Organization for Animal Health
WWF	World-Wide Fund for Nature
ZSL	Zoological Society of London

Camels Matter: History, Economics and Culture

1

Noha Kamel Moussa, Ibtisam Al-Wahaibi,
and Bernard Faye

Abstract

This chapter introduces the history of the camel, highlighting its importance to human civilization since its domestication. Thanks to its unique ability to thrive in harsh environments, camels have played a crucial role in human trade, migration and expansion to new territories. The chapter also considers the deep cultural significance of camels to pastoral societies, social identity, traditions, religion and livelihoods.

Keywords

Camelid domestication · Transport animals · Trade routes · Camel cavalry · Desertification · Camel breeding · Cultural symbolism · Camels in Islam · Bedouins · Reflexive modernity

1.1 Introduction

The camel is much more than a traditional animal in arid environments. It is an inherent, dynamic economic and cultural asset that continues to shape innovation, sustainability and policy transformation processes. From archaeological, anthropological

N. Kamel Moussa (✉)
Independent Researcher in Arabian and Islamic Archaeology, Cairo, Egypt
e-mail: n.mousa@ucl.ac.uk

I. Al-Wahaibi
Business Communication Unit, College of Economics and Political Science, Sultan Qaboos University, Muscat, Sultanate of Oman
e-mail: ibtisamw@squ.edu.om

B. Faye
CIRAD-SELMET (French Agricultural Research Centre for International Development), Montpellier, France

S. A. Gordine et al. (eds.), *Camels and One Health*,
https://doi.org/10.1007/978-981-92-2246-9_1

and literary evidence, this chapter examines the evolving role the camel has played from bronze age societies to the post-oil era, its value and perception shifting through time and space.

This chapter traces the transformation of the camel from domestication, where it was central to iron age transportation, economy and settlement patterns, to the emergence of their influence in pre-Islamic poetry, Islamic thought and Bedouin narratives, where it symbolizes honour, communal belonging and resilience. Finally, the chapter explores the embedding of the camel economy within social relationships and kinship values in premodern Arabia. The chapter highlights the orientalist misconception of camels as a static marker of backwardness and environmental decline. Camels, in the past, present and future, are an economic source as well as a symbol of identity.

In a contemporary context, this chapter situates the camel in the framework of reflexive modernity, highlighting how Gulf states are strategically reinterpreting inherited traditions to build contemporary identities alongside modern economies. These initiatives expand the camel's importance beyond economic and cultural value, positioning it within sustainability agendas, One Health approaches and long-term development strategies in which camels are now a key player in national branding.

1.2 The Domestication of Large Camelids

While the ancestors of all camelids, large and small, come from North America, where they became extinct a few million years ago, the true history of camels (one-humped dromedary camels, *Camelus dromedarius*, and two-humped Bactrian camels, *Camelus bactrianus*) begins with their domestication. Although it is still the subject of debate among specialists, it is agreed that the main focal point of domestication is the south Arabian Peninsula for the dromedary and central Asia for the Bactrian.

The domestication of the dromedary occurred more recently than the Bactrian, probably in the second half of the second millennium BCE, making it the most recently domesticated large herbivore species, just after the yak (Digard 2009; Faye 2022). The direct wild ancestors of the dromedary, characterized by their large size, have all disappeared. They occupied the entire area of hot deserts from East Africa to the Near and Middle East as well as the Arabian Peninsula during the Pleistocene (2.6 million to 11,700 years BP). At the end of this period, this range was reduced to the Arabian Peninsula (Vigne 2025). During the Neolithic period, it appears the wild dromedary was still present in the north of the peninsula. This is demonstrated by the monumental high reliefs of Sakaka (Fig. 1.1), dating from 5200 to 5600 BCE, which seem to indicate dromedaries were hunted for their meat (Guagnin et al. 2022).

However, archaeological excavations carried out on many sites in the peninsula suggest that the domestication of the dromedary began in southern Arabia (present-day Emirates, Sultanate of Oman and southern Saudi Arabia), around the great desert of Rub-Al-Khali (Uerpmann and Uerpmann 2002). The occupants of this region, fishermen or small ruminant herders practicing semi-nomadism, began the domestication process with a small number of animals, as population genetics and ancient DNA have recently confirmed (Fitak et al. 2020). As with other species, however,

Fig. 1.1 Camel carved in high relief from the Camel Site near Sakaka (Saudi Arabia). (© B. Faye)

during the slow and long process of domestication, back-crossing with dromedaries that remained wild have made it possible to establish a high genetic diversity within domestic lines, as shown today by the great diversity of camel breeds in Saudi Arabia and neighbouring countries (Faye et al. 2011). It is this initial genetic diversity that has allowed dromedaries to acclimatize during their geographical expansion, which occurred for more than 3000 years (Almathen et al. 2016; Faye 2022; Vigne 2025). All wild dromedary forms having disappeared, both in the Arabian Peninsula and in the other arid regions of the old world, we can assert that all the dromedaries of the world today originated from this first focus of domestication. However, epigraphic arguments seem to indicate that another focus of domestication may have emerged in connection with that of the Bactrian camel (Heide and Peters 2021).

Indeed, the history of the domestication of the Bactrian camel is older, taking place at least a millennium before that of the dromedary. For quite a long time, scientists thought that its domestication took place in the Gobi Desert between present-day Mongolia and northern China. In favour of this argument is the presence of the wild double-humped camel (*Camelus ferus*), which John Hare (2015) calls the Tartary camel and which was long falsely considered the wild ancestor of the domestic Bactrian. We now know, thanks to genetic studies that it is in fact a "cousin". Clearly, the two species have a common ancestor that has disappeared, and their genetic divergence dates back about 700,000 years (Ji et al. 2009; Fitak et al. 2020). In addition, the first attestations of the domestic Bactrian camel actually come from northern China, about 1800 BCE (Salma 2022).

However, this hypothesis has recently been challenged yet again by new archaeological discoveries in the more western part of Central Asia. Indeed, the wild ancestor of the Bactrian seems to have been present in the Kyzylkum desert (present-day Uzbekistan) as early as the sixth millennium BCE (Ming et al. 2019). Its domestication is thought to have taken place during the third millennium BCE, in an area located between present-day Turkmenistan, Afghanistan and Iran (Vigne 2025). Bones of domestic camels (attested by their size) dated to 2500–2700 BCE have been found in Iran, and figurines dated between 2500 and 1600 BCE representing Bactrian camels pulling chariots have been unearthed in Turkmenistan (Sala 2017). With the development of economic exchange in the heart of Central Asia, the Bactrian camel then migrated to the eastern part of the continent, to Kazakhstan, Mongolia and China. The name Bactrian actually refers to a region (former kingdom conquered by Alexander the Great) located between Afghanistan, Iran and Kazakhstan (Burger et al. 2019).

The domesticated Bactrian camel was present in Mesopotamia at the beginning of the second millennium BCE (Heide and Peters 2021), while the dromedary was still wild. At that time, the Bactrian was used as a pack animal for trade with the great empires, especially during the development of the city of Babylon. According to one hypothesis, this presence of the Bactrian camel, particularly in the Tigris and Euphrates delta (Chatt-El-Arab), would have contributed to the domestication of dromedary camels in this area by hybridization. This crossing between domestic Bactrian males and wild female dromedaries would have been the fastest and most natural way to initiate domestication of dromedaries for the needs of the pastoral populations of the area. This hypothesis, proposed by Walz in 1954, is potentially supported by the presence of two different genotypes between the north and south of the peninsula. This would imply that there were two centres of dromedary domestication, the second giving rise to camel lines originally having some Bactrian genes (Fitak et al. 2020).

In any case, the process of camel domestication originally took place in small geographical areas (in the southern Arabian Peninsula and the Central Asian desert) which were quite distant from each other, implying a clear spatial separation of the two species. However, the high mobility of these animals has facilitated their encounter (Faye and Konuspayeva 2012) and the possibility of crossing between the two species, a practice that is probably very ancient.

The question is also to understand why and how dromedaries and Bactrians moved from these restricted centres of original domestication into the areas they occupy today. This is of course linked to its ability to carry heavy loads (100–300 kg depending on the animals and the distances to be covered) under very limited resource conditions (water and fodder) and in the climatic conditions of the desert (Faye 2013). From this point of view, the camel is very competitive compared to the horse, which has hooves that sink into the sand, a poor ability to space out watering and fragility in the face of "heat stroke". To a certain extent, in the desert, the camel is unrivalled, and thanks to its role in transcontinental trade and military conquests, it may have been the origin of great civilizations.

1.3 The Expansion of the Camel After Domestication: The Role of Trade

One may ask the reason that prompted humanity to domesticate camels. The question is still debated, but it is reasonable to presume that the animal's ability to carry people and goods was the main motivation for domesticating the species, given that the domestic horse was not yet very common (at least in the Arabian Peninsula). Yet even in Central Asia, the cradle of horse domestication (Vigne 2025), the Bactrian camel may have had an advantage over the horse for the transport of goods, the former's carrying capacity being greater.

Indeed, the oldest epigraphic records of large camelids refer to their prominent role in the development of long-distance trade. The existence of trade routes following the development of the Sumerian empire in the Middle East was already mentioned above, but one of the oldest caravan routes in the Arabian Peninsula dating back to the twelfth century BCE and lasting until the second century AD, during the Roman period, was the Incense Roads. These unified the south of Arabia to the Mediterranean Sea and to Mesopotamia thanks to caravans of dromedaries (Finkelstein 1988). This trade route allowed the transport of myrrh and frankincense from the forests of *Commiphora myrrha* of Dhofar (now the Sultanate of Oman) or Hadramaut (now Yemen and southern Saudi Arabia) through the Hejaz mountains, to Egypt or the Levant. In Egypt it was widely consumed for the embalming of the dead, whereas in the Levant Romans used the incense in their temples.

Mentioned in the legend of the Queen of Sheba from where the royal caravan left laden with precious goods (gold, spices and incenses) to King Solomon's palace in Jerusalem, the caravan trade in the Arabic Peninsula probably reached its peak towards the end of the first millennium BCE and made the fortune of the Nabataean Kingdom (Salma 2022). Numerous rock carving representations left by the actors of this period (Faye et al. 2024) testify to the place of the dromedary in these commercial links between oases (Fig. 1.2).

Along these caravan routes, flourishing cities sprang up, such as Petra (present-day Jordan), Hegra and Tayma (Saudi Arabia), Palmyra (Syria) and even Mecca before it became the cradle of Islam. This shows what these cities owe to the dromedary, the main motor of their economic development, and, later, Islamic civilization (see Box 1.1). After all, Prophet Mohammed—peace be upon him

Fig. 1.2 Rock carving representing a camel caravan on the Nabataean route between Hegra and Tayma. (© B. Faye)

Box 1.1: Camel and Islam

The Arabian Peninsula being the cradle both of the camel and of Islam necessarily links their reciprocal destinies. The development of Arab civilization was not only based on the Muslim religion, but also on camels for the connection of spaces across deserts. This connectivity brought not only goods of different kinds, but also ideas and cultures, of which the Muslim religion is a part. This is demonstrated, for example, by the presence of the camel in the sacred scripture of Muslims: the Qur'an. It is interesting to note, from this point of view, the diversity of denominations in the Qur'an. In English translations, there is only one word to designate the camel (for example, the male is a "camel bull", or the young is a "camel calf", i.e. terms used for cattle). In Arabic, however, there are many words describing the camel depending on its age, status, herd size or coat colour. Thus, in the Qur'an, the camel is present under 16 different names (https://www.kamelito.com/le-chameau-dans-le-coran/): (1) In Surah 7 (verse 40), the termٱلْجَمَلُ **(*Al-Djamal*)** designates the large camelids in general. It can be noted that this word (*Al-Djamal*) corresponds to the third letter of the Sumerian alphabet (*gamel*), which gave *gamma* in Greek and to the root of the word "camel" which is found in all Indo-European languages (e.g. chameau, kamelen, Camello). (2) In Surah 6 (verse 44) and 88 (verse 17), we find the word ٱلْإِبِل **(*Al-ibil*)** a generic term which designates adult male and female dromedaries. (3) The word النَّاقَة **(*Al-Naaqa*)** appears seven times in the Qur'an and designates the female camel, including the female camel of Allah; it is also said in the Hadith that the Prophet (PBUH) had a white she-camel named القصواء (*Al-Qaswa'*) which he used during his journey through the Hijra. (4) In Surah 12 (verses 65 and 72) appears the word بَعِيرٍ **(*Al-Ba'ir*)** which designates the mature camel (male or female) on which goods can be transported; i.e. it corresponds to pack camel. (5) In Surah 77 (verse 33), the word جِمَٰلَتٌ **(*jimâlat*)** appears, which is an ambiguous term sometimes translated as yellow camels or black camels in the plural. In the nomenclature of camel breed names in the Arabian Peninsula, the black camel is named المجاهيم (*Al-Majaheem*) (Faye et al. 2011). (6) In Surah 22 in relation to the pilgrimage ٱلْحَجّ (Al-Hajj), in verse 36, the *word* ٱلْبُدْنَ **(*Al-boudn*)** means "mature camel, in good health, with beautiful appearance". (7) In Surah 12 (verses 70, 82 and 94), the word ٱلْعِيرُ **(*Al-cîr*)** refers to pack animals in a caravan. (8) In one Surah (56, verse 55) appears the word ٱلْهِيمِ **(*Al-hîm*)** that can be translated as a camel that is so thirsty it drinks non-stop to the point of risking death. (9) The riding camel رِكَابٍ **(*rikâb*)** appears in Surah 59 (verse 6) in a context of fighting in war. (10) In the Surah 81 (verse 4) appears the word ٱلْعِشَارُ **(*Al-ichâr*)** which refers to the adult female reaching the tenth month of gestation, i.e. the end of pregnancy, which is 13 months

(continued)

Box 1.1 (continued)

in camels. (11) Appearing also once in Surah 22 (verse 27), the term ضَامِرٍ (***Al-Dâmir***) means the camel which has become thinner after the course of a long journey, i.e. the emaciated camel. (12) The word الأَنْعَام (***Al-Ancâm***) (generic term for "livestock" including camels) is present in 22 different surahs. (13) Also appearing once (Surah 5, verse 103), the word بَحِيرَةٍ (***Al-bâhîrah***) means the camel with split ears, i.e. belonging to pagans who split the ears of their camel after delivering its fifth calf. This word, as well as the following, refers to pagan traditions which are not instituted by Allah. (14) The word سَآئِبَةٍ (***Al-sâ'ibah***) in the same verse means the camel which is allowed to be free, i.e. not to be used for riding or as a pack animal. (15) It is also the case of the camel وَصِيلَةٍ (***Al-wasîlah***) in the same verse, which, according to pagan tradition, is the camel that has given birth to a female baby two consecutive times, which is then let free. (16) Finally, still in the same verse appears the breeding camel consecrated to pagan idols, called لاَ حَامٍ (***Al- hâmî***).

There are many other designated words for camels, and the richness of this vocabulary attests to the Arabs' interest in this animal. The same thing can be observed in the Kazakh language, for example, to refer to the Bactrian camel.

(PBUH)—belonged to the tribe of the Quraysh, a merchant tribe of Mecca renowned for its camel caravan trade.

The expansion of the dromedary into Africa was relatively late. The oldest testimonies attest to the presence of the dromedary in Sinai and the eastern part of the Nile Valley during the first millennium BCE, mainly for the transport of goods and agricultural work, as beasts of burden (Agut-Labordère and Redon 2020). However, belonging mainly to Arab tribes, the dromedary camel was not an important element of Egyptian culture and economy (Barnard 2012). The dromedary was never present in hieroglyphics or sacred illustrations, unlike cattle and other domestic livestock. The dromedary was not part of the bestiary of ancient Egypt (Ikram 2017).

The first camels probably reached the Horn of Africa by boat, coming from the Arabian coast of the Red Sea during the first BCE millennium as part of trade (Salma 2022). The dromedary appeared in the north of Africa and in central parts of the Sahara slightly before the Roman Empire, as documented by petroglyphs in the Libyan desert. Its strongest expansion occurred later with the arrival of Islam (seventh to eighth century ACE) and in the fifteenth century after the capture of Constantinople and Mecca by the Ottoman Empire (Vigne 2025). However, the expansion of the dromedary camel in the Sahara was facilitated by its aridification which would have begun at the beginning of the fourth millennium BCE, leading to a favourable environment for the animal (Jung et al. 2004). Thus, the crossing of this immense barrier of sand and arid lands to connect the shores (*Sahel*) of this desert could only be done with the dromedary, justifying the camel's nickname of "ship of desert," given by the Arab people. Thanks to this animal, many trans-Saharan trade routes contributed to the link between the north and south margins of the Sahara and to the economic development of oases, on both sides of this immense desert. Thus, cities such as Sigilmassa (present-day Morocco), Chingetti (Mauritania) and Timbuktu (Mali) in the western part of the Sahara, or Ghadames (Libya), Bilma and Agadez (Niger) in the central part, or Khoufra (Libya), Aswan (Egypt) and Dongola (Sudan) in the eastern part, are among the many oases that have enabled Morocco to connect with the Niger river, Tunisia with Lake Chad and the Maghreb with the upper Nile Valley, carrying gold, dates, salt, copper, ivory and slaves (Austen 2024; Scheele 2024).

In Asia, the expansion of the Bactrian camel was also strongly linked to trade, notably with the emergence and the development of the Silk Road. This was an important network of routes connecting China to the Mediterranean ports for the trade of not just silk, but also porcelain, lapis lazuli and cotton (Frankopan 2015). This trade route is believed to have been active for more than one millennium. Indeed, the Chinese started using the route during the Han dynasty, from the third century BCE, its apogee under the Tang dynasty (seventh to tenth century BCE) and during the *Pax Mongolica* in the thirteenth century under the impetus of Genghis Khan (Franklin 2024).

It is probably also in connection with these caravan routes across the Asian continent that the practice of hybridization developed, an inappropriate term because it is rather a cross-breeding between the dromedary and the Bactrian camel. The hybrids, which have different names in Turkish-Mongolian traditions (Dioli 2020) depending on the type of crossing, are indeed animals that combine the strength of the Bactrian with the endurance of the dromedary, qualities that are very useful in caravan activity across the desert. It can be noted that the practice of hybridization has continued to the present day, especially in Kazakhstan, to obtain hybrids producing abundant (dromedary effect) and richer milk (Bactrian effect) (Faye et al. 2008) and in Türkiye to participate in camel wrestling festivals, which are very popular in Anatolia (Yilmaz 2017).

Of course, caravan activity gradually declined with competition from sea lanes in Asia and the Arabian Peninsula and, then later, the motorization of transport, particularly in the Sahara. However, caravans are perpetuated over short distances, particularly for the transport of non-perishable products (salt, cereals) or contraband objects in areas that are mostly inaccessible by truck and generally over short distances (Alary and Faye 2016).

Thus, dromedary and Bactrian camels contributed to territorial connectivity between Asia, Africa and even Europe, as evidenced by the numerous archaeological sources attesting to the presence of large camelids in Roman times, along the roads linking the great cities of the empire (Pigière and Henrotay 2012). They moved huge quantities of wealth over thousands of kilometres from the Chinese Empire to the Roman Empire, from the *Arabia Felix* to the Levant, from the Mediterranean banks to the Sudan (Fig. 1.3). However, the expansion of the camel through the continents is also linked to another human activity, namely, war (Faye 2022).

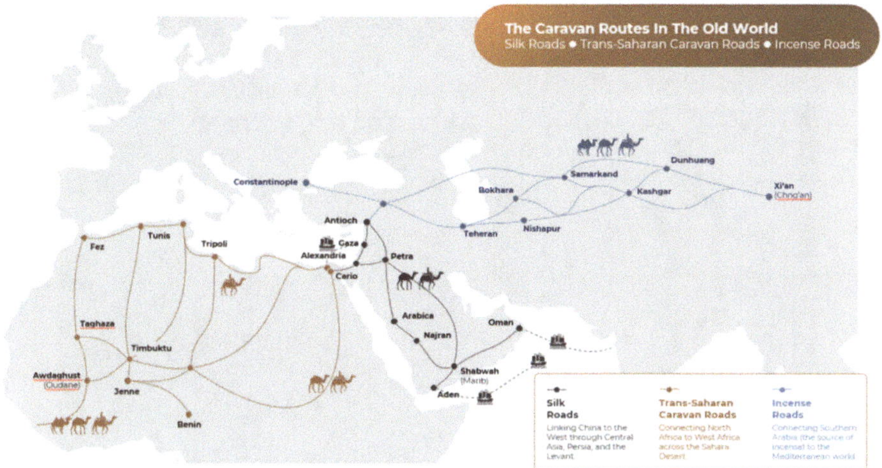

Fig. 1.3 The caravan routes in the Old World (Silk Roads, Trans Saharan Caravan Roads and Incense Roads). (Source: Authors' own work)

1.4 The Expansion of the Camel After Domestication: The Role of War

Although mainly in the hands of Arab tribes, dromedaries seem to have been widely used in ancient conflicts managed by other populations. For example, the introduction of the dromedary into Sinai would have been the consequence of the Assyrian invasions at the beginning of the seventh century BCE. Later, at the end of the sixth century BCE, the Persian invasions used Arab cameleers as auxiliary forces (Cousin 2020). Indeed, if one of the main reasons for the domestication of the dromedary was its use as a pack or riding animal, these functions were consequently highly useful for a mobile army. Some archaeological and epigraphic sources show dromedaries carrying fighters or weapons, as early as the second millennium BCE, as can be seen in a Sumerian fresco dating from this period (Fig. 1.4).

The arrival of the dromedary camel in North Africa with the expansion of Islam has been mentioned above. However, in addition to this peaceful expansion of camel territory, Arab troops brought them to Spain during their conquests (Insoll 1996). The military vocation of the dromedary has been perpetuated to the present day, with the use of camel cavalry and the creation of dedicated camel corps. The first of these was probably created by General Bonaparte during the Egyptian campaign in 1799 (Cvikel and Goren 2008). Widely present during the colonial conquest in the Sahara and the Levant, the Méharist camel companies of the French (Perez 2003) and British colonial administrations (Wilson 2016) traditionally used dromedary camels as a military auxiliary. Nowadays, despite the use of vehicles in modern conflicts, some Méharist companies are still operational for border control in desert countries such as Mauritania and India (Choudhary 2019).

Fig. 1.4 Sumerian frescoes showing a camel raid by Arab tribes in the second millennium BC in the Euphrates valley (museum of Doumat-Al-Jandal, Saudi Arabia). (© B. Faye)

If the dromedary, thanks to its speed, was valued by both the military and rebels for use in direct combat or raids, the heavier Bactrian camel was rather confined to the transport of goods or weapons, even by the Soviet army during the second world war (Mighty 2023).

The movements of conquest by Genghis Khan in the thirteenth century were achieved thanks to the remarkable Mongol horse rather than to the Bactrian camel. However, ancient petroglyphs in China attest to the use of the Bactrian camel for hunting antelopes (Fig. 1.5). Camel riding is also practised by enthusiasts of falcon or eagle hunting in the Middle East and Central Asia. However, riding camels for war or hunting required faster, slenderer (longilinear) animals, while the pack camels required robust, more wiry animals (brevilinear) able to carry heavy loads. As a result, very soon after domestication, we believe a certain selection was put in place to formalize two types of animals: for packing and for riding. Some people consider that the term *dromedary* should only refer to animals adapted to racing (the Greek term *dromos* which means *to run* is found in words such as hippodrome), while pack camels should be called *Arabian camel* distinguished from the Asian (Bactrian) camel.

It should also be noted that the commercial and military mobility of large camelids contributed to the strong dispersal and mixing of their genes. We can also trace these flows to a large extent from the study of genes (Almathen et al. 2016) and consider that their mixing across ages and spaces has led to a panmictic population, i.e. globally not very diverse, compared to other domestic species, at least until very recent times (Burger et al. 2019). If human activities such as trade and war strongly contributed to the expansion of camel rangelands, desertification also played an important role in the conquest of new territories.

Fig. 1.5 Petroglyphs representing the hunt of antelopes by camel riders (Alashan desert, Inner Mongolia, China). (© B. Faye)

1.5 Camels, Desertification and Climatic Changes

It can be presumed that the presence of large camelids accompanied the aridification of the environments in the old world. The aridification of the Sahara facilitated its occupation by the dromedary from the beginning of the Christian era and the spread of the Bactrian in the cold deserts of Mongolia and China at the end of the second millennium BCE (Yam and Khomeiri 2015; Faye 2022). Since the great droughts that affected the Sahelian regions in Africa at the beginning of the twentieth century and especially since the 60s, the expansion of camel farming systems has continued beyond the traditional camel countries (Mauritania, Mali, Niger, Chad, Sudan, Ethiopia, Somalia) into neighbouring countries (Senegal, Burkina Faso, Nigeria, Cameroon, Uganda and Tanzania among others) (Faye et al. 2012; Wilson 2017; Asiimwe et al. 2020). In Ethiopia and Kenya, camel farming expanded into mountainous areas formerly devoted to cattle and small ruminants (Tefera and Abebe 2012; Volpato and King 2019). Overall, camel farming is considered to have migrated more than 500 km further south in sub-Saharan Africa, less because of the migration of traditional camel drivers than because of the adoption of the animal by traditional cattle breeders to secure their production system in the face of climatic hazards (Faye 2020).

The large camelids have been the pillars of many nomadic groups (Mongols, Kazakhs, Arabian Bedouins, Somalis, Toubous, Tuaregs, Moors), some of whom have given birth to great empires and civilizations (Mongol Empire, Arab-Muslim world). Because of their adaptation to resource-poor environments, the slow expansion of camel breeding beyond the arid zones has led to their gradual integration into agricultural systems and their sedentarization in some places of the world. More generally, current climate changes and economic globalization are leading to profound changes in camel breeding around the world, particularly regarding the increasing presence of large camelids in the production and leisure economies. Long an auxiliary to intercontinental trade, camels were rarely themselves the object of trade, apart from the purchase and sale of pack animals for caravans. However, the history of the camel does not end with the decline of large caravans and with urban development in arid countries, as is exemplified by the current major trends in camel production systems.

1.6 The Camel: Between Orientalist Representations and Indigenous Perception

The word *camel* conjures a specific mental image based on what has been socially ingrained in the readers' subconscious. Depending on cultural and environmental backgrounds, camels can either represent the *other* or can play a part in the representation of self-identity.

The Western image of the Orient as a land of dunes and camel caravans was a constructed Orientalist narrative. As Said (1978: 1) argues, it was the Orient that helped Europe and the West define its image and personality as contrasting to oriental societies and cultures. Camels and nomads were viewed as the unchanging, premodern Orient, compared with Europe's self-image as progressive and dynamic. The environmentalist Diana Davis (2007) presents the concept of environmental Orientalism, through which the West embarked on its colonial projects for irrigation and improvement, based on the perception of the landscapes in the Middle East and North Africa as barren, degraded and in need of "Western" restoration. In this representation, camels have double symbolism, being both the subject of romantic magic but also an indication of environmental backwardness (Davis 2007). Moreover, images in colonial environmental discourse reinforced utilitarian and civilizational hierarchies, as control over nature became seen as a measure of progress. In this representation, the camel embodied inertia compared to Western dynamism (Appuhn 2005).

While in the past the Western understanding of the environment was limited to control over nature as a measure of progress (Appuhn 2005), new studies reaffirm the importance of integrating the traditional *hema* system for regenerating rangelands to avoid desertification. The *hema* system involves rotational grazing, and is cited in the Qur'an, supported by Shari' ah (Islamic law), Hadith (sayings of the Prophet) and Sunnah (traditions of the Prophet). The *hema* system was possibly developed as a way of conserving and protecting natural pastoral resources, representing the earliest known record of rotational range management in history (Chaudhary and Le Houérou 2006 in Breulmann et al. 2007).

Unsurprisingly, the Arabian perception of the camel is therefore dramatically different as is expressed through original artistic and literary traditions. Camels were depicted in literature, poetry and art as tools for movement, war and social prestige. This does not represent inertia, as perceived by the West, but rather dynamic and effective agency (Jung 1994). It is necessary, therefore, to re-evaluate colonial preconceptions of camels and Arab societies within a new framework of sustainability, innovation, cultural identity and technological adaptation.

1.7 Camels and Arabian Cultural Development

Camels serve multiple functions in desert societies, providing economic benefit as well as cultural symbolism. Their attributes are highly valued by those who inhabit the unforgiving desert environment, who perceive the camel a representation of resilience and strength, revering its capacity to survive under challenging circumstances. In the Arab region, camels are treated with respect and even a sense of worship due to their significant contribution to household wealth, their perceived sacred nature and tribal honour. The Qur'an sanctifies the camel as a divine sign of creation and reflection: "أَفَلَا يَنْظُرُونَ إِلَى الْإِبِلِ كَيْفَ خُلِقَتْ" "Do they not look at the camels, how they are created?" (Qur'an 88: 17), while the Bedouin describe the camel as عطايا الله "Ata Allah" or God's gift.

The camel occupies the Arabian mind and soul, having become one of the iconic symbols of honour and pride. In pre-Islamic poetry, in the *Muʿallaqa* or odes [*Muʿallaqat* are the hanging poems, which are famous long pre-Islamic Arabic poems that were believed to have been hung on the Kaaba] of the poets Ṭarafa ibn al-ʿAbd and Labīd ibn Rabīʿa, an image of accepted and negotiated social values is depicted through the slaughter of the she-camel. Ṭarafa ibn al-ʿAbd in his words.

فمرّتْ كَهاةٌ ذاتُ خَيْفٍ جُلالةٌ عقِيلَةُ شَيْخٍ كالوبِيل يَلنْدَد
يقولُ، وقد تَرّ الوَظِيفُ وساقُها : ألَسْتَ ترى أنْ قد أتَيْتَ بمُؤيِد؟
وقال: ألا ماذا تَرَونَ بِشاربٍ . شديدٍ علينا بَغْيُهُ، مُتَعَمِّدِ ؟
(Nasir al-Din 2002: 28)

Arberry (1957: 88) translates the verses as:

"Many's the kneeling, sleeping camel—the fear of me stalking with naked blade has oft startled the runaways; then some ancient she-camel with flaccid udders, huge, the pride of an elder thin as a stick, quarrelsome, has passed me by and he remarking to me (for her pastern and shank were slit) "Don't you see what ruination you've brought on me now? What think you," this to the tribesmen, "we should do with a drunkard whose wickedness presses hard on us, a willful sinner?""

Tarafa visualizes the camel as his inner life where its slaughter reflects the inner conflict between pride, generosity and social obligation and the individual acceptance of sacrificing something the person does not own. On the other hand, Labīd ibn Rabīʿa's visualizes his generosity and his sense of communal belonging by his voluntary sacrifice of his own camels by saying:

وجَزُورِ أَيْسَارٍ دَعَوْتُ لِحَتْفِهَا بِمَغَالِقٍ مُتَشَابِهٍ أَجْسَامُهَا.
"أَدْعُو بِهِنَّ لِعَاقِرٍ أَوْ مُطْفِلٍ بُذِلَتْ لِجِيْرَانِ الجَمِيْعِ لِحَامُهَا
فَالضَّيْفُ والجَارُ الجَنِيْبُ كَأَنَّمَا هَبَطَا تَبَالَةَ مُخْصِباً أَهْضَامُهَا
(Al-Hasani 2012: 75)

which is translated in Arberry (1957: 146–47) as:

> I've disowned the wrong, and boldly maintained the right as I saw it, and none of those noble gentry could glory over me. And many a time I've called for the gambling-arrows, so like each to each in shape, to kill a gambler's slaughtering-beast, called for the arrows to choose a barren or bearing camel whose flesh was distributed to the poor relations of all; and the guest and the poor stranger must have thought themselves come down upon Tabla, whose valleys are ever green.

The sacrifice of the person's own camel reflects the sense of communal belonging, while the slaughter of another's camel signifies social separation—a contrast in which camels played a dual role as both a sacred and social mediator between self, tribe and universe (Sells 1986).

In addition, there is the pre-Islamic tale of the forbidden love between Antarah ibn Shaddad, the son of a black slave, and his cousin Abla bint Malik. This story, which has a significant influence in formulating the Arabian understanding and collective memory of love and romance, documents the excessive dowry asked of Antarah to marry Abla. This dowry took the form of one thousand red she-camels, rare and only possessed by the vassal of the Lakhmid King al-Mundhir, leading Antarah to make a dangerous journey to present-day Iraq to bring the dowry (Devic 1898).

In his biographical literature that recounts the life of prophet Mohamed (PBUH), Ibn Hishām (1937) narrates about when Abraha intended to destroy the Kaaba. Abd al-Mutalib, the grandfather of prophet Mohamed (PBUH), met with Abraha and asked him to return 200 camels he had seized. Abraha was astonished that he cared more about the camels than the Kaaba. Abd al-Mutalib replied:

إِنِّي أَنَا رَبُّ الإِبِلِ، وَلِلْبَيْتِ رَبٌّ يَحْمِيهِ

which translates as:

> I am the lord (owner) of the camels, and the Kaaba has a Lord who will protect it.

The Qur'ān also narrates the story of prophet Salih and his people the Thamud, who asked their prophet for a miracle. Allah sent them a she-camel (Surah Al-A'raf 7: 73):

هَٰذِهِۦ نَاقَةُ ٱللَّهِ لَكُمْ آيَةً

which translates as:

> This is the she-camel of Allah, a sign for you.

Allah asked them not to harm the she-camel and to give her a turn to drink, whereby the camel should drink one day and the Thamud take their turn to drink the next. The Thamud killed the she-camel and in consequence Allah punished them by a thunderous blast, killing them all (according to Surah Hud 11: 68).

Fig. 1.6 A vibrant scene of a camel gathering in a desert area preparing for Ardha Race (festival race), featuring a man riding a camel in the foreground among a crowd of people in traditional attire. (Alwahaibi 2024)

Throughout these accounts, the camel was governed by sets of rules and resources which changed across time and space. Drawing on Giddens (1984) and Hobsbawm and Ranger (2012: 1) posit that tradition, understood as a set of practices accepted by communal rules and resources, solidifies cultural and religious ties with the camel. In the past, these rules included codes of honour, ownership, owners' communal obligations, the exchange of camels for a bride's dowry and accepted norms governing the care of camels. Such rules were demonstrated by binding honour, kinship and obligation towards the community. In modern times this tradition is expressed differently, through practices such as camel racing festivals (Fig. 1.6) and modern Gulf poetry.

The camel-based trade routes created networks of exchange and alliances, linking settled farmers and nomads. Excavations at the *Camel Site* in the Negev reveal that obsidian, shells and copper were traded through multi-level trading systems (Finkelstein 1988; Rosen 2011). This exchange network created what Rosen (2011: 209–210) terms "multi-resource nomadism—a dynamic economy of interdependence between desert and town".

The tent and the camel were considered to be a new architectural and technological innovation which provided another form of Bedouin life that combined mobility and stability (Rosen and Saidel 2010). Early inhabitants of the Arab peninsula organized their camps to revolve around the camel corrals, as the physical presence of camels shaped the geometry and rhythm of Bedouin life. This coexistence of architecture and animal husbandry determined the integration of environment, economy, and kinship in early Arab societies (Haimi and Rosen 2011).

Camels also served the ambition of the emerging elite class in ancient South Arabia in opening the way for trans-regional trade (such as to the South Arabian kingdoms) and new environmental zones for regional trade (such as Muweilah) (Kamel Moussa 2019). As previously mentioned, camel figures were depicted on rocks over almost all of Saudi Arabia, sometimes accompanied by ancient Arabian inscriptions. Such depictions were located as far as Yemen and Jordan, where camels were sacrificed as offerings to the deities and gods with camel bones found buried in the temples at Al-Fao in Saudi Arabia (Khan 2022). The early South Arabian inscriptions and grave stelae depicting armed camel riders associated the animal with honour, mobility and warfare, confirming its centrality to moral and political life across the peninsula (Jung 1994).

The camel emerged as a unit of social and moral value, which embodied kinship, prestige, and ethical relations. Camel ownership served the political and moral order for the pastoral community, which valued lineage identity and social reciprocity (Köhler-Rollefson 1993). Arabian tribes mark their camels with their tribal sign (a practice called *Wasum)* to prove tribal ownership (Khan 2022). As noted in Arabic literature, camels functioned as *mahr* (dowry) in marriage alliances and as *diyah* (blood compensation). Aktham ibn Sayfi Al-Tamimi wrote in his will to the Tayy tribe: “Do not place the necks of camels except where they are deserved, for in them lies the price of the noble steed (dowry), the payment of blood money (diyah), their milk is an offering to the distinguished, and a nourishment to the young” (KSA, Ministry of Culture 2024: 23).

These practices demonstrate how the camel economy extended beyond materialism to support moral ethics and systems of justice. This concept is defined by Scott (1976) as the Moral Economy, that is, an economy governed by collective socially accepted ethics and obligations. Polanyi’s (2001) economic concept of embeddedness in pre-modern societies explains that the camel economy in Arabian societies was embedded in social relationships and values. The camel thus embedded kinship, economy and social values.

1.8 Camels and Bedouin Identity

Among the Arab Bedouins, a camel can symbolize dignity and honour as well as affection, psychological stability and health (Khan 2022; Alwahaibi et al. 2024). This symbology is well ingrained in their culture and resounds in the values and beliefs of these people. This symbolism should be acknowledged when engaging with the Arab Bedouin people because it provides an insight into their culture of existence. By better understanding the camel, it is possible to discern more about the values and customs of the peoples in these regions.

In Bedouin societies across the Arabian Peninsula and Oman, the ownership of purebred camels serves as a significant marker of social prestige and wealth, with camel herders historically occupying the highest status positions within tribal hierarchies due to their control of these valuable animals essential for desert survival and trade (Lancaster 1981; Meeker 1979; Rosen and Saidel 2010; Alwahaibi et al. 2024). The acquisition and breeding of racing camels and pageant camels, in particular, elevates an owner's standing within the prestigious community of camel owners, conferring social weight and cultural recognition that extends beyond mere economic value to encompass identity, tribal affiliation and community (Le Magoariec 2025; Alwahaibi et al. 2025).

While camels are most commonly recognized for their profound loyalty and occasional jealousy towards their handlers, they also exhibit a less prevalent but distinct sense of pride in their demeanour. This complex emotional intelligence challenges historical misconceptions of them as merely stubborn beasts of burden, revealing a deeply intuitive and sentient nature (Maitra et al. 2025). Bedouins cherish the time they spend with their camels, rearing, coaching and feeding them. For camel owners and their children, engaging with camels is part of their daily routine (Eades et al. 2013; Lancaster 1981; Meeker 1979; Rosen and Saidel 2010; Alwahaibi et al. 2024). Children in the Arabian Gulf engage with camels at a very young age (Fig. 1.7). This continuous proximity to and interaction with camels tends to result in a strong bond with their animal companions.

Fig. 1.7 A group of camels standing in a sandy desert area, with two boys in traditional attire holding their reins, preparing them for their daily run exercise. (Alwahaibi 2024)

1.9 The Camel as a Commodity

The perception of camels has undergone a modern transformation from an iconic local symbol to a global economic commodity. As discussed in Appadurai's (1986) "Social life of things", the social and cultural value of camels is expressed through activities like racing festivals, beauty contests and cultural tourism, while their commodity value is realized through the export of milk, meat, infant nutrition, cosmetics and more.

Using camels to produce milk, meat as well as wool is probably as old as domestication (Yam and Khomeiri 2015). As the wild camels of the Neolithic period were hunted, it can be assumed that the consumption of camel meat was perpetuated from the beginning of domestication. Similarly, it is likely that in the absence or almost absence of cattle in the areas of origin of the domestic camel, humans consumed camel milk. Bactrian camel milk consumption is attested in the archaeological site of Ayakagytma in the Kyzylkum desert, in Uzbekistan, 3000–4000 BCE (Suska-Malawska et al. 2022). The fact that nowadays the dromedary, particularly in the Arabian Peninsula, is considered a dairy animal with high potential leads us to consider that the consumption of their milk was common from the beginning of domestication. However, most of this milk was self-consumed or given to passing guests and poor people, so much so that for a long time, the sale of camel milk was considered taboo. It was not until the end of the twentieth century that a real market for this product appeared (Faye and Corniaux 2024), first nationally and then on the international market with the development of camel milk powder (Konuspayeva et al. 2022). For a long time camel milk was only available in nomad camps, while today camel milk is available in the supermarket. Big camel dairy farms are implemented in Gulf countries, Saudi Arabia and central Asia to provide processed camel milk to the international market, including Europe (Ait El Alia et al. 2025).

In traditional societies, camel meat was consumed only on special occasions such as weddings, funerals, or a chief's installation, because the animal was large enough to feed many people. For a long time, it was common to have camels slaughtered only by specialized butchers for local markets. Butchery practices involving Bactrian camels were identified in different archaeological sites from China during the Han dynasty (206 BCE to 220 ACE) (Ding et al. 2026). Camel meat consumption has also been mentioned in Bedouin culture for centuries (Schmitt 2025). However, the modernization of this sector will be linked to the implementation of specific slaughterhouses and the development of frozen carcasses for export, rather than the export of live animals for food on trucks, as is the dominant way for other meat (Boukrouh et al. 2025). Moreover, for both milk and meat, production modernization is accompanied by an important diversification in camel products.

In the past, camel wool and camel leather had little market value and were limited to the local handicraft sector for traditional use. Currently though, these products are seeing noteworthy interest on international markets for the luxury sector, notably in China and Mongolia for wool and in the Emirates for leather.

Besides the use of camels in production, the camel has also entered the world of sports and leisure, especially in the Middle East since the oil boom. While camel racing is undoubtedly very old, directly resulting from the raids organized by nomadic societies during tribal conflicts, a real race economy has been emerging in recent decades with racing competitions, markets with racing accessories, racing stables, selection programs and the fabulous prizes of champions. On the other hand, former Bedouins who have now predominately become sedentary city dwellers have turned their enthusiasm for camel beauty into organized camel beauty contests, the winners of which are worth gold (Bridi and De Man 2025). These events have now expanded far beyond traditional Bedouin communities, attracting a wide range of participants such as camel breeders, entrepreneurs and investors from across the Gulf region. Larger camel events such as the King Abdulaziz Camel Festival include both beauty (Mazayen) and racing competitions, attracting international participation and large investments.

Another new camel-based economy is emerging in the deserts of the world and beyond, through the evolution of the traditional trade caravans as a modern tourist caravan (Maitra et al. 2025). On a private scale, the concept of "the camel for the weekend" is arising in countries of the Arabian Peninsula (Faye 2016b). This concept sees stressed city dwellers keeping a small herd of camels cared for by a shepherd on the outskirts of cities, which the owners come to spend the weekend with in a comfortably furnished Bedouin tent as a way of reconnecting with their pastoral past.

1.10 Modern Transformations: Vision 2030, Tourism and National Branding

Long regarded as an animal of the past, the camel has entered modernity, especially in the Middle East (Faye 2016a). Camels serve as an important element in post-oil strategies across the GCC, with the Kingdom of Saudi Arabia, the UAE and Qatar investing heavily in developing their value. In their national visions, heritage is designed to play a central role in identity formation, in national branding and as a tool for diversified economies. Local cultural symbols, such the camel embed soft power, where they link the community to their ancestors and act as a safeguard for the national identity. The sociologist Anthony Giddens (1991: 39) provides a theoretical framework for understanding how the 2030 national visions reinterpret tradition. Giddens presents the concept of reflexive modernity, in which "social practices are constantly examined and reformed in the light of incoming information" from scientists or experts. Innovations in pastoralism, therefore, demonstrate a focus on adaptation rather than a decline in practices (Köhler-Rollefson 1993).

The UNESCO Doha Office report entitled "The Camel: From Tradition to Modern Times" (Breulmann et al. 2007) highlights the camel's role in combating desertification and maintaining ecological balance in drylands. The camel also has the potential to contribute to climate-resilient, low-carbon development through camel dairy innovation that is well suited to arid regions (Faye 2016a). The dairy industry within the GCC alone was valued at USD 765 million in 2024 and is projected to reach USD 1.08 billion by 2033 (IMARC 2024). Similarly, the use of camel wool in fashion, compared to other textiles such as cotton or synthetic fibres, saves vast amounts of water and requires less energy to produce (Ministry of Culture 2024).

To realize this potential, Saudi Arabia, the UAE, Qatar and other Gulf countries are leading initiatives in research and development for stretching the camel's value chain while promoting the reinterpreted camel tradition. For example, Saudi Arabia provides a grant for Camel Studies which covers the historical, cultural, social, economic, environmental and health aspects of camels. The country is also investing in sequencing the camel genome (Ministry of Culture 2024). The UAE is investing in camel cloning using biotechnological innovations (Leigh-Hewitson 2023) and Qatar has organized the world's first camel technology conference, "Camel Tech", in 2023 (Ministry of Culture 2024). These same countries are also promoting the camel tradition by a series of events, media, festivals, races and contests. For example, Saudi Arabia designated 2024 as the *year of the camel*. The film industry in the Kingdom produced six Saudi films centred around the camel such as ناقة *The She-Camel* and هجّان *Hajjan*, along with modern Saudi visual artists who created art pieces influenced by the unique features and aesthetics of the camel (Ministry of Culture 2024).

These countries also organize camel festivals, races and heritage markets where sport, technology and tourism come together. Organizing a series of events such as King Abdulaziz Camel Festival and the AlUla Camel Cup creates benefits that extend beyond cultural preservation to act as a tool for cultural diplomacy. For example, the King Abdulaziz Camel Festival attracts thousands of international tourists and fosters cultural exchange between breeders, experts and researchers from around the world. The festival also promotes innovation in the camel industry by showcasing new techniques in breeding, care and training (Ministry of Culture 2024). Camel events also contribute to the economic development by increasing the demand for goods and services associated with camel breeding, creating new job opportunities in fields like veterinary care and logistics. These initiatives represent forms of heritage diplomacy, where cultural authenticity is mobilized as a diplomatic and economic resource (Allen 2023). This reinvention of animal symbolism under postcolonial regimes demonstrates how heritage operates as a vehicle of modernization, linking national identity to global engagement (Appuhn 2005).

1.11 Reframing the Camel: From Orientalist Symbol to Modern Agency

Camels, both dromedaries and Bactrians, share a long common history with humans. They shared the harsh conditions of life in the deserts together and jointly contributed, particularly through caravan activities, to the emergence of great civilizations from the Neolithic period until the late Middle Age. Through this long-distance mobility, they have not only exchanged goods, but also ideas and cultures and have contributed greatly to the spread of Islam. At the turn of the last century, one might have thought that the end of caravan activity, the motorization of desert spaces, the oil revolution and rapid urbanization would signal decline for a species competing with other more prolific and productive domestic animals. However, this view does not account for the resilience of a species that seems to be adapting to the constraints of a modern world. It is likely that the evolving history of the large camelids still has surprises in store for us.

Through the lens of Giddens' (1984, 1991) concept of reflexive modernity and Hobsbawm and Ranger's (2012) concept of invented tradition, the camel's reemergence in biotechnology, eco-tourism and policy frameworks illustrates how cultural symbols evolve into instruments of modernization and global engagement. Once a signifier of exotic immobility in Orientalist art, the camel now embodies biotechnological potential, climate adaptation and heritage diplomacy that marks a paradigm shift from representation to participation. The camel, once a marker of desert marginality, now stands at the intersection of biotechnology, tourism and sustainability.

References

Agut-Labordère D, Redon B (2020) Dromadaires et chameaux de l'Asie centrale au Nil dans les mondes anciens (IVe millénaire av. J.-C.-premiers siècles de notre ère). In: Les vaisseaux du désert et des steppes: les camélidés dans l'Antiquité (Camelus dromedarius et Camelus bactrianus). Archéologie(s) 2. MOM Éditions, Lyon, pp 9–20

Ait El Alia O, Zine-Eddine Y, Chaji S, Boukrouh S, Boutoial K, Faye B (2025) Global camel milk industry: a comprehensive overview of production, consumption trends, market evolution, and value chain efficiency. Small Rumin Res 243:107441. https://doi.org/10.1016/j.smallrumres.2025.107441

Alary V, Faye B (2016) The camel chains in East Africa—importance of gaps between the data and the apparent reality. J Camelid Sci 9:1–22

Al-Hasani MA (ed) (2012) Mu'allaqat Labid ibn Rabi'ah, 1st edn. Abu Dhabi Tourism & Culture Authority, National Library, Abu Dhabi

Allen C (2023) Soft power and heritage economies in Saudi Vision 2030. Middle East Policy 30(1):77–95

Almathen F, Charruau P, Mohandesan E, Mwaharo JM, Orozco-Terwengel P, Pitt D, Abdussamad A, Uerpmann M, Uerpmann HP, Decupere B, Magee P, Alnaqeeb M, Salim B, Raziq B, Dessie T, Abdelhadi O, Banabazi M, El-Eknah M, Walzer C, Faye B, Hofreiter M, Peters J, Hanotte O, Burger P (2016) Ancient and modern DNA reveal dynamics of domestication and cross-continental dispersal of the dromedary. PNAS 113(24):6707–6712. https://doi.org/10.1073/pnas.1519508113

Alwahaibi I (2024) God's gift to the Bedouin: the camel traditions of the People of Oman. Nomad Publishing. ISBN: 9781-917045-06-3

Alwahaibi I, Dauletova V, Wels H (2024) Camel ownership as a passage to adulthood in Omani Bedouin culture. Anthrozoös 37(2):197–211

Alwahaibi I, Khan L, Dauletova V (2025) So much more than the ship of the deserts! How naming of camels preserves identity, cultural heritage, and well-being. Soc Identities 31(2):169–186. https://doi.org/10.1080/13504630.2024.2439793

Appadurai A (ed) (1986) The social life of things: commodities in cultural perspective. Cambridge University Press, Cambridge

Appuhn K (2005) Beasts of many burdens. Environ Hist 10(4):667–669

Arberry AJ (1957) The seven odes. George Allen & Unwin, London

Asiimwe R, Ainembabazi JH, Egeru A, Isoto R, Aleper DK, Namaalwa J, Diiro GM (2020) The role of camel production on household resilience to droughts in pastoral and agro-pastoral households in Uganda. Pastoralism 10(1):5. https://doi.org/10.1186/s13570-020-0160-x

Austen RA (2024) Trans-Saharan slave trade. In: Oxford research encyclopedia of African history. https://doi.org/10.1093/acrefore/9780190277734.013.838

Barnard H (2012) From Adam to Alexander (500,000–2,500 years ago). In: Barnard H, Duistermaat K (eds) The history of the peoples of the Eastern Desert. Regents of the University of California, Oakland

Boukrouh S, Ait El Alia O, Faye B (2025) Worldwide camel meat and products: an extensive analysis of production, consumption patterns, market evolution, and supply chain effectiveness. Meat Sci 228:109882. https://doi.org/10.1016/j.meatsci.2025.109882

Breulmann M, Böer B, Wernery U, Wernery R, El-Shaer H, Al-Hadrami GA, Gallacher J, Peacock J, Chaudhary SA, Brown G, Norton J (2007) The camel from tradition to modern times: a proposal towards combating desertification via the establishment of camel farms based on fodder production from indigenous plants and halophytes. UNESCO, Doha

Bridi RM, De Man A (2025) From heritage to high-tech: the impact of technology on camels in the United Arab Emirates. Heritage 8(5):165. https://doi.org/10.3390/heritage8050165

Burger P, Ciani E, Faye B (2019) Old World camels in a modern world—a balancing act between conservation and genetic improvement. Anim Genet 50:598–612. https://doi.org/10.1111/age.12858

Choudhary M (2019) Changing area, shrinking spaces and struggling species: history of camels. Hist Sociol South Asia 13(2):93–111. https://doi.org/10.1177/2230807521 1043285

Cousin L (2020) Le dromadaire (Camelus dromedarius) dans le Proche-Orient ancien au Ier millénaire av. J.-C.: présentation de la documentation épigraphique et retour sur quelques documents iconographiques. In: Les vaisseaux du désert et des steppes: les camélidés dans l'Antiquité (Camelus dromedarius et Camelus bactrianus). Archéologie(s) 2. MOM Éditions, Lyon, pp 65–80

Cvikel D, Goren H (2008) Where are Bonaparte's siege cannon? An episode in the Egyptian campaign. Mediterr Hist Rev 23(2):129–142. https://doi.org/10.1080/09518960802528803

Davis DK (2007) Imperialism, orientalism, and the environment in the Middle East: history, policy, power, and practice. In: Burke E, Davis DK (eds) Environmental imaginaries of the Middle East and North Africa. Ohio University Press, Athens

Devic M (1898) Antar: poème héroïque arabe/d'après la traduction de Marcel Devic; illustrations en couleurs de E. Dinet. Available at https://catalogue.bnf.fr/ark:/12148/cb332429289

Digard JP (2009) L'homme et les animaux domestiques. Fayard, Paris

Ding Y, Zhu J, Ma J, Festa M (2026) Following camels between bone and culture: camel–human interactions in China from the Neolithic to the late Imperial period. Animals 16(5):772. https://doi.org/10.3390/ani16050772

Dioli M (2020) Dromedary (*Camelus dromedarius*) and Bactrian camel (*Camelus bactrianus*) crossbreeding husbandry practices in Turkey and Kazakhstan: an in-depth review. Pastor Res Policy Pract 10:6. https://doi.org/10.1186/s13570-020-0159-3

Eades D, Watson JCE, Ahmad al-Mahri M (2013) Camel culture and camel terminology among the Omani Bedouin. J Semit Stud 58(1):169–186. https://doi.org/10.1093/jss/fgs044

Faye B (2013) 'La dune et la bosse', in 'Vivre le sable ! Corps, matière et sociétés'. Technol Cult 61(2):60–75. https://doi.org/10.4000/tc.7222

Faye B (2016a) The camel, new challenges for a sustainable development. Trop Anim Health Prod 48:689–692. https://doi.org/10.1007/s11250-016-0995-8

Faye B (2016b) Des dromadaires et des hommes au Moyen-Orient: identité et modernité. Anthropol Middle East 11(1):51–65

Faye B (2020) How many large camelids in the world? A synthetic analysis of the world camel demographic changes. Pastor Res Policy Pract 10:25. https://doi.org/10.1186/s13570-20-00176-z

Faye B (2022) Is the camel conquering the world? Anim Front 12(4):8–16. https://doi.org/10.1093/af/vfac034

Faye B, Corniaux C (2024) Le lait de chamelle au risque de l'économie politique: de l'économie du don à l'économie marchande. Rev Elev Med Vet Pays Trop 77:1–7. https://doi.org/10.19182/remvt.37263
Faye B, Konuspayeva G (2012) The encounter between Bactrian and dromedary camels in Central Asia. In: Knoll EM, Burger P (eds) Camels in Asia and North Africa: interdisciplinary perspectives on their past and present significance. Austrian Academy of Sciences Press, Vienna, pp 27–33
Faye B, Konuspayeva G, Messad S, Loiseau G (2008) Discriminant milk components of Bactrian camel (*Camelus bactrianus*), dromedary (*Camelus dromedarius*) and hybrids. Dairy Sci Technol 88(6):607–617. https://doi.org/10.1051/dst:2008008
Faye B, Abdallah H, Almathen F, Harzallah B, Al-Mutairi S (2011) Camel biodiversity: camel phenotypes in the Kingdom of Saudi Arabia. FAO/Camel Breeding, Protection and Improvement Center, Project UTF/SAU/021/SAU, Riyadh
Faye B, Chaibou M, Vias G (2012) Integrated impact of climate change and socioeconomic development on the evolution of camel farming systems. Br J Environ Clim Change 2(3):227–244. https://doi.org/10.13140/2.1.2172.4164
Faye B, Nehmé L, Norris J (2024) Les représentations et les zoonymes du dromadaire Camelus dromedarius Linnaeus (1758) dans l'art rupestre et l'épigraphie le long de la route caravanière nabatéenne entre Hégra et Taymāʾ (Arabie Saoudite). Anthropozoologica 59(12):183–204. https://doi.org/10.5252/anthropozoologica2024v59a12
Finkelstein I (1988) Arabian trade and socio-political conditions in the Negev in the twelfth–eleventh centuries BCE. J Near East Stud 47(4):241–252. https://doi.org/10.1086/373318
Fitak RR, Mohandesan E, Corander J, Yadamsuren A, Chuluunbat B, Abdelhadi O, Raziq A, Nagy P, Walzer C, Faye B, Burger PA (2020) Genomic signatures of domestication in Old World camels. Commun Biol 3(1):316. https://doi.org/10.1038/s42003-020-1039-5
Franklin K (2024) Archaeology of the Silk Road: challenges of scale and storytelling. J Archaeol Res 32(2):263–308. https://doi.org/10.1007/s10814-023-09188-w
Frankopan P (2015) The silk roads: a new history of the world. Bloomsbury, London
Giddens A (1984) The constitution of society: outline of the theory of structuration. University of California Press, Berkeley
Giddens A (1991) The consequences of modernity. Polity Press, Cambridge
Guagnin M, Charloux G, AlSharekh AM, Crassard R, Hilbert YH, Andreae MO, AlAmri A, Preusser F, Dubois F, Burgos F, Flohr P, Mora P, AlQaeed A, AlAli Y (2022) Life-sized Neolithic camel sculptures in Arabia: a scientific assessment of the craftsmanship and age of the Camel Site reliefs. J Archaeol Sci Rep 42:103165. https://doi.org/10.1016/j.jasrep.2021.103165
Haimi Y, Rosen SA (2011) The organization of space at the Camel Site. In: Rosen SA, Abadi-Reiss Y, Avni Y, Mayer DBY, Gottesman M, Hermon S, Haimi Y, Ryan P, Saidel BA, Segal I, Tykot RH, Vardi J, Weisskopf A (eds) An Investigation into early desert pastoralism: excavations at the Camel Site, Negev. Cotsen Institute of Archaeology Press at UCLA, Los Angeles. https://doi.org/10.2307/j.ctvhhhfv6
Hare J (2015) The lost camels of tartary: a quest into forbidden China. Abacus, London
Heide M, Peters J (2021) Camels in the biblical world. Pennsylvania State University Press, University Park
Hobsbawm E, Ranger T (eds) (2012) The invention of tradition. Cambridge University Press, Cambridge
Ibn Hishām, ʻAbd al-Malik (1937) Sīrat al-Nabī, vol 1, ʻAbd al-Ḥamīd (ed). Maṭbaʻat Ḥijāzī, M.M. Cairo
Ikram S (2017) Animals in ancient Egyptian religion. In: The Oxford handbook of zooarchaeology. Oxford University Press, Oxford
IMARC Group (2024) GCC camel dairy market report and forecast 2024–2033 [Report ID: SR112025A956]. Available at https://www.imarcgroup.com/gcc-camel-dairy-market
Insoll T (1996) The archaeology of Islam in sub-Saharan Africa. J World Prehist 10(4):439–504. https://doi.org/10.1007/BF02221202

Ji R, Cui P, Ding F, Geng J, Gao H, Zhang H, Yu J, Hu S, Meng H (2009) Monophyletic origin of domestic Bactrian camel (*Camelus bactrianus*) and its evolutionary relationship with the extant wild camel (*Camelus bactrianus ferus*). Anim Genet 40(4):377–382. https://doi.org/10.1111/j.1365-2052.2008.01848.x

Jung M (1994) On representations of camels and camel-riders in the rock art of North Yemen. East West 44(2/4):231–248

Jung SJ, Davies GR, Ganssen GM, Kroon D (2004) Stepwise Holocene aridification in NE Africa deduced from dust-borne radiogenic isotope records. Earth Planet Sci Lett 221(1–4):27–37. https://doi.org/10.1016/S0012-821X(04)00095-0

Kamel Moussa N (2019) The contribution of the domesticated camel and advanced irrigation techniques to the Iron Age economy and settlement patterns of the Oman Peninsula and Arabia. J Gen Union Arab Archeol 4(1):87–115

Khan M (2022) Camel: the symbol of honor, pride and love of Arab Bedouins. In J Archaeol 10(1):20–23

Köhler-Rollefson I (1993) Camels and camel pastoralism in Arabia. Biblic Archaeol 56(4):180–188

Konuspayeva G, Faye B, Duteurtre G (2022) Online camel milk trade: new players, new markets (update). Rev Elev Med Vet Pays Trop 75(4):95–101. https://doi.org/10.19182/remvt.37041

Lancaster W (1981) The Rwala Bedouin Today. Cambridge: Cambridge University Press

Leigh-Hewitson N (2023) Why camel cloning is big business in Dubai. CNN, March. Available at https://edition.cnn.com/travel/article/camel-cloning-dubai-spc-intl-scn

Le Magoariec R (2025) The camel in the shadow of Vision 2030. Arabian Gulf Strategic Institute (AGSI). Retrieved from https://agsi.org/analysis/the-camel-in-the-shadow-of-vision-2030/

Maitra A, Iglesias Pastrana C, Faye B, Burger P, Ciani E (2025) Breeding of camels in Europe: between continuity and innovation. Appl Sci 15(3):1644. https://doi.org/10.3390/app15031644

Meeker ME (1979) Literature and Violence in North Arabia. Cambridge: Cambridge University Press

Mighty T (2023) How the Red Army used camels to fight the Nazis at Stalingrad and on to Berlin. We Are the Mighty. Available at https://www.wearethemighty.com/articles/red-army-camels/

Ming L, Yuan L, Yi L, Ding G, Hasi S, Chen G, Jambl T, Hedayat-Evright N, Batmunkh M, Badmaevna MK, Gan-Erdene T, Batskh TS, Zhang W, Zulipikaer A, Erdemt H, Natyrov A, Mamay P, Narenbatu, Meng G, Narangerel C, Khongorzul O, He J, Hai L, Lin W, Sirendalai, Sarentuya, Aiyisi, Li Y, Wang Z, Jirimutu (2019) Whole-genome sequencing of 128 camels across Asia provides insights into origin and migration of domestic Bactrian camels. bioRxiv:656231. https://doi.org/10.1101/656231

Ministry of Culture, Kingdom of Saudi Arabia (2024) Camel-economy report. Available at https://share.google/MU0ACGgA6fy352egy

Nasir al-Din MM (2002) Diwan of Ṭarafah ibn al-ʻAbd. Dar al-Kutub al-ʻIlmiyyah, Beirut

Perez J (2003) Les compagnies méharistes au Levant, 1921-1941. Revue historique des Armées 233(4):79–96. https://doi.org/10.3406/rharm.2003.5545

Pigière F, Henrotay D (2012) Camels in the northern provinces of the Roman Empire. J Archaeol Sci 39(5):1531–1539. https://doi.org/10.1016/j.jas.2011.11.014

Polanyi K (2001) The great transformation. Farrar & Rinehart, New York

Rosen SA (2011) The camel site in perspective. In: An investigation into early desert pastoralism, UCLA Cotsen Institute of Archaeology Press Monographs, No. 69. Cotsen Institute of Archaeology Press, Los Angeles, pp 199–215

Rosen SA, Saidel BA (2010) The camel and the tent: an exploration of technological change among early pastoralists. J Near East Stud 69(1):63–77

Said EW (1978) Orientalism. Pantheon, New York

Sala R (2017) The domestication of camel in the literary, archaeological and petroglyph records. J Arid Land Stud 26(4):205–211. https://doi.org/10.14976/jals.26.4_205

Salma MB (2022) Economic relations between the regions of the Arabian Peninsula in the ancient history. Int J Econ Perform 5(2):170–194. Available at https://www.asjp.cerist.dz/en/article/206519

Scheele J (2024) Trans-Saharan trade. In: Oxford research encyclopedia of African history. https://doi.org/10.1093/acrefore/9780190277734.013.176
Schmitt J (2025) Tasty animals: animal edibility in premodern Arabic zoological writings. Alif J Comp Poet 45(1–4):37–57. https://doi.org/10.1163/3050760x-20240014
Scott JC (1976) The moral economy of the peasant: rebellion and subsistence in Southeast Asia. Yale University Press, New Haven
Sells M (1986) The "Muʿallaqa" of Ṭarafa. J Arab Lit 17:21–33
Suska-Malawska M, Kot M, Gręzak A, Mętrak M, Khudjanazarov M, Szymczak K (2022) Potential impact of Holocene climate changes on camel breeding practices of Neolithic pastoralists in the Central Asian drylands: a preliminary assessment. The Holocene 32(11):1132–1143. https://doi.org/10.1177/09596836221111428
Tefera M, Abebe G (2012) The camel in Ethiopia. Ethiopian Veterinary Association, Addis Ababa
Uerpmann H, Uerpmann M (2002) The appearance of the domestic camel in southeastern Arabia. J Oman Stud 12:235–260
Vigne JD (2025) Aux origines de la domestication animale. Quae, Versailles
Volpato G, King EG (2019) From cattle to camels: trajectories of livelihood adaptation and social-ecological resilience in a Kenyan pastoralist community. Reg Environ Chang 19(3):849–865. https://doi.org/10.1007/s10113-018-1438-z
Wilson RT (2016) The one-humped camel in the Anglo-Egyptian military campaigns in Sudan, 1885–1926. J Camel Pract Res 23(2):193–205. https://doi.org/10.5958/2277-8934.2016.00034.5
Wilson RT (2017) The one-humped camel in Uganda. J Camel Pract Res 24(1):1–7. https://doi.org/10.5958/2277-8934.2017.00001.7
Yam BAZ, Khomeiri M (2015) Introduction to camel origin, history, raising, characteristics, and wool, hair and skin: a review. Res J Agric Environ Manag 4(11):496–508
Yilmaz O (2017) History of camel wrestling in Turkey. Int J Livest Res 7(10):235–239. https://doi.org/10.5455/ijlr.20171021041317

[illegible]

Camels and the Concept of One Health 2

Pamela Burger and Bernard Faye

Abstract

This chapter outlines the evolution of the One Health concept from early veterinary approaches to a modern, integrated framework linking human, animal, and ecosystem health. It highlights key principles such as transdisciplinarity and sustainability. Additionally, it examines their relevance to camels in the context of changing farming systems, disease dynamics, and zoonotic risks, particularly MERS-CoV. The chapter emphasises the need for holistic research, improved management, and sustainable development within camel production systems.

Keywords

Ecopathology · Agroecosystem health · Camel farming · Sustainable development · Animal welfare

2.1 Paving the Way for the One Health Concept

The concept of One Health was developed within the scientific community in the early 2000s in response to the need for a holistic approach to deal with new diseases that are common to both humans and animals. However, holistic approaches to health, particularly in the field of veterinary medicine, had been established long before this. The development of intensive livestock farming in Europe after the Second World War, alongside social, technical, and economic changes, led to a

P. Burger (✉)
Research Institute of Wildlife Ecology, University of Veterinary Medicine, Vienna, Austria
e-mail: pamela.burger@vetmeduni.ac.at

B. Faye
CIRAD-SELMET (French Agricultural Research Centre for International Development), Montpellier, France

S. A. Gordine et al. (eds.), *Camels and One Health*,
https://doi.org/10.1007/978-981-92-2246-9_2

"productivist" management system that aimed to maximise output through the efficient use of resources (Cronin et al. 2014). The "Pasteurian" approach to disease management, based on a linear relationship between disease causation (microorganism, parasite, toxic, deficiency) and its treatment or prevention, successfully controlled many diseases within this developing productivist system. At this time, many fatal or economically devastating diseases declined in many western countries. These included tuberculosis, rabies, foot-and-mouth disease, and brucellosis in cattle (Moutou 1990; Aubert et al. 1994; Benet 1994); Aujeszky's disease, influenza, and swine erysipelas in pigs (Van Oirschot 1994); and avian influenza and Newcastle diseases in poultry (Alexander 1995).

With the intensification of livestock production, however, new, multifactorial pathologies began appearing on farms. It was no longer possible to attribute disease to a single cause, but rather "a spider web of causes" (Schwabe 1993). Health disorders like mastitis, infertility, lameness, behavioural problems, lack of productivity, and loss of appetite could not be explained by the traditional linear approach. In this context, the concept of "ecopathology" was developed, followed by the concept of "animal hygiene". However, these two ideas were based on different approaches. The field of animal hygiene aims to prevent disease by promoting animal health and welfare (definition of the International Society of Animal Hygiene [ISAH]). This scientific approach is analytic and experimental, addressing areas such as the impact of animal housing (Goodger et al. 1994), the microbiome (Saltijeral et al. 1994), feeding (Sazzad and Barreto 1994), and breeding management (Böhm 1994) on the health and productivity of farm animals.

The concept of ecopathology was precisely a much more holistic approach than the classical Pasteurian approach (one disease, one pathogen agent). Ecopathology is the "study in the livestock farms of the factors which, in interrelationships in the biological, physical, human and economical of the animals, are susceptible to induce a pathological state or to affect their productivity and the quality of their products" (Ganière et al. 1991). The epistemological choices of this approach are the complexity of the interactions between the risk factors, the globality of the farming system (including the farmer regarded as the "pilot" of the system), and the study in real environment, excluding the experimental trials (Landais 1991). The ecopathological approach therefore consists of integrating animal health into the functioning of the livestock system (Landais 1992) and, therefore, considering the disease as a disorder of this system, which includes the decisions of the breeder, his management practices, his technical and economic constraints, his resources, and the parameters of the biological environment with which he is confronted (Bernues et al. 1994; Faye 1995).

This ecopathology concept was first applied in fish farming systems (Tuffery et al. 1971), followed by pig (Tillon 1987; Madec and Tillon 1988), dairy cattle (Barnouin and Brochart 1981; Faye and Barnouin 1987), and poultry farming (Sigognault et al. 1994), and later transferred to tropical countries (Faye and Lefèvre 1993; Faye et al. 1994; Bengoumi et al. 2004). The ecopathology concept was taken up by professional organisations as part of the creation of the Animal Ecopathology Center (Calavas and Rosner 1997). This centre was active for several years, achieving many ecopathological surveys in sheep, goat, and meat cattle farming. For camels, some initiatives were developed in North Africa and in Saudi Arabia (see Fig. 2.1).

An Ecopathological Diagnosis on a Camel Farm

The Problem

A large camel farm (1200 head) in northern Saudi Arabia suffered a high mortality rate (more than 10%) of its adult female camels during the winter. The camels were noted to be significantly emaciated. In response to this, the vet took blood samples, expecting to find an infectious cause for these deaths. Alternative potential causes were also investigated by conducting an "eco-pathological" visit of the farm.

The Findings

The farm was composed of dromedary camels that had been transferred the previous July from Jeddah, which is located in the south-west part of the country, where the winters are mild and humid. These camels arrived in the north of the country, spending their first winter in very cold (sometimes below 0 °C) and dry conditions. The vet observed that almost all the camels were affected by mange (significant hair loss). All the camels were housed in small enclosures in very high density (groups of around 100 camels) with only 1 or 2 feeders available. The feed was composed of good alfalfa hay, but, due to the lack of feeders, was largely wasted because it was scattered on the ground and soiled by animal urine and excrement. The ecopathological assessment is summarised in the figure below:

In response to these findings, the following treatment recommendations were made: (1) decrease the stocking density in paddocks, (2) allow the camels spend at least 2 h each day in natural desert pastures, (3) treat the animals for mange, and (4) introduce more feeders. After the plan was implemented, the problem was solved within 3 months.

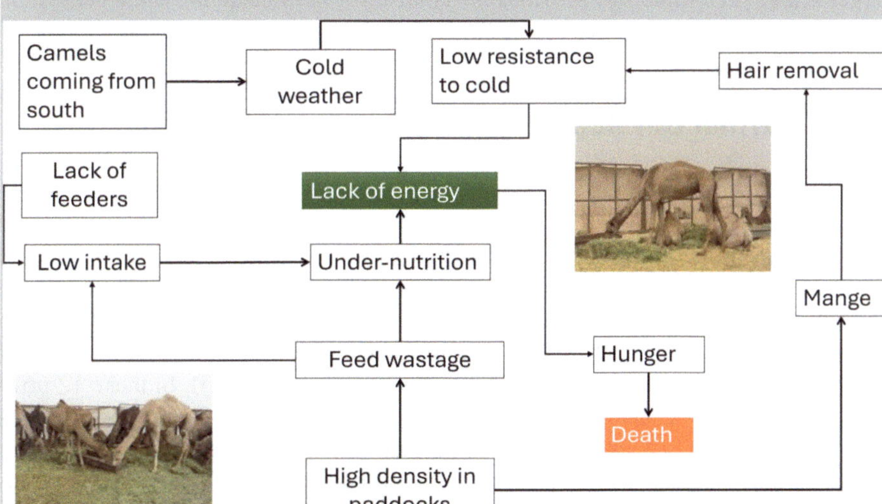

Fig. 2.1 An ecopathological diagnosis on a camel farm

Despite the potential of the ecopathology approach for understanding the relationship between multifactorial risk factors and disease, its use has declined since the 2000s for various reasons. In his critical analysis of the ecopathology approach, Fortané (2023) concluded that ecopathology has failed because (i) it contributes to "*the adaptation of animal health management to the intensification process*" when a social demand for another approach was emerging and (ii) the appropriation of ecopathological knowledge and techniques by farmers and other agricultural actors "*contributes to preventing veterinary medicine from breaking away from its clinical and infectious straitjacket, and from developing agricultural expertise*".

Another concept was emerging towards the end of the 1990s—the concept of "Agroecosystem Health" (Waltner-Toews 1996). An agroecosystem (AES) is defined as plants and animals interacting with their physical and chemical environments to produce food, fibre, and other agricultural products for humans (Conway 1987). AES include people, domestic plants and animals, and living and non-living components of the soil, as well as wildlife and natural vegetation. An AES is a collective human creation aiming to achieve nutritional and socio-economic goals. Thus, it has not only an environmental dimension but also considers socio-economics and public health. In the AES entity, farms, watersheds, and agricultural spaces are inextricably linked in a socio-ecological web (Giampetro 1994). However, under pressure (whether external or internal), the AES can fail to address problems within the system, leading to negative impacts such as environmental degradation, emerging diseases, and economic stagnation (Gallopin 1995). The transition from the concept of ecopathology to AES corresponded to changes of scale: from animal disease management in the context of "Farmer, Herd, Resource" to disorder management in the context of "human population, animal population, AES resources". If in ecopathology the herd was considered to be a community of microorganisms and animals and the farm was a human-made ecosystem, the concept of AES expanded the approach by including agricultural, health, epidemiologic, and social sciences at a larger scale. "*Ecopathology provides the context for herd medicine while ecosystem health provides the context of ecopathology*" (Faye et al. 1999).

2.2 Further Development of the One Health Concept

In 2004, a symposium at the Rockefeller University chaired by the Wildlife Conservation Society (WCS; www.wcs.org) developed 12 recommendations "for establishing a more holistic approach to preventing epidemic/zoonotic diseases and for maintaining ecosystem integrity for the benefit of humans, their domesticated animals and foundational biodiversity". These are known as the Manhattan Principles on One World, One Health (https://oneworldonehealth.wcs.org/about-us/mission/the-manhattan-principles.aspx; accessed on 12.12.2025). In these 12 principles, the world's leaders, the global health community, and scientific institutions are called upon to recognise the essential link between human and animal (both domestic and wildlife) health, as well as the importance of biodiversity for maintaining functioning ecosystems. For the 15 years following the announcement of these recommendations, efforts have focussed on reconnecting the components of human, animal, and ecosystem health. This has achieved mixed success as shown by the COVID-19 pandemic, which was foretold, but unfortunately not prevented.

To revisit the Manhattan Principles in light of newly emerging disease, increased climate change, and biodiversity loss, the German Federal Foreign Office and WCS launched a conference in 2019 to bring together the One Health community with researchers advocating similar concepts such as Ecopathology, Eco Health, and Planetary Health. Hence, experts from diverse fields prepared a Call to Action, the "Berlin Principles of One Health". This aims to reintegrate the components of human, animal, and ecosystem health while addressing other pressing issues, like newly emerging pathogens and their spillover, climate change, and antimicrobial resistance. These ten Berlin Principles (see Fig. 2.2) aim to "overcome the most

The Berlin principles of One-Health

1. Recognize and take action to retain the essential health links between humans, wildlife, domesticated animals and plants, and all nature; and ensure the conservation and protection of biodiversity which, interwoven with intact and functional ecosystems, provides the critical foundational infrastructure of life, health, and well-being on our planet;
2. Take action to develop strong institutions that integrate understanding of human and animal health with the health of the environment, and invest in the translation of robust science-based knowledge into policy and practice;
3. Take action to combat the current climate crisis, which is creating new severe threats to human, animal, and environmental health, and exacerbating existing challenges;
4. Recognize that decisions regarding the use of land, air, sea, and freshwater directly impact health and well-being of humans, animals, and ecosystems and that alterations in ecosystems paired with decreased resilience generate shifts in communicable and non-communicable disease emergence, exacerbation and spread; and take action to eliminate or mitigate these impacts ;
5. Devise adaptive, holistic, and forward-looking approaches to the detection, prevention, monitoring, control, and mitigation of emerging/resurging diseases and exacerbating communicable and non-communicable diseases, that incorporate the complex interconnections among species, ecosystems, and human society, while accounting fully for harmful economic drivers, and perverse subsidies;
6. Take action to meaningfully integrate biodiversity conservation perspectives and human health and well-being when developing solutions for communicable and non-communicable disease threats;
7. Increase cross-sectoral investment in the global human, livestock, wildlife, plant, and ecosystem health infrastructure and international funding mechanisms for the protection of ecosystems, commensurate with the serious nature of emerging/resurging and exacerbating communicable and non-communicable disease threats to life on our planet;
8. Enhance capacity for cross-sectoral and trans-disciplinary health surveillance and clear, timely information-sharing to improve coordination of responses among governments and non-governmental organizations, health, academia and other institutions, the private sector and other stakeholders;
9. Form participatory, collaborative relationships among governments, NGOs, Indigenous Peoples, and local communities while strengthening the public sector to meet the challenges of global health and biodiversity conservation;
10. Invest in educating and raising awareness for global citizenship and holistic planetary health approaches among children and adults in schools, communities, and universities while also influencing policy processes to increase recognition that human health ultimately depends on ecosystem integrity and a healthy planet.

Fig. 2.2 The Berlin principles of One Health (Gruetzmachet et al. 2021)

important systemic policy and societal barriers to transform and enable the world community to tackle increasing health threats at a global scale" (Gruetzmacher et al. 2021).

In 2021 four global partners established the interdisciplinary *One Health High-Level Expert Panel (OHHLEP)* (https://www.who.int/groups/one-health-high-level-expert-panel) to enhance their cross-sectoral collaboration. The panel consists of:

- The Food and Agriculture Organization (FAO)
- The World Organization for Animal Health (OIE)
- The United Nations Environment Programme (UNEP)
- The World Health Organization (WHO)

The first priority of the OHHLEP was to consolidate the different perceptions of One Health and to develop a consensus for a working definition as a solid basis to support a common understanding among the panel members and the partner organisations.

2.3 Current Definition of One Health and Its Key Underlying Principles

One Health is an integrated, unifying approach that aims to sustainably balance and optimise the health of people, animals, and ecosystems. It recognises that the health of humans, domestic and wild animals, plants, and the wider environment (including ecosystems) is closely linked and interdependent (One Health High-Level Expert Panel et al. 2022).

To bring One Health from theory to practice by implementing this definition, the "4 Cs" are highlighted: communication, coordination, collaboration, and capacity building (Fig. 2.3). The key underlying principles of One Health include:

1. *Equity* between sectors and disciplines.
2. Sociopolitical and multicultural *parity* (the doctrine that all people are equal and deserve equal rights and opportunities) and inclusion and engagement of communities and marginalised voices.
3. Socioecological *equilibrium* that seeks a harmonious balance between human–animal–environment interactions and acknowledges the importance of biodiversity, access to sufficient natural space and resources, and the intrinsic value of all living things within the ecosystem.
4. *Stewardship* and the responsibility of humans to change behaviour and adopt sustainable solutions that recognise the importance of animal welfare and the integrity of the whole ecosystem, thus securing the well-being of current and future generations.
5. *Transdisciplinarity* and multisectoral collaboration, which includes all relevant disciplines, both modern and traditional forms of knowledge and a broad representative array of perspectives.

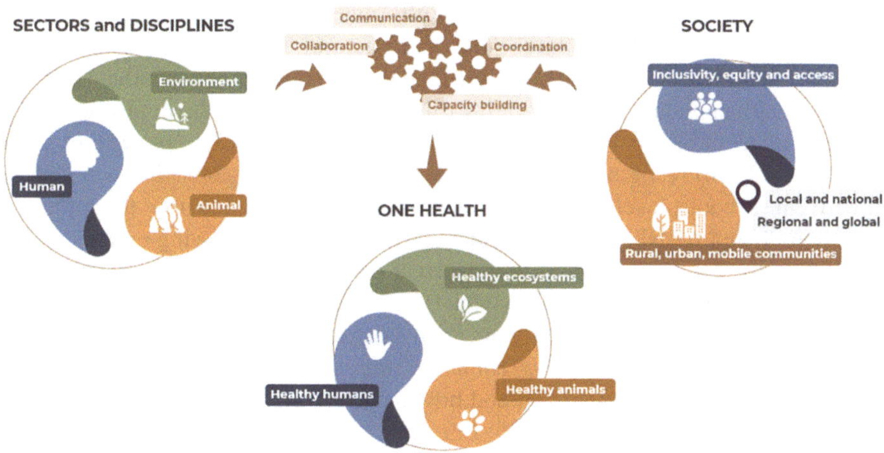

Fig. 2.3 A schematic overview of One Health including its key underlying principles and 4 Cs for effective implementation (One Health High-Level Expert Panel et al. 2022). (Reprint of this figure made available under the Creative Commons CC0 public domain dedication)

This newly defined One Health concept should be perceived as an overarching concept to be brought into action by different stakeholders and sectors such as policymakers, food and water security, energy, animal, human, and ecosystem health. It is important to note that the One Health approach goes beyond zoonotic disease and antimicrobial resistance, but aims to address a broader concept of health by incorporating disease prevention and health improvement as well as detection, response, and recovery from epidemic/-zoonotic health crises. The collaboration between sectors and disciplines on regional, national, and international levels, including the knowledge of local communities, is imperative to improve health for humans, animals, and the ecosystem.

2.4 Further Development and a Systemic Approach of the One Health Concept

In light of today's challenges of a recent pandemic, biodiversity loss, and global climate change, we recognise that these issues are "systemic", concerning not only a part of society or a health sector but affecting all of us on the planet. Thus, we need to address the challenges in a systemic, holistic, and innovative way. The One Health concept is a holistic approach, which aims to sustainably balance and optimise the health of people, animals, and ecosystems (One Health High-Level Expert Panel et al. 2022).

The concept of Systems Science is a multi- and transdisciplinary approach that is principally concerned with the complexity of natural, social, and other systems. It can provide a theoretical and practical framework to develop sustainable solutions to One Health challenges by integrating different fields of cognitive, physiological, social, and technological systems (Burger 2024). *General Systems Theory* (von Bertalanffy 1956) recognised that specific principles like organisation, wholeness, directiveness, control, self-regulation, and differentiation are intrinsic to biological, behavioural, and social systems. As such, they are essential in living organisms and social groups. This new approach induced a paradigm shift from perceiving the world as consisting of independent objects towards an understanding that systems are adaptive and constantly changing, with interdependent components (Rosseau 2017). This new way of *Systems Thinking* aims to characterise the fundamental parts and components of a system and how they interact with each other (*Systems Dynamics*; Homer and Hirsch 2006; Leischow and Milstein 2006). For example, this might involve identifying key players and factors that play a role in health and disease in a human or animal population within their respective environments or ecosystems (One Health). To bridge the gap between theory and reality, *Systems Designs* applies multiple transdisciplinary approaches, putting scientific solutions into practice (Banathy 1996) (Fig. 2.4).

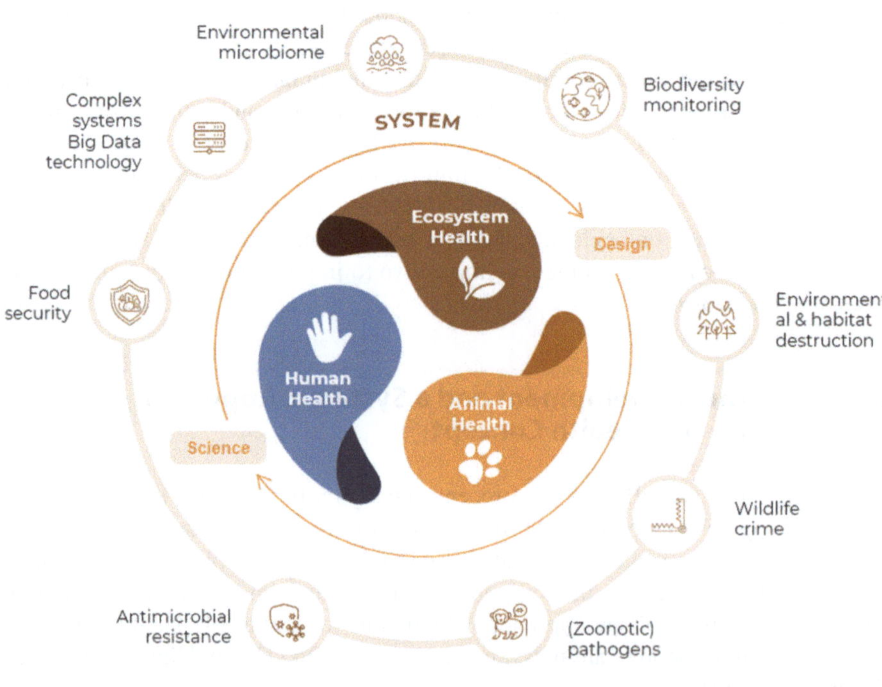

Fig. 2.4 Graphical summary of a Systems Science One Health approach, which uses real-time genomic and epidemiological big data from the three systems of human, animal, and ecosystem health and their interacting subsystems (displayed as circles) to develop sustainable solutions for challenges like zoonotic pathogens, antimicrobial resistance, biodiversity monitoring, and food security

To develop sustainable solutions in One Health, big data from the three systems of ecosystem, human, and animal health need to be integrated with a participative approach to follow the key principles of parity, equity, and transdisciplinarity. The development of new technologies in the field of genomics (Urban et al. 2023), proteomics, and the exposome (all environmental influences on an organism), together with complexity theory and network modelling, enables us to address complex challenges in a multidisciplinary way, respecting local knowledge and building capacity for One Health actions. Placing camel research in the middle of the One Health context will form new links between traditional camel management, sustainable transformation of camel production, and an omics-based health surveillance of animals, humans, and their environment.

2.5 Relevance of the One Health Concept for Camels

2.5.1 Changes to Farming Systems

Traditionally extensive, camel farming is changing in many countries in response to three main drivers: (i) the market integration of camel products in a globalised and more urbanised world. This is especially important for the camel milk sector which has evolved within a few decades from a "gift economy to market economy" (Faye and Corniaux 2024). (ii) Climatic changes are leading to a scarcity of natural resources. This has made it necessary to intensify food production, resulting in a global intensification of camel production processes (Nagy et al. 2022). (iii) There is growing international interest in camel products. The perceived dietetic and medicinal properties of camel milk and meat are driving increasing consumption of these products (Alavi et al. 2017; Kadim et al. 2022), and the market value of camel leather and wool is also increasing.

The intensification of camel farming is not limited to the implementation of "modern" dairy farms, feed-lots, or racing stables as can be observed in Gulf countries. Indeed, in the Sahel countries in Africa, recurrent episodes of drought since the 1960s have pushed pastoralists to move closer to cities, resulting in the peri-urbanisation of camel farming systems (Faye 2019; Amine et al. 2024). In these new constrained pastoral environments, camel farmers have better access to veterinary input and feed supplements, but high competition for space (for example, from horticulture or buildings). This encourages them to intensify their production systems, particularly as regards feeding (Faye 2016). The concentration of camel herds around cities in arid countries results in a more or less permanent sedentary lifestyle, and sometimes only the productive animals (such as lactating camels) remain close to urban areas, resulting in them being separated from the rest of the camel flock, which remains in pastoral areas (Faye 2018).

The intensification of camel farming can result in important changes in housing (indoor systems with limited mobility), supplementary feeding (with feeds rich in protein and energy or in rapidly-released energy for racing camels), use of reproduction technologies (artificial insemination, embryo transfer), milking management (machine milking, early weaning), and health care (intensive use of medicine). Such changes could have a significant impact on the health status of the animal. Among the changes induced by modern farming systems, the following have been observed:

- A change in feeding systems: When in pastoral ranges, the camel grazes a wide variety of plants, including halophytes, grasses, bushes, and trees, and spends more than 8 h/day grazing. In intensive farming, camels are settled in one place and are fed with a monotonous diet (for example, alfalfa hay with barley or commercial concentrates) that is distributed twice a day (Faye 2020). Consequently, these camels spend less time eating but increase their feed intake, from approximately 1.8–2.2 kg DM/100 kg LW (dry matter per 100 kilograms of live weight) up to 2.5–3.0 kg/DM/100 kg LW (Faye 2016).
- A decrease (and sometimes complete cessation) of mobility, leading to a lack of exercise in these animals that are accustomed to spending all day walking in search of food, traveling up to 30–40 km/day (Amsidder et al. 2021).
- The early weaning of the camel calf for milking. Conversely, in pastoral areas, the camel calf is rarely weaned before 6 months or even 12 months (Awoke and Ali 2015).
- The development of feed-lots for the rapid fattening (within 3 months) of young males before slaughtering. While, on average, the daily weight gain in pastoral areas is around 300–500 g/day and 500–800 g in intensive farming, it can reach 1.5 kg in feed-lots (Mushtaq et al. 2015).
- A high stocking concentration, in a species that is not used to being closely housed. Additionally, the recommended space per camel in indoor housing systems (15–20 m^2) is not respected in some cases, resulting in the poor welfare status of some animals (Faraz et al. 2024).
- The use of milking machines designed for cows that is not necessarily adapted for camels, particularly with regards to stimulation time, vacuum level, and pulsation rate (Marnet et al. 2016; Ayadi et al. 2018).

2.5.2 The Impact of Environment and Management Practices on Disease Patterns

Farming intensification impacts either animals themselves (metabolism, health, physiology) or the farm environment. These changes can result in the modification of classical disease patterns that are observed in pastoral environments. For example, regarding animals, changes to feeding systems can impact the digestive physiology of camels (nitrogen recycling, intestinal transit, rumen flora). The distribution of a high proteo-energetic diet to camels, who are generally accustomed to eating grasses that are low in nutrients, can lead to excess fat deposition (Al-Jassim 2024) and obesity. Similarly, the use of milking machines requires good knowledge of lactation physiology (Kaskous 2018) and reproduction biotechnology to understand the physiological and hormonal constraints of milking (Skidmore 2011). The main consequences of intensification on disease patterns can be summarised as follows:

- The increasing risk of metabolic diseases in overfed animals, leading to liver disorders (hepatic lipidosis, hepatitis, cirrhosis, cholestasis, biliary epithelial hyperplasia, hepatic necrosis) or renal failure (necrosis, tubular nephrosis, glomerular congestion) without specific visible symptoms before a sudden death (Salem and Hassan 2011; Belina et al. 2015).
- The emergence of "diseases of production" such as infertility (Ali et al. 2010), mastitis (Iyer et al. 2014), or genital disorders (Benaissa et al. 2015; Khalafallah et al. 2017) as well as illness that is linked to lack of mobility, high stocking density, and misuse of machine milking.
- The increasing prevalence of parasitic diseases in camel farms with high housing density, facilitating the dissemination of external parasites such as mange (Chhabra and Khurana 2007) or internal parasites including zoonoses (Sazmand et al. 2019).
- A potential deleterious effect on the camel products. Indeed, a diet with limited plant diversity could impact the true or supposed health effect of camel milk (Konuspayeva et al. 2004; Kaskous 2015). Nomads attribute health benefits of milk to its composition, which is linked to the desert plants. Several studies show that farming systems (traditional or intensive) and the type of feed could have significant effects on composition and functional properties of camel milk (Benmohamed et al. 2018) and meat (Kadim 2013).
- The intensification of camel farming could also explain the onset of emerging diseases such as MERS-coronavirus or prion disease (Babelhadj et al. 2018), with a risk of transboundary contamination through the export of live camels to regional markets (Younan et al. 2016). The virus responsible of MERS is proposed to have mutated and spread due to changes in camel management (Hemida et al. 2017). Camel farming intensification in the Arabian Peninsula has also been suggested to be a risk factor for the transmission of this virus from camels to humans (Funk et al. 2016).

Faced with these changes, veterinarians have needed to develop investigations and more holistic approaches for managing the health of the camels under their care. The One Health approach has been utilised in several instances of zoonotic diseases on intensive camel farms (Mohamed et al. 2024), including a MERS-CoV outbreak (Hemida and Alnaeem 2019).

Regarding the impact of farming intensification on the environment, one of the most important concerns relates to water management. Indeed, irrigated forage replaces desert plants in camel indoor systems, with considerable impact on water usage. In a study conducted in Saudi Arabia comparing Bedouin (high camel mobility, low inputs, pastoral feeding) to intensive farming systems (irrigated feedstuffs, low camel mobility, high market integration), Faye (2013) reported an increase in water consumption from 3000 to 35,000 m^3/ha with a biomass productivity increasing from 5 to 18 tons/ha (Faye 2013). Alternative forage irrigated with salty water has been proposed as a potential solution for a better balance between natural resources and camel production (Abu-Alrub et al. 2018).

2.5.3 MERS-CoV: A Change in the Epidemiological Research

The Middle Eastern Respiratory Syndrome Corona Virus (MERS-CoV) belongs to the family of betacoronaviruses, which include also SARS-CoV-1 and SARS-CoV-2. MERS-CoV shares 50% sequence identity with SARS-CoV-2 and all the three viruses cause similar symptoms in humans (fever and respiratory problems) with a mortality rate for MERS of up to 35%. In dromedaries, however, the infection has a mild course, ranging from asymptomatic to mild nasal discharge (Hemida et al. 2014; Hemida et al. 2017; Hu et al. 2021). Although only detected in 2012 in a patient from Saudi Arabia, MERS-CoV-specific antibodies have retrospectively been identified as early as 1992 in dromedary samples from more than 25 African and Asian countries (Gossner et al. 2016) and more recently also in Bactrian camels (Lau et al. 2020). There are multiple routes of MERS-CoV transmission from camels to humans, including droplet infection, camel urine, and food-borne transmission through the consumption of unpasteurised milk and raw meat (Gossner et al. 2016).

Although humans act as transient and terminal hosts, the human-to-human transmission rate is low (Dudas et al. 2018). As the consumption of camel milk and meat is on the rise and as camel products gain access to wider markets, the impact of camel-associated zoonotic diseases on public health and on economies will also grow with increasing urbanisation. Based on whole camel genomes, which include detailed information on immune response regions (Lado et al. 2020; Ming et al. 2020), candidate genes have been identified that have important functions in the adaptive immune response (such as *HLA-A-24*-like and *HLA-DPB1*-like) and the innate immune response (such as *PTPN4*, *MAGOHB*, and *DNAH7*). Some of these genes have previously been associated with viral replication in SARS-CoV-1/-2 in humans, while others play an important role in the movement of bronchial cilia (Lado et al. 2021). These results suggest similar host genetic pathways associated with these betacoronaviruses, which again highlights the relevance of camels in the One Health context.

2.6 Future Camel Research and Development in the Context of One Health

Future interdisciplinary research, including ethnoveterinary and cultural aspects, parity, and local capacity building into camel and human health, is needed. As an example, wild camel conservation research projects in the Mongolian Great Gobi a Strictly Protected Area (GGASPA) implement genetic monitoring, behavioural observations, zoonotic pathogen screening, and the detection of hybridisation with domestic Bactrian camels in the buffer zone of the national park. The importance of wild camel conservation and its perception amongst camel herders is evaluated, and local capacities for wild camel conservation as well as the management of domestic Bactrian camels are improved.

It will be necessary to develop a holistic approach for veterinary medicine as well as animal breeding, welfare, and management training in the camel sector. As an example, the European Association of Establishments for Veterinary Education (EAEVA; https://www.eaeve.org/) included One Health as a competence in their "Day One Competences" of veterinarian graduates. This refers to the knowledge, skills, and attitudes acquired throughout veterinary studies and involves further applying relevant knowledge with the confidence and ability to transfer this knowledge in a variety of contexts. Veterinary graduates "use their professional capabilities to contribute to the advancement of veterinary knowledge and the One Health concept, in order to improve animal health and welfare, the quality of animal care and veterinary public health".

Placing camel research in the middle of the One Health context (Fig. 2.5) and maintaining sustainable camel management contributes to the Sustainable Development Goals (SDGs) announced by the United Nations in 2015 (Chikha and Faye 2025). Respecting camel welfare is important for animal and human health in view of the camel's role in zoonotic disease and the specificities of the camel immune system, i.e. nanobodies (see Chap. 6). Camels are of strong social and cultural importance to many people, and many livelihoods depend on camelid products and services. Greater support to the camelid sector can accelerate progress towards achieving Zero Hunger (SDG 2) and other SDGs (SDG 1, 3, 5; Fig. 2.5). Ultimately, camels can contribute to the sustainable use of terrestrial ecosystems and help to build resilience to climate change (SDG 15).

Fig. 2.5 Placing camel research in the middle of the One Health context and aligning it with the sustainability goals (SDGs) of the United Nations. (Source: Matilda Burger)

References

Abu-Alrub I, Marcum KB, Kabir N, Aran A, Al Hammadi M (2018) Productivity and nutritional value of four forage grass cultivars compared to Rhodes grass irrigated with saline water. Aust J Crop Sci 12(02):203–209

Al Jassim R (2024) Good feeding: nutrition and feeding of the Arabian camel (*Camelus dromedarius*). In: Padalino B, Faye B (eds) Dromedary camel behaviour and welfare—camel friendly management practice, Animal welfare 24. Springer, pp 55–70

Alavi F, Salami M, Emam-Djomeh Z, Mohammadian M (2017) Nutraceutical properties of camel milk. In: Watson RR, Collier RJ, Preedy VR (eds) Nutrients in dairy and their implications on health and disease, Nutrients in dairy and their implications on health and disease. Academic, pp 451–468

Alexander DJ (1995) The epidemiology and control of avian influenza and Newcastle disease. J Comp Pathol 112(2):105–126

Ali A, Al-sobayil FA, Tharwat M, Al-Hawas A, Ahmed AF (2010) Causes of infertility in female camels (*Camelus dromedarius*) in middle of Saudi Arabia. J Agric Vet Sci 2(2):59–68

Amine MAM, Koussou MO, Moulin CH, Duteurte G (2024) The current state of camel herding in Chad. J Camelid Sci 18:37–54

Amsidder L, Alary V, Sraïri TM (2021) An empirical approach of past and present mobility management in the desert societies of camel breeders in South Eastern Morocco. J Arid Environ 189:104501

Aubert M, Barrat J, Masson E (1994) L RAGE EN France et en Europe en 1993. Epidemiol Santé Anim 26:131–144

Awoke K, Ali SM (2015) Traditional husbandry practices and major challenge of young stock (camel calf) in Fafen Zone, Ethiopian Somali Regional State, Ethiopia. J Environ Anal Toxicol 5(6):1000321

Ayadi M, Musaad A, Aljumaah R, Matar A, Konuspayeva G, Abdelrahman M, Abid I, Bengoumi M, Faye B (2018) Machine milking parameters for an efficient and healthy milking in dairy camels (*Camelus dromedarius*). J Camel Pract Res 25(1):81–88

Babelhadj B, Di Bari MA, Pirisinu L, Chiappini B, Gaouar SBS, Riccardi G, Marcon S, Agrimi U, Nonno R, Vaccari G (2018) Prion disease in dromedary camels, Algeria. Emerg Infect Dis 24(6):1029–1036

Banathy BH (1996) Designing social systems in a changing world. Plenum Press, New York

Barnouin J, Brochart M (1981) L'enquête écopathologique continue: une nouvelle méthode de travail en pathologie chez les ruminants. Bull Techn CRZV-Theix INRA 46:11–14

Belina D, Giro B, Ashenafi H, Demissie T, Muktar Y (2015) Review on camel liver pathology and its major diagnostic approaches. Glob J Vet Med Res 3(1):68–79

Benaissa MH, Faye B, Kaidi R (2015) Reproductive abnormalities in female camel (*Camelus dromedarius*) in Algeria: relationship with age, season, breed and body condition score. J Camel Pract Res 22(1):67–73

Benet JJ (1994) Epidémiologie de la tuberculose bovine en France: état des connaissances et perspectives. Le Point Vet 26(159):13–25

Bengoumi M, Berrada J, Hidane K, Faye B (2004) Diarrhées du chamelon: écopathologie, physiopathologie et prévention. In: Lhoste F (ed) Lait de chamelle pour l'Afrique: atelier sur la filière laitière camélin en Afrique. FAO Publ, Rome, pp 137–146

Benmohamed C, Siboukeur O, Edoud A (2018) Influence of feeding on some physicochemical and biochemical characteristics of camel milk (*Camelus dromadarius*). Emir J Food Agric 30(4):251–255

Bernues A, Manrique E, Maza MT (1994) Farming systems and sanitary problems in mountain cattle farms. Vet Res 25(2–3):305–312

Böhm R., 1994. Environmental effects of cleaning and disinfection in animal husbandry. First results of ecotoxicological testing. In: Diesch (ed) Proceedings of the VIIIth International Congress of the International Society for Animal Hygiene, 12–16 September 1994, Saint-Paul, USA, pp 28–31

Burger PA (2024) Integrating one health into systems science. One Health (Amsterdam, Netherlands) 18:100701. https://doi.org/10.1016/j.onehlt.2024.100701

Calavas D, Rosner G (1997) Institutionnalisation d'une recherche-action en santé animale: l'expérience du Centre d'Ecopathologie Animale. Études et recherches sur les systèmes agraires et le développement 30:91–113

Chhabra MB, Khurana KL (2007) Parasitisms of camels revisited: 3-ectoparasites. J Camel Pract Res 14(1):1–8

Chikha M, Faye B (2025) Camel milk: white gold and its contribution to the sustainable development goals—a review. Outlook Agric:1–13. https://doi.org/10.1177/00307270251315472

Conway GR (1987) The properties of agroecosystems. Agric Syst 24(2):95–117

Cronin GM, Rault JL, Glatz PC (2014) Lessons learned from past experience with intensive livestock management systems. Rev Sci Tech 33(1):139–151

Dudas G, Carvalho LM, Rambaut A, Bedford T (2018) MERS-CoV spillover at the camel-human interface. eLife 7:e31257. https://doi.org/10.7554/eLife.31257

Faraz A, Hussain SM, Iglesias PC, Zappaterra M (2024) Good housing: camels and their interaction with the environment. In: Padalino B, Faye B (eds) Dromedary camel behaviour and welfare—camel friendly management practice, Animal welfare 24. Springer, pp 71–109

Faye B (1995) L'éleveur, son environnement, ses pratiques et la santé de son troupeau. Approche écopathologique des troubles de santé en élevage bovin laitier. PhD thesis report of the University of Paris XII (France), 227 pp

Faye B (2013) Camel farming sustainability: the challenges of the camel farming system in the XXIth century. J Sustain Dev 6(12):74–82
Faye B (2016) The camel, new challenges for a sustainable development. Trop Anim Health Prod 48(4):689–692
Faye B (2018) What future for camel pastoralism in the world? In: Sghiri A, Kichou F (eds) Proceedings of the 5th conference ISOCARD "Recent advances in camelids biology, health and Production", 12–15 November 2018, Laâyoune, Morocco, IAV Hassan II, pp 32–38
Faye B (2019) La marchandisation du lait de chamelle et la « périurbanisation » de l'élevage camelin: quel modèle de développement ? In: 3ème Renc Int sur «le lait, vecteur de développement», Dakar (Sénégal), 12–13 June 2019, 7p. https://www.inra.fr/lait-vecteur-developpement/
Faye B (2020) Chapter 17: interaction between camel farming and environment. In: AlHaj O, Faye B, Agrawal RD (eds) Handbook of research on health and environmental benefits of camel products. IGI Global, Hershey, pp 363–378
Faye B, Barnouin J (1987) Motivations et objectifs de l'enquête écopathologique Bretagne. Bull Techn CRZV-Theix, INRA 69:9–13
Faye B, Corniaux C (2024) Le lait de chamelle au risque de l'économie politique : de l'économie du don à l'économie marchande. Rev Elev Med Vet Pays Trop 77:37263. https://doi.org/10.19182/remvt.37263
Faye B, Lefèvre PC (1993) Introduction à l'écopathologie. Le développement des enquêtes écopathologiques en milieu tropical: contraintes et perspectives. Rev Elev Méd Vét Pays Trop 46(3):481–483
Faye B, Lefèvre PC, Quirin R, Lancelot R (1994) Ecopathologie animale: méthodologie, applications en milieu tropical. Ed. INRA-CIRAD, Versailles, 119 pp
Faye B, Waltner-Toews D, McDermott J (1999) From "ecopathology" to "agroecosystem health". Prev Vet Med 39:111–128
Fortané N (2023) Rise and fall of ecopathology: the thwarted development of a preventive approach to animal health (France, 1970s-1980s). In: Taboas DL, Martiin C, Prieto LF, Herment L (eds) From breeding & feeding to medicalization: animal farming, veterinarization and consumers in twentieth-century Western Europe. Brepols Publication, Turnhout, pp 337–360
Funk AL, Goutard FL, Miguel E, Bourgarel M, Chevalier V, Faye B, Peiris JSM, Van Kerkhove MD, Roger FL (2016) MERS-CoV at the animal–human Interface: inputs on exposure pathways from an expert-opinion elicitation. Front Vet Sci 3:88
Gallopin G (1995) The potential of agroecosystem health as guiding concept for agricultural research. Ecosyst Health 1:129–140
Ganière JP, André-Fontaine G, Drouin P, Faye B, Madec F, Rosner G, Fourichon C, Wang B, Tillon JP (1991) L'écopathologie: une méthode d'approche de la santé en élevage. INRA Prod Anim 4(3):247–256
Giampetro M (1994) Sustainability and technological development in agriculture. Bioscience 44:677–689
Goodger W, Franck G, Nordlund KV, Hilbelink B (1994) Environmental mastitis, suboptimal production, reproductive inefficiency and lameness caused by an inadequate housing system resulting in a limital financial return: four-year case study in a 160 cow Wisconsin dairy herd. In: Diesch (ed) Proceedings of the VIIIth International Congress of the International Society for Animal Hygiene, 12–16 September 1994, Saint-Paul, USA, pp 78–81
Gossner C, Danielson N, Gervelmeyer A, Berthe F, Faye B, Kaasik Aaslav K, Adlhoch C, Zeller H, Penttinen P, Coulombier D (2016) Human–dromedary camel interactions and the risk of acquiring zoonotic Middle East respiratory syndrome coronavirus infection. Zoonoses Public Health 63:1–9. https://doi.org/10.1111/zph.12171
Gruetzmacher K, Karesh WB, Amuasi JH, Arshad A, Farlow A, Gabrysch S, Jetzkowitz J, Lieberman S, Palmer C, Winkler AS, Walzer C (2021) The Berlin principles on one health—bridging global health and conservation. Sci Total Environ 764:142919. https://doi.org/10.1016/j.scitotenv.2020.142919

Hemida MG, Chu DK, Poon LL, Perera RA, Alhammadi MA, Ng HY, Siu LY, Guan Y, Alnaeem A, Peiris M (2014) MERS coronavirus in dromedary camel herd, Saudi Arabia. Emerg Infect Dis 20(7):1231–4. https://doi.org/10.3201/eid2007.140571

Hemida MG, Alnaeem A (2019) Some One Health based control strategies for the Middle East respiratory syndrome coronavirus. One Health 8:100102

Hemida MG, Elmoslemany A, Al-Hizab F, Alnaeem A, Almathen F, Faye B, Chu DKW, Perera RAPM, Peiris M (2017) Dromedary camels and the transmission of Middle East respiratory syndrome coronavirus (MERS-CoV). Transbound Emerg Dis 64:344–353

Homer JB, Hirsch GB (2006) System dynamics modeling for public health: background and opportunities. Am J Public Health 96:452–458. https://doi.org/10.2105/AJPH.2005.062059

Hu B, Guo H, Zhou P, Shi ZL (2021) Characteristics of SARS-CoV-2 and COVID-19. Nat Rev Microbiol 19:141–154. https://doi.org/10.1038/s41579-020-00459-7

Iyer AP, Albaik M, Baghallab I (2014) Mastitis in camels in African and Middle East countries. J Bacteriol Parasitol 5(3):1000188

Kadim I (2013) A review of the nutritive value and meat quality characteristics of the dromedary (*Camelus dromedaries*) camel meats. CAMEL Int J Vet Sci 1(2):129–156

Kadim IT, Al-Amri IS, Alkindi AY, Haq QM (2022) Nutritional values and health benefits of dromedary camel meat. Anim Front 12(4):61–70

Kaskous S (2015) Importance of camel milk for human health. Emir J Food Agric 28(3):158–163

Kaskous S (2018) Physiology of lactation and machine milking in dromedary she-camel. Emir J Food Agric 30(4):295–303

Khalafalla AI, Al Eknah MM, Abdelaziz M, Ghoneim IM (2017) A study on some reproductive disorders in dromedary camel herds in Saudi Arabia with special references to uterine infections and abortion. Trop Anim Health Prod 49(5):967–974

Konuspayeva G, Loiseau G, Faye B (2004) La plus-value « santé » du lait de chamelle cru et fermenté: L'expérience du Kazakhstan. Rencontre Recherches Ruminants 11:47–50

Lado S, Elbers JP, Rogers MF, Melo-Ferreira JKM, Yadmsuren A, Corander J, Horin P, Burger PA (2020) Nucleotide diversity of functionally different groups of immune response genes in Old World camels based on newly annotated and reference-guided assemblies. BMC Genomics 21:606. https://doi.org/10.1186/s12864-020-06990-4

Lado S, Elbers JP, Plasil M, Loney T, Weidinger P, Camp JV, Kolodziejek J, Futas J, Kannan DA, Orozco-terWengel P, Horin P, Nowotny N, Burger PA (2021) Innate and adaptive immune genes associated with MERS-CoV infection in dromedaries. Cells 10(6):1291. https://doi.org/10.3390/cells10061291

Landais E (1991) Ecopathologie et systémique. Etudes et recherches sur les systèmes agraires et le développement. INRA-SAD 21:5–11

Landais E (1992) Les trois pôles des systèmes d'élevage. Cah Rech Dev 32:3–5

Lau SKP, Li KSM, Luk HKH, He ZR, Teng JLL, Yuen KY, Wernery U, Woo PCY (2020) Middle East respiratory syndrome coronavirus antibodies in Bactrian and hybrid camels from Dubai. Msphere 5:e00898–e00819. https://doi.org/10.1128/mSphere.00898-19

Leischow SJ, Milstein B (2006) Systems thinking and modeling for public health practice. Am J Public Health 96:403–405. https://doi.org/10.2105/AJPH.2005.082842

Madec F, Tillon JP (1988) Ecopathologie et facteurs de risque en médecine vétérinaire. Analyse rétrospective (1977-1987) de l'expérience acquise en élevage porcin intensif. Recl Med Vet 164(8–9):607–616

Marnet PG, Atigui M, Hammadi M (2016) Developing mechanical milking in camels? Some main steps to take…. Trop Anim Health Prod 48(5):889–896

Ming L, Wang Z, Yi L, Batmunkh M, Liu T, Siren D, He J, Juramt N, Jambl T, Li YX, Jirimutu (2020) Chromosome-level assembly of wild Bactrian camel genome reveals organization of immune gene loci. Mol Ecol Resour 20. https://doi.org/10.1111/1755-0998.13141

Mohamed MYI, Lakshmi GB, Sodagari H, Habib I (2024) A One Health perspective on camel meat hygiene and zoonoses: insights from a decade of research in the Middle East. Vet Sci 11(8):344

Moutou F (1990) Prévision du risque de Fièvre Aphteuse en Europe. Description et analyse. Epidemiol Santé Anim 18:45–50

Mushtaq MH, Hussain D, Hussain F, Warraich MA (2015) Challenges and prospects of camel meat production and marketing in Pakistan. In: Al-Tariki Y, Al-Abdullrahman A, Al-Naeem A (Eds) Proceedings of the "Conference of sustainability of camel population and production (CSCPP)", 17–20 February 2013, Al-Ahsa University, Saudi Arabia, pp 126–129

Nagy PP, Skidmore JA, Juhasz J (2022) Intensification of camel farming and milk production with special emphasis on animal health, welfare, and the biotechnology of reproduction. Anim Front 12(4):35–45

One Health High-Level Expert Panel (OHHLEP), Adisasmito WB, Almuhairi S, Behravesh CB, Bilivogui P, Bukachi SA, Casas N, Cediel Becerra N, Charron DF, Chaudhary A, Ciacci Zanella JR, Cunningham AA, Dar O, Debnath N, Dungu B, Farag E, Gao GF, Hayman DTS, Khaitsa M, Koopmans MPG et al (2022) One Health: a new definition for a sustainable and healthy future. PLoS Pathog 18(6):e1010537. https://doi.org/10.1371/journal.ppat.1010537

Roussou D (2017) Systems research and the quest for scientific systems principles. System 5:25. https://doi.org/10.3390/systems5020025

Salem SI, Hassan AM (2011) Clinicopathological, cytological and histopathological studies on liver and kidney affections in camels. Glob Vet 7(6):557–571

Saltijeral JA, Gonzalez C, Arias L (1994) Dynamic aerial bacterial population in a Holstein calf production farm utilizing a centrifugal ambient detector. In: Diesch (ed) Proceedings of the VIIIth International Congress of the International Society for Animal Hygiene, 12–16 September 1994, Saint-Paul, USA, pp 94–97

Sazmand A, Joachim A, Otranto D (2019) Zoonotic parasites of dromedary camels: so important, so ignored. Parasit Vectors 12(1):610

Sazzad MH, Barreto SLT (1994) Effects of various levels of available phosphorus during peak production on performance, shell quality and phosphorus retention in two strains of laying hens. In: Diesch (ed) Proceedings of the VIIIth International Congress of the International Society for Animal Hygiene, 12–16 September 1994, Saint-Paul, USA, pp 63–66

Schwabe CW (1993) The current epidemiological revolution in veterinary medicine. Part II. Prev Vet Med 18:3–16

Sigognault A, Drouin P, Toux JY, Guittet M, Bennejean G (1994) Les lésions cutanées d'origine traumatique chez le pintadeau de chair: résultats d'une enquête écopathologique préliminaire. Sci Techn Avic 8:19–25

Skidmore JA (2011) Reproductive physiology in female Old World camelids. Anim Reprod Sci 124(3–4):148–154

Tillon JP (1987) Facteurs de risque et pathologie d'élevage: études des correspondances en élevage porcin intensif. Epidemiol Santé Anim 12:95–105

Tuffery G, Beaudois M, Dubois-Darnaudpeys A (1971) L'écopathologie des systems piscicoles. Semain d'hydrobiologie INRA, Biarritz, 10 p

Urban L, Perlas A, Francino O, Martí-Carreras J, Muga BA, Mwangi JW, Boykin Okalebo L, Stanton JL, Black A, Waipara N, Fontsere C, Eccles D, Urel H, Reska T, Morales HE, Palmada-Flores M, Marques-Bonet T, Watsa M, Libke Z, Erkenswick G, van Oosterhout C (2023) Real-time genomics for One Health. Mol Syst Biol 19:e11686. https://doi.org/10.15252/msb.202311686

Van Oirschot JT (1994) Vaccination in food animal populations. Vaccine 12(5):415–418

von Bertalanffy L (1956) General system theory. Gen Syst 1:1–10

Waltner-Toews D (1996) Ecosystem health: a framework for implementing sustainability in agriculture. Bioscience 46(9):686–689

Younan M, Bornstein S, Gluecks IV (2016) MERS and the dromedary camel trade between Africa and the Middle East. Trop Anim Health Prod 48(6):1277–1282

Camel Zoonoses: The Link to Human Disease

3

Abdelmalik Ibrahim Khalafalla

Abstract

Zoonotic diseases transmitted from camels to humans are an emerging public health concern, particularly in regions with close human-camel interactions. This chapter reviews key camel-associated zoonoses, including MERS-CoV, brucellosis, plague, and Rift Valley fever, and examines their transmission through direct contact, consumption of raw products, and cultural practices. It highlights the role of camels as reservoirs, the influence of environmental and socioeconomic factors, and the challenges in surveillance and control. Emphasis is placed on the importance of integrated One Health approaches to improve prevention, monitoring, and preparedness for future outbreaks.

Keywords

MERS-CoV · Brucellosis · Rift Valley fever · Disease transmission · Public health

3.1 Introduction

Zoonotic diseases are those that spread from animals to people. They have exerted significant influence on society throughout history. The role of camels as reservoirs and spreaders of disease is becoming more widely recognized, particularly since the 2012 emergence of the Middle East Respiratory Syndrome coronavirus (MERS-CoV). The spread of zoonotic diseases from camels to humans is driven by a complex mix of biology, ecology, and cultural practices in camel-rearing societies.

Known as the "ships of the desert," camels have played a significant role in the economies of the desert environments of North Africa, Asia, and the Middle East for

A. I. Khalafalla (✉)
Department of Microbiology, Faculty of Veterinary Medicine, University of Khartoum, Khartoum, Sudan

S. A. Gordine et al. (eds.), *Camels and One Health*,
https://doi.org/10.1007/978-981-92-2246-9_3

thousands of years. They are evolutionarily adapted to arid climates, which makes them a traditional source of food and transportation in those parts of the world. Unfortunately, that same utility and close contact with people makes camels obvious candidates for becoming carriers and reservoirs of zoonotic diseases.

Direct contact remains the primary mode of disease transmission from camels to people, and individuals who frequently come into contact with camels are most at risk. By the nature of their jobs, camel owners, farmers, herders, veterinarians, and slaughterhouse employees have the highest incidence of direct contact with camels.

Other disease vectors include the customary consumption of undercooked camel meat or unpasteurized camel milk. For example, sick animals that are about to die are slaughtered for food in rural communities in Africa and Asia. Numerous cases of zoonotic illnesses associated with camel meat consumption have been documented, including anthrax (Musa et al. 1993) and plague (Yersinia pestis). The latter has been reported in Libya since the 1970s and more recently documented in several studies (Khalafalla 2023). Due to long-standing cultural practices, camel milk is widely consumed raw. A recent study (Holloway et al. 2025) found that rural Arab communities in Jordan tested positive for Brucella antibodies, likely due to the daily consumption of raw camel milk.

Cultural practices also greatly influence the possibility of transmission. For example, Middle Eastern countries have traditions of breeding and racing camels, along with the aforementioned consumption of camel meat and milk. When coupled with inadequate biosecurity measures, these behaviors allow zoonotic diseases to spread across species boundaries. Camel races and beauty (Mazayna) pageants are significant cultural events, held throughout the Persian Gulf region, primarily in the UAE, Saudi Arabia, and Qatar, from October to February. These large gatherings attract participants and tourists from around the world, with the high concentration of both people and camels creating an ideal environment for the rapid spread of infectious pathogens. The camel festival risk assessment (Bieh et al. 2019) emphasized that an effective strategy for mass gatherings must be comprehensive, include all risks, and involve collaborative solutions to improve existing capabilities like disease surveillance.

The upcoming sections provide a detailed analysis of major zoonotic diseases that are transmitted from camels to humans, especially focusing on the threats posed by MERS-CoV, brucellosis, plague, and RVF. Our examination will highlight how these pathogens are spread due to the practices previously discussed, resulting in considerable global health challenges.

3.2 Key Zoonotic Diseases Transmitted from Camels to Humans

The main camel-associated zoonotic disease outbreaks involve MERS-CoV infection, brucellosis, plague caused by *Y. pestis*, camelpox, hepatitis E, and anthrax (Khalafalla 2023). More recently, a new zoonotic disease transmission linked to dromedary camels has been reported in the UAE, specifically a case of

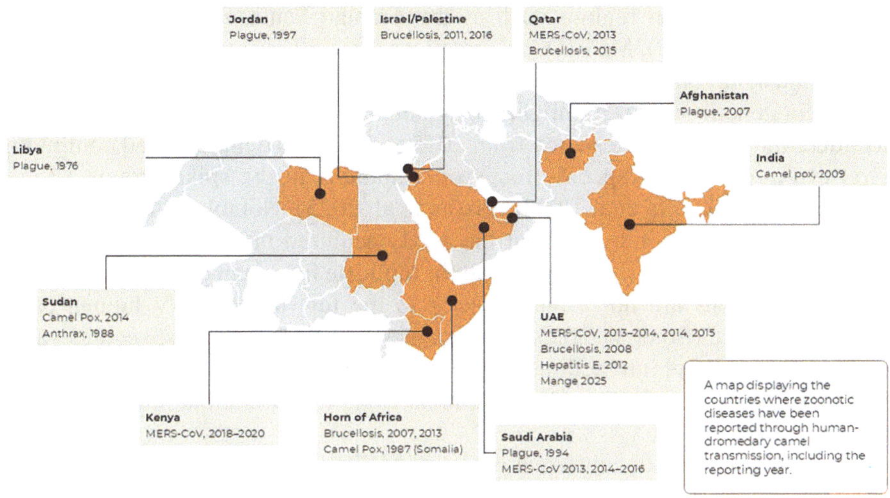

Fig. 3.1 A map displaying the countries where zoonotic diseases have been reported through human-dromedary camel transmission, including the reporting year

pseudoscabies. This case involved a rapid transmission of sarcoptic mange, a highly contagious skin disease caused by the mite *Sarcoptes scabiei*. The disease spread from one dromedary camel to a group of other dromedaries and to humans, even those wearing personal protective equipment (Christiana et al. 2025).

Although there is no direct proof of a transmission event, camels have also been linked to the spread of several other zoonotic diseases, such as toxoplasmosis, Rift Valley fever (RVF), tuberculosis (TB), Crimean-Congo hemorrhagic fever (CCHF), and Q fever caused by the bacterium *Coxiella burnetii* (Khalafalla 2023). Reports of these zoonotic diseases have been growing, due to the increased contact between humans and animals, from increasing global trade and travel, as well as environmental changes. Of note, the period from 2010 to the present accounts for approximately 60% of all reported cases.

Figure 3.1 illustrates cases of zoonotic disease spread from camels to humans in various countries. The map depicts the infectious diseases by country and the year they first appeared. These countries, which extend from Libya in the west to India in the east, include seven Asian countries and six African countries and are part of the Camel Belt.

3.2.1 MERS-CoV Infection

Middle East Respiratory Syndrome (MERS), a viral respiratory disease, is caused by the MERS coronavirus (MERS-CoV) and first appeared in Saudi Arabia in 2012. While bats are considered the virus' ultimate origin, dromedary camels are the primary reservoir (WHO 2022; Al-Tawfiq et al. 2022). MERS-CoV often causes mild or asymptomatic infections in camels and readily spreads among camel herds. This allows it to persist in regions such as the Arabian Peninsula, and according to Al-Tawfiq et al. (2022), MERS-CoV infections can spill over to humans. In contrast to the mild symptoms in camels, human cases result in severe illness.

Recent genetic studies highlight the dynamic evolution of MERS-CoV through its infection of camels. Research from 2015 to 2025 has demonstrated continuing viral evolution, with a specific focus on mutations in the spike protein (Azhar et al. 2023; Khalafalla et al. 2023; Hassan et al. 2025). Notably, a study of recent MERS-CoV isolates indicates that the newly acquired mutations do not significantly enhance the virus' ability to enter or replicate in human cells. Further, spike protein mutations are not a major factor in the few MERS-CoV human cases observed to date (So et al. 2025). This suggests that other factors, such as variation in human-camel interactions or herd immunity in camels, may have a greater impact on the fluctuating number of human cases than the mutations themselves (So et al. 2026).

Studies have identified different clades of MERS-CoV. African clade C viruses, prevalent in East African dromedaries, do not replicate in human cells as readily as Middle Eastern clade B viruses (Karani et al. 2025). This variation is partly attributed to differences in the spike protein and deletions in accessory proteins such as ORF4b and ORF3. These factors indicate that the virus may be adapting to camels differentially in different regions, which could affect the likelihood of successful human spillover events in those regions (Al-Tawfiq et al. 2022). The viral adaptations observed in East African dromedaries indicate that the virus may be specializing to its animal host, which could increase the barrier for transmission to humans.

It is generally accepted that camels are the primary source of MERS-CoV infections, and that human cases often represent "evolutionary dead ends" due to the limited potential for sustained human-to-human transmission (Dudas et al. 2018).

3.2.1.1 Camel-to-Human Transmission

Although MERS-CoV infections in people can arise through both direct and indirect contact with infected dromedary camels, the primary cause of zoonotic transmission is close interaction with camels, which includes activities such as handling, training, milking, and coming into contact with their feces. According to epidemiological studies, individuals who work with camels, such as camel shepherds and slaughterhouse workers, show a higher prevalence of MERS-CoV antibodies (Khudhair et al. 2019; Aljasim et al. 2020; Holloway et al. 2024). Further supporting this direct link, identical or nearly identical MERS-CoV genetic sequences have been identified in linked human and camel cases (Al Hammadi et al. 2015; Muhairi et al. 2016; Paden et al. 2018), providing strong molecular evidence for direct zoonotic transfer and further highlighting the importance of the one health approach in identifying disease spill-over from camels.

3.2.1.2 Transmission Mechanisms from Camels

Direct or close physical contact with infected dromedaries or their products continues to be the main way that MERS-CoV is spread from camels to humans. These covers come into contact with respiratory droplets, saliva, camel feces, and nasal secretions. There is also a risk of consuming raw camel products, such as unpasteurized milk or undercooked meat (Al-Tawfiq et al. 2023; CDC 2024).

There is limited evidence for transmission through camel products other than the milk or asymptomatic MERS patients. However, current epidemiological research suggests direct contact with live camels or individuals with symptomatic MERS plays a significant role in transmission (Killerby et al. 2020).

3.2.1.3 Clinical Signs in Dromedary Camels

Dromedary camels are particularly effective silent carriers of MERS-CoV because they often experience subclinical infections (WHO 2025; ECDC 2025). When clinical signs do manifest, they typically affect the upper respiratory tract and are nonspecific and mild (Alnaeem et al. 2020, 2021; WOAH 2021; WHO 2025). Clinical manifestations may include lacrimation and moderate nasal discharge (rhinorrhea) (Khalafalla et al. 2015; Alnaeem et al. 2020, 2021). Figure 3.2 illustrates the clinical signs of nasal discharge and increased lacrimation in a young camel calf diagnosed with MERS-CoV through PCR testing.

Although infected young dromedary camels may not show overt signs of illness, research indicates that they can harbor significant MERS-CoV viral loads and viral RNA concentrations, as measured in nasal swab samples (Wernery et al. 2015; WOAH 2021). Some reports also indicate a brief and moderate increase in body temperature (Alnaeem et al. 2020).

MERS-CoV infections in camels are difficult to track epidemiologically, due to the significant differences between the disease's severe pathology at the level of tissues and cells, while the infected animals generally appear asymptomatic or with mild symptoms (Alnaeem et al. 2020, 2021). This is indicative of a unique relationship between the host and pathogen, allowing the virus to replicate and damage tissue without causing the camels to show serious symptoms (WHO 2024, 2025). As a result, an animal's positive infectious state cannot be ruled out through visual inspection (WHO 2024, 2025; WOAH 2021). This inability to readily identify which animals are infected makes disease surveillance and control of zoonotic

Fig. 3.2 Illustration of the clinical signs of MERS-CoV in young camel calf

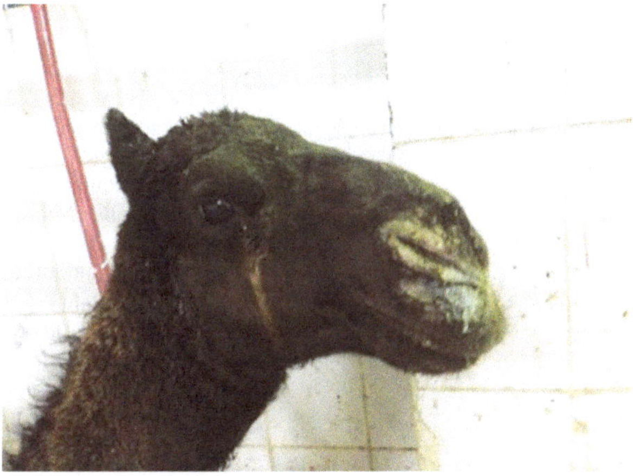

transmission particularly challenging because infected animals often appear healthy. Consequently, laboratory testing, such as PCR of nasal swabs, is essential for accurately identifying infected animals and the reservoir of the virus.

3.2.1.4 Clinical Signs in Humans

Until the outbreak of MERS-CoV in Saudi Arabia in 2012, most human coronavirus infections were not regarded as particularly pathogenic. As noted previously, and shown in Table 3.1, MERS-CoV infections typically present with more severe clinical symptoms in humans than in camels. Human infections with MERS-CoV can manifest in a variety of ways, ranging from mild respiratory symptoms to severe acute respiratory disease and death. Common symptoms include coughing (83%), shortness of breath (72%), and fever (observed in 98% of patients in one review) (WHO 2024, 2025; CIDRAP 2025; CDC 2024). Pneumonia is frequently observed (WHO 2024, 2025; CDC 2024). Other commonly reported symptoms include headache, chills or rigors, and myalgia (32%) (CDC 2024). Gastrointestinal symptoms

Table 3.1 Comparing the clinical signs of MERS-CoV in humans and dromedary camels

Clinical presentation	Dromedary camels	Humans (common/mild)	Humans (severe/complications)
Asymptomatic	In most cases	Possible, especially in close contact	Rare, usually precedes severe progression
Respiratory signs/symptoms	Mild, if present: nasal discharge, lacrimation, mild fever	Fever, cough, shortness of breath, nonproductive cough, sore throat	Severe pneumonia, acute respiratory distress syndrome (ARDS)
Gastrointestinal signs/symptoms	Not commonly reported	Diarrhea (26%), vomiting (21%), abdominal pain (17%), nausea	May precede respiratory symptoms; can contribute to multi-organ dysfunction
Systemic signs/symptoms	Mild fever	Chills/rigors, headache, myalgia, dizziness	Septic shock, multi-organ failure (e.g., kidney failure), disseminated intravascular coagulation (DIC), pericarditis
Pathological findings	Microscopic: ciliary loss, epithelial hyperplasia, interstitial pneumonia, renal degeneration	Macroscopic/radiological: pneumonia, lung infiltrates. Microscopic: diffuse alveolar damage, inflammation	Extensive lung damage, kidney injury, systemic inflammation leading to organ dysfunction
Risk factors for severity	Not available (usually mild or asymptomatic in camels)	Not available (for common/mild presentation)	Older age, weakened immune systems, chronic diseases (renal, cancer, lungs, diabetes)
Case fatality rate	Negligible	Not available (for mild cases)	35–36% of laboratory-confirmed cases

also occur regularly, with frequency of vomiting, diarrhea, and abdominal discomfort reported in 21%, 26%, and 17% of cases, respectively (WHO 2025; CDC 2024). Further, atypical presentations, such as diarrhea occurring before the onset of pneumonia, have also been reported (CDC 2024).

Acute respiratory distress syndrome (ARDS) can develop rapidly from severe illness, often necessitating intensive care unit (ICU) support and mechanical ventilation (WHO 2025; ECDC 2025; CDC 2024). Additionally, serious complications such as pericarditis, disseminated intravascular coagulation (DIC), and renal failure may occur (Willman et al. 2019). Approximately 35% of laboratory-confirmed cases result in death. The high incidence of severe respiratory and kidney issues is attributed to the virus' affinity for the DPP4 receptor, which is widely expressed in both lung and kidney cells (Willman et al. 2019). MERS in humans differs from many other respiratory viruses in that it affects the lungs and the kidney in sequence, significantly increasing the mortality rate and complicating patient management.

The wide range of symptoms in humans also presents a significant diagnostic challenge. These symptoms can range from asymptomatic to severe, as well as non-specific and atypical presentations (WHO 2025; CDC 2024). This issue has been observed in non-endemic areas or during periods when other respiratory viruses circulate widely, as MERS-CoV infection may be overlooked or misdiagnosed when clinical symptoms are nonspecific or moderate. Delays in detection can postpone the isolation of infected individuals, hindering the timely implementation of infection prevention and control measures, with direct consequences for public health (WHO 2024; CDC 2024). Without isolation, the risk of human-to-human transmission increases, especially in medical facilities where close contact with critically ill patients is common (CDC 2024, 2025; WHO 2025; Willman et al. 2019). Further, the potential for asymptomatic transmission complicates control measures (CDC 2024).

Transmission in healthcare settings has made controlling the spread of MERS-CoV extremely difficult. The 2015 outbreak in South Korea, for example, started with a single infected returned traveler from an endemic area and quickly spread to other hospitals, resulting in 186 infections and 38 fatalities (Hui 2016).

One consistent observation is that the disease is more severe in certain vulnerable groups, including older adults and those with chronic conditions, such as compromised immune systems, diabetes, cancer, renal disease, and chronic lung disease (CDC 2024, 2025; CIDRAP 2025; Willman et al. 2019). Individuals with comorbidities are significantly more likely to suffer from severe cases of MERS-CoV (Willman et al. 2019; CDC 2025). For these high-risk populations, targeted preventative measures and increased awareness are needed.

3.2.2 Brucellosis

Brucellosis is another zoonotic disease that can be transmitted from camels to humans and presents a significant public health risk in regions such as the Middle East and parts of Africa, where close contact with animals and traditional farming practices are common. The disease is caused by gram-negative bacteria from the genus *Brucella*, with *B. abortus* and *B. melitensis* of biggest concern to humans (Radwan 2014).

3.2.2.1 Transmission Mechanisms from Camels

The disease spreads to humans primarily through contact with infected animals in two ways: direct contact with the animals or with their bodily fluids. These infection paths are relevant for those dealing directly with camels, including camel handlers, herders, veterinarians, and slaughterhouse workers. Additional infection risk comes from consuming contaminated products, including raw or unpasteurized camel dairy products like milk and cheese. Other biological materials of particular concern due to high concentrations of the bacteria include aborted fetuses, placental tissue, or vaginal discharges from infected camels (Gwida et al. 2012).

Evidence of disease transmission from camels to humans has been documented in several regions, including the United Arab Emirates (UAE), the Horn of Africa, Qatar, and Israel/Palestine. Schulze zur Wiesch et al. (2010) reported a notable case in the UAE where a veterinarian in a laboratory was diagnosed with acute brucellosis. This case highlighted the high infection potential, including airborne transmission, of brucellosis in veterinary settings.

In another study, Shimol et al. (2012) described an outbreak of human brucellosis in Israel/Palestine, stemming from the consumption of camel milk. Subsequent whole-genome sequencing in Israel/Palestine further linked the *B. melitensis* infections to uncontrolled livestock trading and the sale of raw camel milk (Bardenstein et al. 2021). Similarly, Rhodes et al. (2016) reported cases of human brucellosis infections in travelers and immigrants from countries in the Horn of Africa (Somalia, Ethiopia, and Djibouti), again determining that source of infection was the consumption of raw camel milk.

Garcell et al. (2016) reported an outbreak of human brucellosis in a Qatari family: 14 family members who owned camels and sheep were affected. Collectively, such studies add to our understanding of camel-associated brucellosis and underscore the significance of camels as a source of zoonotic diseases, with the consumption of raw camel milk as a significant infection pathway.

3.2.2.2 Epidemiological Management to Reduce Zoonotic Spread

Preventing the transmission of brucellosis from camels requires a multipronged strategy. Educational campaigns encouraging people to avoid consuming raw milk and unpasteurized dairy products are effective. Additionally, implementing animal health programs, such as "test-and-slaughter" policies and vaccinating young camels, is crucial for eradicating the disease at its source.

3.2.3 Camel Plague (*Yersinia pestis* Infection)

All camelids, including Dromedary and Bactrian camels and New World camelids, are susceptible to *Y. pestis*, and plague cases have been reported in these animals in different regions. Arab traditions held that widespread camel deaths signaled an impending human plague, likely due to past transmission events from camels to humans. Early twentieth-century cases of *Y. pestis* spreading from infected camels to people were documented in the former USSR.

The consumption of raw or undercooked camel meat and organs formed the most common transmission route in recent outbreaks, while direct exposure to infected camels or their bodily fluids is another route. Four recent studies describe a link between *Y. pestis* infection and contact with dromedary camels. An investigation into a 1997 plague outbreak in Jordan found that the patients, who experienced fever and swollen cervical lymph nodes, had consumed raw or cooked camel meat (Arbaji et al. 2005). A similar event in Saudi Arabia resulted in five plague cases with severe pharyngitis and submandibular lymph node swelling, all linked to the consumption of raw camel liver (Bin Saeed et al. 2005). Leslie et al. (2011) reported an outbreak of plague in Afghanistan that resulted in 17 deaths, and in which the patients presented with rare gastrointestinal symptoms, associated with consuming or handling camel meat. Another incident in Libya in 1976 involved the consumption of meat from a slaughtered sick camel (Christie et al. 1980).

Unlike the classic bubonic plague transmitted by flea bites, plague from camels often presents with different clinical manifestations, such as severe pharyngitis (inflammation of the throat) and gastrointestinal symptoms.

Controlling and preventing plague transmission from camels requires a comprehensive approach that combines public health and veterinary measures. Public health initiatives focus on educating communities about the associated risks, promoting the thorough cooking of camel meat, and establishing surveillance systems to detect outbreaks early. Additionally, veterinary measures are essential; these include actively monitoring camel herds, quarantining sick animals, and ensuring the proper disposal of carcasses to prevent the spread of disease.

3.2.4 Rift Valley Fever

Rift Valley fever (RVF) is a significant mosquito-borne viral zoonosis that affects both animals and humans, predominantly across Africa and the Arabian Peninsula (Kwasnik et al. 2021). The disease is caused by the RVF virus (RVFV), which belongs to the Bunyaviridae family and is primarily transmitted by mosquitoes. Since its first discovery in Kenya in 1930, the disease has periodically spread to cattle and small ruminants. The disease has also spread to humans in sub-Saharan Africa and the Arabian Peninsula as a result of these outbreaks.

The mosquitos that carry RVF are expanding their ecological range due to rising global temperatures and increasingly unpredictable precipitation patterns, significantly altering the disease's epidemiology in camel populations across Africa and the Arabian Peninsula. The disease is often associated with "abortion storms" and high mortality rates in sheep and cattle, while the dromedary camel's significance in the virus's spread is recently becoming increasingly recognized (Zhu et al. 2019). The presentation of RVF in camels is complicated and can alternate between silent, asymptomatic circulation and severe reproductive losses (Abdallah et al. 2015).

In 2010, northern Mauritania experienced a severe Rift Valley fever (RVF) outbreak, causing mass abortions in livestock and 63 human cases with 13 deaths (El Mamy et al. 2011). This event was unique as camels displayed advanced clinical symptoms aside from abortion, including hemorrhagic septicemia and severe respiratory distress. Disease seroprevalence in camels was high during this initial period, reaching up to 38.5% across the region. In the 2022 outbreak, there were 47 confirmed human cases and a high fatality rate of nearly 50% among animal breeders (WHO 2022). These data suggest that camels are important vectors of RVF transmission, playing a role as significant as sheep and goats in regional outbreaks.

3.2.4.1 Transmission, Clinical Signs, and Epidemiology

Like in other susceptible species, RVFV is primarily transmitted to camels through the bites of infected mosquitoes in the *Aedes* and *Culex* genera (Kwasnik et al. 2021). Heavy rainfall and flooding trigger RVF outbreaks by inducing the hatching of *Aedes* mosquito eggs that were transovarially infected. In Sudan, Mauritania, and Egypt, camels act as central amplification hosts for the virus (El Mamy et al. 2011). Camels serve as long-distance biological "bridge hosts": Their extensive grazing patterns and role in trade across national boundaries mean that infected camels can spread the virus across arid regions into new areas where suitable mosquito vectors are present. Their high mobility and cross-border trade facilitated the introduction of RVFV into new territories like the Arabian Peninsula. The presence of competent *Culex pipiens* vectors in Europe and North Africa, along with rising temperatures, signals a significant risk of the virus spreading to new camel-rearing regions (Napp et al. 2018).

Rift Valley fever (RVF) often remains subclinical or presents only mild symptoms in adult dromedaries, which allows the virus to circulate undetected for long periods (El Mamy et al. 2011). In contrast, pregnant camels can experience devastating "abortion storms," leading to herd loss rates of up to 100% (Kwasnik et al. 2021). Young calves are particularly susceptible to severe infections that are characterized by high fever and rapid mortality.

3.2.4.2 Zoonotic Risk and Control

RVF in camels poses a direct zoonotic threat to herders, veterinarians, and slaughterhouse workers through contact with infected fluids or tissues or consumption of raw camel milk (Khalafalla 2023). Managing the disease is difficult because outbreaks are sporadic and vaccines may not be safe for pregnant camels (Abdallah et al. 2015; Napp et al. 2018). Following a "One Health" approach, weather monitoring and vector control can be used to predict and mitigate risks.

3.3 Potential Pandemic Preparedness for Zoonoses Transmitted by Camels

Preparedness strategies must encompass a One Health approach, integrating human, animal, and environmental health surveillance systems, particularly in isolated or resource-constrained environments where camels are raised and regularly come into close contact with humans. A notable example of effective collaboration is the early warning and response system in Abu Dhabi, UAE. This system was established by the Department of Health and the Abu Dhabi Agriculture and Food Safety Authority, which focus on human and animal health, respectively. The system is designed with efficient bilateral notification processes and web-based surveillance systems to facilitate the rapid sharing of epidemiological data, laboratory results, and risk assessments among relevant authorities. This capability allows authorities to develop and implement targeted, evidence-based control strategies, such as regulating the movement of camels, which has been identified as an effective approach to controlling the spread of MERS-CoV, and is documented in several publications (Al Hammadi et al. 2015; Muhairi et al. 2016; Paden et al. 2018).

International coordination is also crucial. The WHO, FAO, and the World Organization for Animal Health (WOAH) can also facilitate standardized reporting, rapid response mechanisms, and the sharing of epidemiological data through their collaborating and reference centers.

Limited diagnostic capabilities, a lack of knowledge among communities and healthcare professionals on camel diseases and their zoonotic impact, and insufficient veterinary services are some of the challenges to controlling the spread of zoonotic diseases from camels. In most countries where camels are raised, rates of diagnosis were low, suggesting the need for increasing the capacity for diagnostic testing and treatment among high-risk rural communities. To mitigate future outbreaks, investment in diagnostic and research infrastructure in camel-raising regions should also be prioritized.

The informal trading and transportation of camels often bypass health inspections and can introduce diseases into new areas. To reduce the risk of transmitting infections to humans, quarantine measures should be enhanced and testing for zoonotic diseases in camels conducted, particularly for MERS-CoV and brucellosis. This testing should be performed at borders and before camels participate in races and beauty festivals.

Monitoring camel populations, molecular diagnostics to identify active infections, and serological surveys to determine exposure levels are all part of the enhanced surveillance efforts. By integrating this data with human health information, hotspots can be detected early and possible epidemics prevented. Recent advancements in molecular epidemiology and genetics have enabled researchers to identify previously unknown diseases in camels, as well as hidden disease reservoirs. For instance, in 2024, researchers successfully linked the Wesselsbron virus to significant outbreaks in Ethiopian dromedary camels by utilizing the pan-PCR technique followed by genome sequencing (Ishag et al. 2025). In parallel, the scientific understanding of these diseases has evolved, from initial observations to advanced genetic and diagnostic techniques.

Public health education aimed at populations that closely interact with camels—such as herders, traders, and slaughterhouse workers—is essential. Educational campaigns should promote practices that minimize exposure to camel secretions and advocate for the safe handling of camel products. The awareness messages should emphasize the importance of boiling raw camel milk before consumption or milking brucella-free she-camels after testing to reduce zoonotic transmission. The camel festival risk assessment (Bieh et al. 2019) emphasized that an effective strategy for mass gatherings must consider all hazards and involve multi-sectoral collaboration to improve existing capabilities like disease surveillance.

Additionally, a mathematical model (Alatawi and Gumel 2024) indicates that MERS-CoV could be eliminated in Saudi Arabia by focusing on camel vaccination and mask use, preferably together, to lower the disease burden in the camel population. Developing and deploying veterinary vaccines for camels could help reduce viral circulation within the animal reservoir and limit spillover events.

A significant challenge in the veterinary field is that vaccination of camels is not a common practice. Currently, there is only a camelpox vaccine specifically developed for camels. While vaccines licensed for other large animals, such as those for rabies, hemorrhagic septicemia, and *Clostridium perfringens,* are sometimes used in camels, they are not specifically approved for this species. A major focus of vaccine research aims to develop a vaccine to prevent the transmission of MERS-CoV from camels to humans, using a One Health approach. Several obstacles hinder the advancement of camel vaccines, including the lack of established guidelines, insufficient comprehensive research on camel diseases and immunology, and logistical challenges in delivering vaccines to pastoral or nomadic camel herds located in remote areas.

References

Abdallah MMM, Adam IA, Abdalla TM, Abdelaziz SA, Ahmed ME, Aradaib IE (2015) A survey of rift valley fever and associated risk factors among the one-humped camel (*Camelus dromedaries*) in Sudan. Ir Vet J 69. https://doi.org/10.1186/s13620-016-0065-6

Al Hammadi ZM, Chu DK, Eltahir YM, Al Hosani F, Al Mulla M, Tarnini W, Hall AJ, Perera RA, Abdelkhalek MM, Peiris JS, Al Muhairi SS, Poon LL (2015) Asymptomatic MERS-CoV infection in humans possibly linked to infected dromedaries imported from Oman to United Arab Emirates. Emerg Infect Dis 12:2197–2200

Alatawi A, Gumel AB (2024) Mathematical assessment of control strategies against the spread of MERS-CoV in humans and camels in Saudi Arabia. Math Biosci Eng 21(7):6425–6470. https://doi.org/10.3934/mbe.2024281. PMID: 39176403

Aljasim TA, Almasoud A, Aljami HA, Alenazi MW, Alsagaby SA, Alsaleh AN, Alharbi NK (2020) High rate of circulating MERS-CoV in dromedary camels at slaughterhouses in Riyadh, 2019. Viruses 12(11):1215. https://doi.org/10.3390/v12111215. PMID: 33120981; PMCID: PMC7692456

Alnaeem A, Kasem S, Qasim I, Al-Doweriej A, Al-Houfufi A, Alwazan A, Albadrani A, Alshaammari K, Refaat M, Al-Shabebi A, Hemida MG (2020) Some pathological observations

on the naturally infected dromedary camels (*Camelus dromedarius*) with the Middle East respiratory syndrome coronavirus (MERS-CoV) in Saudi Arabia 2018-2019. Vet Q 40(1):190–197. https://doi.org/10.1080/01652176.2020.1781350. PMID: 32543343; PMCID: PMC7734115
Alnaeem A, Kasem S, Qasim I, Refaat M, Alhufufi AN, Al-Doweriej A, Al-Shabebi A, Hereba ART, Hemida MG (2021) Scanning electron microscopic findings on respiratory organs of some naturally infected dromedary camels with the lineage-B of the Middle East respiratory syndrome coronavirus (MERS-CoV) in Saudi Arabia-2018. Pathogens 10(4):420. https://doi.org/10.3390/pathogens10040420. PMID: 33916036; PMCID: PMC8065699
Al-Tawfiq JA, Al-Amri AF, Memish ZA (2022) Unresolved questions in the zoonotic transmission of MERS. Pathog Glob Health 116(1):1–11. https://pmc.ncbi.nlm.nih.gov/articles/PMC8734234/
Al-Tawfiq JA, Barry M, Memish ZA (2023) MERS-CoV as an emerging respiratory illness: a review of prevention methods. J Travel Med 30(4):taad059. https://pmc.ncbi.nlm.nih.gov/articles/PMC7110694/
Arbaji A, Kharabsheh S, Al-Azab S, Al-Kayed M, Amr ZS, Abu Baker M, Chu MC (2005) A 12-case outbreak of pharyngeal plague following the consumption of camel meat, in North-Eastern Jordan. Ann Trop Med Parasitol 99(8):789–793. https://doi.org/10.1179/136485905X65161. PMID: 16297292
Azhar EI, Velavan TP, Rungsung I, Traore T, Hui DS, McCloskey B, El-Kafrawy SA, Zumla A (2023) Middle East respiratory syndrome coronavirus-a 10-year (2012–2022) global analysis of human and camel infections, genomic sequences, lineages, and geographical origins. Int J Infect Dis 131:87–94. https://doi.org/10.1016/j.ijid.2023.03.046. Epub 2023 Mar 29. Erratum in: Int J Infect Dis. 2024;149:107261. https://doi.org/10.1016/j.ijid.2024.107261. PMID: 36996998; PMCID: PMC10050196
Bardenstein S, Gibbs RE, Yagel Y, Motro Y, Moran-Gilad J (2021) Brucellosis outbreak traced to commercially sold camel milk through whole-genome sequencing. Israel Emerg Infect Dis 27(6):1728–1731. https://doi.org/10.3201/eid2706.204902. PMID: 34013863; PMCID: PMC8153880
Bieh K, ElGanainy A, Yezli S, Malik M, Jokhdar HA, Asiri A, Alotaibi B (2019) Health risk assessment at mass gatherings: a report of the camel festival in Saudi Arabia. East Mediterr Health J 25(9):647–655. https://doi.org/10.26719/emhj.18.071. PMID: 31625590
Bin Saeed AA, Al-Hamdan NA, Fontaine RE (2005) Plague from eating raw camel liver. Emerg Infect Dis 11(9):1456–1457. https://doi.org/10.3201/eid1109.050081. PMID: 16229781; PMCID: PMC3310619
Center for Infectious Disease Research and Policy (CIDRAP) (2025). Retrieved from https://www.cidrap.umn.edu/mers-cov/who-notes-4-new-mers-cases-2-fatal-saudi-arabia-september
Center for Infectious Disease Research and Policy (CIDRAP). Study: MERS-CoV in Saudi camels matches human isolates. Retrieved from https://www.cidrap.umn.edu/mers-cov/study-mers-cov-saudi-camels-matches-human-isolates
Centers for Disease Control and Prevention (CDC) (2024) Interim guidance for MERS-CoV. Retrieved from https://archive.cdc.gov/www_cdc_gov/coronavirus/mers/interim-guidance.html
Centers for Disease Control and Prevention (CDC) (2025) Update on the epidemiology of Middle East respiratory syndrome coronavirus—worldwide, 2017–2023 | MMWR—CDC. Retrieved from https://www.cdc.gov/mmwr/volumes/74/wr/mm7419a1.htm
Christiana H, Rolf Karl S, Joerg K, Ulrich W (2025) One Health alert: zoonotic scabies from dromedary camels—a case report and call for vigilance. Front Vet Sci 12:1500916. https://doi.org/10.3389/fvets.2025.1500916. PMID: 40144525; PMCID: PMC11938839
Christie AB, Chen TH, Elberg SS (1980) Plague in camels and goats: their role in human epidemics. J Infect Dis 141(6):724–726. https://doi.org/10.1093/infdis/141.6.724. PMID: 7391614
Dudas G, Carvalho LM, Rambaut A, Bedford T (2018) MERS-CoV spillover at the camel-human interface. ELife 7:e31257. https://doi.org/10.7554/eLife.31257. Erratum in: Elife 7:e37324. https:doi.org/10.7554/eLife.37324. PMID: 29336306; PMCID: PMC5777824

El Mamy ABO, Baba MO, Barry Y, Isselmou K, Dia ML, Hampate B, Diallo MY et al (2011) Unexpected Rift Valley Fever outbreak, Northern Mauritania. Emerg Infect Dis 17(10):1894–1896. https://doi.org/10.3201/eid1710.110397

European Centre for Disease Prevention and Control (ECDC) (2025) Middle East respiratory syndrome coronavirus: factsheet. Retrieved from https://www.ecdc.europa.eu/en/middle-east-respiratory-syndrome-coronavirus/factsheet

Garcell HG, Garcia EG, Pueyo PV, Martín IR, Arias AV, Alfonso Serrano RN (2016) Outbreaks of brucellosis related to the consumption of unpasteurized camel milk. J Infect Public Health 9(4):523–527. https://doi.org/10.1016/j.jiph.2015.12.006. Epub 2016 Jan 12. PMID: 26796768

Gwida M, El-Gohary A, Melzer F, Khan I, Rösler U, Neubauer H (2012) Brucellosis in camels. Res Vet Sci 92(3):351–355. https://doi.org/10.1016/j.rvsc.2011.05.002

Hassan AM, Mühlemann B, Al-Subhi TL, Rodon J, El-Kafrawy SA, Memish Z, Melchert J, Bleicker T, Mauno T, Perlman S, Zumla A, Jones TC, Müller MA, Corman VM, Drosten C, Azhar EI (2025) Ongoing evolution of Middle East respiratory syndrome coronavirus, Saudi Arabia, 2023-2024. Emerg Infect Dis 31(1):57–65. https://doi.org/10.3201/eid3101.241030. Epub 2024 Dec 6. PMID: 39641462; PMCID: PMC11682817

Holloway PM, Gibson MD, Holloway TT, van Doremalen N, Munster VJ, Al-Omari B, Letko MC, Nash S, Cardwell JM, Abu-Basha EA, Hayajneh W, Mangtani P, Guitian J (2024) MERS-CoV exposure and risk factors for MERS-CoV ELISA seropositivity among members of livestock-owning households in southern Jordan: a population based cross-sectional study. Lancet Microbe 5(9):100866. https://doi.org/10.1016/S2666-5247(24)00082-X. Epub 2024 Jul 22. PMID: 39053480

Holloway P, Gibson M, Holloway T, Pickett I, Crook B, Cardwell JM, Nash S, Musallam I, Al-Omari B, Al-Majali A, Hayajneh W, Abu-Basha E, Mangtani P, Guitian J (2025) Camel milk is a neglected source of brucellosis among rural Arab communities. Nat Commun 16(1):861. https://doi.org/10.1038/s41467-024-55737-2. PMID: 39833143; PMCID: PMC11756418

Hui DS (2016) Super-spreading events of MERS-CoV infection. Lancet 388(10048):942–943. https://doi.org/10.1016/S0140-6736(16)30828-5. Epub 2016 Jul 9. PMID: 27402382; PMCID: PMC7136991

Ishag H, Tigani El Tigani-Asil E, Zawde W, Khalafalla AI, Albreiki M, Al Zarooni N, Mohammed A, Asgedom H, Abei G, Riufael T, Dabasa G, Birasa D, Debano J, Abdelwahab G, Habeeba S, Belexe M, Abduljami G, Amssalu K, Yusof MF, Ragassa F, Mohamed A (2025) Molecular detection of Wesselsbron virus in dromedary camels, Borana Zone, Ethiopia, 2024. Emerg Infect Dis 31(6):1263–1265. https://doi.org/10.3201/eid3106.250130. PMID: 40439475; PMCID: PMC12123917

Karani A, Ombok C, Situma S, Breiman R, Mureithi M, Jaoko W, Njenga MK, Ngere I (2025) Low-level zoonotic transmission of clade C MERS-CoV in Africa: insights from scoping review and cohort studies in hospital and community settings. Viruses 17(1):125. https://doi.org/10.3390/v17010125. PMID: 39861917; PMCID: PMC11768526

Khalafalla AI (2023) Zoonotic diseases transmitted from the camels. Front Vet Sci 10:1244833. https://doi.org/10.3389/fvets.2023.1244833. PMID: 37929289; PMCID: PMC10620500

Khalafalla AI, Lu X, Al-Mubarak AI, Dalab AH, Al-Busadah KA, Erdman DD (2015) MERS-CoV in upper respiratory tract and lungs of dromedary camels, Saudi Arabia, 2013–2014. Emerg Infect Dis 21:1

Khalafalla AI, Ishag HZA, Albalushi HIA, Al-Hammadi ZMA, Al Yammahi SMS, Shah AAM, Al Muhairi SSM (2023) Isolation and genetic characterization of MERS-CoV from dromedary camels in the United Arab Emirates. Front Vet Sci 10:1182165. https://doi.org/10.3389/fvets.2023.1182165. PMID: 37720473; PMCID: PMC10500840

Khudhair K, Killerby M, Al Mulla M, Abou Elkheir K, Ternanni W, Bandar Z, Weber S, Khoury M, Donnelly G, Al Muhairi S, Khalafalla A, John Watson Gerber S, Al Hosani F, Hall A (2019) Risk factors for MERS-CoV seropositivity among animal market and slaughterhouse workers, Abu Dhabi, United Arab Emirates, 2014–2017. Emerg Infect Dis 25(5):927–935. https://doi.org/10.3201/eid2505.181728

Killerby ME, Biggs HM, Midgley CM, Gerber SI, Watson JT (2020) Middle East respiratory syndrome coronavirus transmission. Emerg Infect Dis 26(2):191–198. https://doi.org/10.3201/eid2602.190697. PMID: 31961300; PMCID: PMC6986839

Kwasnik M, Rożek W, Rola J (2021) Rift Valley fever—a growing threat to humans and animals. J Vet Res 65:7–14. https://doi.org/10.2478/jvetres-2021-0009

Leslie T, Whitehouse CA, Yingst S, Baldwin C, Kakar F, Mofleh J, Hami AS, Mustafa L, Omar F, Ayazi E, Rossi C, Noormal B, Ziar N, Kakar R (2011) Outbreak of gastroenteritis caused by Yersinia pestis in Afghanistan. Epidemiol Infect 139(5):728–735. https://doi.org/10.1017/S0950268810001792. Epub 2010 Jul 22. PMID: 20663260

Muhairi SA, Hosani FA, Eltahir YM, Mulla MA, Yusof MF, Serhan WS, Hashem FM, Elsayed EA, Marzoug BA, Abdelazim AS (2016) Epidemiological investigation of Middle East respiratory syndrome coronavirus in dromedary camel farms linked with human infection in Abu Dhabi Emirate, United Arab Emirates. Virus Genes 52(6):848–854. Epub 2016 Jun 29

Musa MT, Shomein AM, Abd El Razig YM, Meki NT, El Hassan SM (1993) Anthrax in humans and camels in the Sudan with reference to the disease in the country. Rev Elev Med Vet Pays Trop 46:438–439. https://doi.org/10.19182/remvt.9438

Napp S, Chevalier V, Busquets N, Calistri P, Casal J, Attia M, Elbassal R et al (2018) Understanding the legal trade of cattle and camels and the derived risk of Rift Valley Fever introduction into and transmission within Egypt. PLoS Negl Trop Dis 12:e0006143. https://doi.org/10.1371/journal.pntd.0006143

Paden CR, Yusof MFBM, Al Hammadi ZM, Queen K, Tao Y, Eltahir YM, Elsayed EA, Marzoug BA, Bensalah OKA, Khalafalla AI, Al Mulla M, Elkheir KA, Issa ZB, Pradeep K, Elsaleh FN, Imambaccus H, Sasse J, Weber S, Shi M, Zhang J, Li Y, Pham H, Kim L, Hall AJ, Gerber SI, Al Hosani FI, Tong S, Al Muhairi SSM (2018) Zoonotic origin and transmission of Middle East respiratory syndrome coronavirus in the UAE. Zoonoses Public Health 65(3):322–333

Radwan AI (2014) Brucellosis in dromedary camels: a review. Rev Sci Tech Off Int Epiz 33(3):785–796

Rhodes HM, Williams DN, Hansen GT (2016) Invasive human brucellosis infection in travelers to and immigrants from the Horn of Africa related to the consumption of raw camel milk. Travel Med Infect Dis 14(3):255–260. https://doi.org/10.1016/j.tmaid.2016.03.013. Epub 2016 Mar 30. PMID: 27038818

Schulze zur Wiesch J, Wichmann D, Sobottka I, Rohde H, Schmoock G, Wernery R, Schmiedel S, Dieter Burchard G, Melzer F (2010) Genomic tandem repeat analysis proves laboratory-acquired brucellosis in veterinary (camel) diagnostic laboratory in the United Arab Emirates. Zoonoses Public Health 57(5):315–317. https://doi.org/10.1111/j.1863-2378.2009.01258.x. Epub 2009 Nov 18. PMID: 19925579

Shimol SB, Dukhan L, Belmaker I, Bardenstein S, Sibirsky D, Barrett C, Greenberg D (2012) Human brucellosis outbreak acquired through camel milk ingestion in Southern Israel. Isr Med Assoc J 14(8):475–478. PMID: 22977965

So RTY, Lau KKM, Zhou Z, Poon LLM, Peiris M (2025) The impact of clade B lineage 5 MERS coronaviruses spike mutations from 2015 to 2023 on virus entry and replication competence. bioRxiv [Preprint]:2025.06.30.662263. https://doi.org/10.1101/2025.06.30.662263. PMID: 40631265; PMCID: PMC12236716

So RTY, Hui KPY, Ho JCW, Lau KKM, Zhou Z, Chan MCW, Poon, LLM, Peiris M (2026) The impact of clade B lineage 5 MERS coronaviruses spike mutations from 2015 to 2023 on virus entry and replication competence. PLOS Pathog 22(2):e1013336. https://doi.org/10.1371/journal.ppat.1013336

Wernery U, Corman VM, Wong EY, Tsang AK, Muth D, Lau SK, Khazanehdari K, Zirkel F, Ali M, Nagy P, Juhasz J, Wernery R, Joseph S, Syriac G, Elizabeth SK, Patteril NA, Woo PC, Drosten C (2015) Acute Middle East respiratory syndrome coronavirus infection in livestock Dromedaries, Dubai, 2014. Emerg Infect Dis 21(6):1019–1022. https://doi.org/10.3201/eid2106.150038. PMID: 25989145; PMCID: PMC4451903

Willman M, Kobasa D, Kindrachuk J (2019) A comparative analysis of factors influencing two outbreaks of Middle Eastern respiratory syndrome (MERS) in Saudi Arabia and South Korea. Viruses 11(12):1119. https://doi.org/10.3390/v11121119. PMID: 31817037; PMCID: PMC6950189

World Health Organization. Disease outbreak news. Rift Valley Fever—Mauritania. Available at https://www.who.int.United/emergencies/disease-outbreak-news/item/2022-DON417

World Health Organization (WHO) (2022) Middle East respiratory syndrome coronavirus (MERS-CoV). https://www.who.int/news-room/fact-sheets/detail/middle-east-respiratory-syndrome-coronavirus-(mers-cov)

World Health Organization (WHO) (2024) Middle East respiratory syndrome coronavirus—World Health Organization (WHO). Retrieved from https://www.who.int/emergencies/disease-outbreak-news/item/2024-DON506

World Health Organization (WHO) (2025) Middle East respiratory syndrome coronavirus (MERS-CoV). Retrieved from https://www.who.int/health-topics/middle-east-respiratory-syndrome-coronavirus-mers

World Organization for Animal Health (WOAH) (2021) Chapter 3.5.2. Middle East respiratory syndrome (infection of dromedary camels with Middle East respiratory syndrome coronavirus). Retrieved from https://www.woah.org/fileadmin/Home/fr/Health_standards/tahm/3.05.02_MERS-CoV.pdf

Zhu S, Zimmerman D, Deem SL (2019) A review of zoonotic pathogens of dromedary camels. EcoHealth 16:356–377. https://doi.org/10.1007/s10393-019-01413-7

Camel Susceptibility to Disease: Genetic Drivers

4

Carlos Iglesias Pastrana and Nieves Cisneros Rodríguez

Abstract

This chapter explores how camel genetic diversity and immunogenetic adaptations influence susceptibility to zoonotic diseases and reservoir potential. It reviews camel evolution, domestication, and hybridization and examines key immune features linked to host–pathogen interactions, including responses to MERS-CoV and other zoonotic agents. The chapter also considers how behavior, pastoralist systems, environmental stressors, and human–camel contact shape disease dynamics. Finally, it highlights the importance of genomics for camel conservation, disease surveillance, selective breeding, and One Health strategies.

Keywords

Genetic diversity · Heavy-chain antibodies · Hybridization · Host-pathogen interactions · Camel genomics

4.1 Introduction

Camels have long been integral to the socio-economic and cultural fabric of arid and semiarid regions across North Africa, the Middle East, and Central and South Asia. These are some of the planet's most ecologically challenging regions, and human habitation there are only possible due to camels' remarkable physiological and

C. Iglesias Pastrana (✉)
Andalusian Institute for Agricultural, Fishing, Food and Ecological Production Research and Training (IFAPA), Centro Alameda del Obispo, Alameda del Obispo, Córdoba, Spain

N. Cisneros Rodríguez
Department of Genetics, Faculty of Veterinary Sciences, University of Córdoba, Córdoba, Spain

S. A. Gordine et al. (eds.), *Camels and One Health*,
https://doi.org/10.1007/978-981-92-2246-9_4

anatomical adaptations. These adaptations include efficient water conservation, thermotolerance, and the ability to thrive on low-quality forage (Burger et al. 2019).

In recent years, there has been a surge of international interest in camels outside their traditional pastoralist settings. Among the three living species, the domesticated one-humped dromedary (*Camelus dromedarius*) and the two-humped Bactrian camel (*Camelus bactrianus*) are particularly attractive, due to their use in sustainable meat and milk production. Their resilience and efficient resource consumption make them valuable alternatives to conventional livestock. As a result, camel husbandry is expanding in regions increasingly affected by ecological stress and diversifying through new breeding, dairy, and transport applications (Faye and Konuspayeva 2012).

However, this geographic expansion also raises concerns regarding zoonotic disease transmission. Camels are potential amplifiers and reservoirs of zoonotic pathogens due to their close contact with humans, mobility across vast territories and national boundaries, and participation in complex agro-pastoral systems (Faye 2019). A prominent example is the role of *C. dromedarius* as the primary animal reservoir of Middle East Respiratory Syndrome Coronavirus (MERS-CoV) and their involvement in recurrent human outbreaks since its emergence in 2012 (Müller 2014; Sabir et al. 2016). In addition to MERS-CoV, camels have been implicated in spreading several other deadly zoonotic diseases, such as brucellosis, plague, hepatitis E, camel pox, Rift Valley fever, and anthrax. These pathogens can be transmitted through direct contact, aerosol exposure, raw milk consumption, or vector-borne routes, particularly in pastoralist environments with significant contact between camels and humans (Khalafalla 2023).

In traditional pastoralist and transhumant societies, camels are not merely economic assets but integral to household survival, providing transport, nutrition, labor, and cultural identity (Volpato 2025). This high degree of human–animal interdependence, coupled with frequent close contact and shared environments, increases the likelihood of zoonotic spillover and complicates outbreak response. Unfortunately, access to veterinary infrastructure is often limited in such societies, and surveillance systems for emerging diseases are under-resourced or absent (Kena 2022).

The One Health framework is particularly useful for understanding how genetic, ecological, and socio-economic variables combine to shape patterns of disease emergence, host susceptibility, and transmission potential.

This chapter explores how camels' genetic variability influences their susceptibility to zoonotic pathogens and their potential role as disease reservoirs. We draw on insights from population genetics, immunogenetics, and behavioral ecology to examine the effects of domestication, selective breeding, and environmental stressors on the genetic makeup of camel populations. We also discuss how this knowledge can inform public health surveillance, conservation efforts, and sustainable livestock development in regions increasingly reliant on camelid resources. In regions where wild and domestic camel ranges overlap, we emphasize the critical need to consider both species-level genetic diversity and the implications of hybridization. Hybridization refers to interbreeding between individuals from different species or genetically distinct populations, and introgression describes the subsequent incorporation of genetic material from one group into the gene pool of another through repeated backcrossing. In Mongolia and China, the proximity of *C. bactrianus* to the endangered wild camel (the two-humped camel *C. ferus*) has already led to documented cases of hybridization and introgression, threatening the evolutionary integrity of the latter (Burger et al. 2019). Beyond conservation concerns, hybridization among domestic camel species may also have implications for disease susceptibility. Hybridization may modify immunogenetic profiles in ways that affect host susceptibility and pathogen reservoir dynamics (Balard and Heitlinger 2022).

Genomic studies offer valuable tools for identifying genetic variants related to disease resistance or susceptibility, which are essential not only to understanding zoonotic transmission but also for designing conservation and selective breeding strategies. These insights may ultimately support the evolutionary rescue and long-term viability of *C. ferus*, the last existing truly wild camel species (Lin et al. 2025).

4.2 Camel Evolutionary Origins, Speciation, and Hybridization

Old World camels have a long and complex evolutionary history. Modern camel species belong to the order Artiodactyla, suborder Tylopoda, and family Camelidae, which is divided into two tribes: Camelini (Old World camelids) and Lamini (New World camelids). Fossil and molecular evidence suggests that the ancestors of camelids originated in North America during the Eocene epoch (~45 Mya). They subsequently migrated to Eurasia through the Bering land bridge approximately 6.5–7.5 million years ago. These ancestral populations diversified into the three extant species: the one-humped dromedary (*Camelus dromedarius*), the two-humped Bactrian camel (*Camelus bactrianus*), and the critically endangered wild camel (*Camelus ferus*) (Gentry et al. 2004; Wu et al. 2014). A clear and accessible visual summary of camelid migration and diversification can be found in Burger et al. (2019).

The divergence between dromedaries and Bactrian camels is estimated to have occurred approximately 4.4 million years ago (Wu et al. 2014), while the split between Bactrian and wild camels is much more recent (~1.1 million years ago) (Ji et al. 2009; Mohandesan et al. 2017). Despite the phenotypic similarities between the two-humped camel species, mitochondrial and nuclear DNA analyses confirm that *C. ferus* is genetically distinct from *C. bactrianus*, supporting its status as a true wild species rather than a feral descendant of domesticated camels (Ji et al. 2009; Ming et al. 2022; Mohandesan et al. 2017; Sequencing and Consortium 2013; Yadamsuren et al. 2012; Yi et al. 2017). The wild camel is now present in just four isolated regions in China (Taklamakan desert, Gashun Gobi desert and Arjin Mountains in the Lop Nur Lake region) and Mongolia (Great Gobi Strictly Protected Area "A"). This species is critically endangered, with estimates of 900–1600 individuals remaining (Lei et al. 2012; Yadamsuren et al. 2012).

Domestication has had a significant effect on camel genetic diversity. Bactrian camels are thought to have been domesticated in the late fourth to early third millennium before the common era (BCE) (Ming et al. 2022), through one of two main hypotheses. The first locates it in northeastern Iran and the Kopet Dagh region, supported by early (2500–3000 BCE) archaeological remains (Khomeiri 2015) and the genetic diversity of Iranian camel populations (Jirimutu and Ming 2018; Ming et al. 2020b). The second hypothesis favors a Central Asian origin, backed by mitochondrial haplotypes from remains in Uzbekistan and Siberia (Trinks et al. 2012). However, these competing views could both be valid, as the current genetic landscape may be the result of multiple founder events, secondary migrations, or introgressive, post-domestication hybridization. Further analysis of ancient DNA is needed to fully resolve the question of domestication in this camel species.

After domestication, Bactrian camels spread widely across Central Asia, with major populations in Mongolia, China, Kazakhstan, northeastern Afghanistan, Russia, Crimea, and Uzbekistan and smaller groups in Iran, Pakistan, Turkey, and India (Mirzaei 2000; Vyas et al. 2015). China currently has the largest camel populations, mainly in the Chinese provinces of Inner Mongolia, Xinjiang, Qinghai, and Gansu. Phylogeographic and genome-wide studies reveal shared mitochondrial haplotypes and gene flow across those regions (Jirimutu and Ming 2018; Ming et al. 2017; Silbermayr et al. 2010). This suggests long-standing genetic links across them, facilitated by transboundary movements and traditional herding practices.

The dromedary camel was domesticated later, likely during the late second millennium BCE on the southeastern coast of the Arabian Peninsula (Almathen et al. 2016; Magee 2015; Uerpmann and Uerpmann 2002). Evidence supports a restocking model, where wild individuals were repeatedly added to domestic herds. Mitochondrial haplotyping indicates that six maternal lineages were incorporated during domestication, with one dominant lineage still present in ~70% of modern dromedaries (Almathen et al. 2016). Following domestication, dromedaries dispersed widely across Africa, Asia, and, very recently, Australia, facilitated by ancient trade networks (i.e., the incense and silk routes). This dispersion contributed to a largely panmictic global population—that is, a population in which individuals from different regions interbreed freely—leading to relatively low genetic diversity across geographies (Lado et al. 2018).

Hybridization between dromedaries and Bactrian camels has occurred for centuries in some countries in Western and Central Asia where the geographic distributions of both species overlap. This takes place both through intentional, human-mediated crossbreeding and through unmanaged interbreeding (Burger et al. 2019). Historically, hybridization was promoted to improve endurance and climatic tolerance. According to reports, first-generation (F1) hybrids from Middle Eastern and Central Asian countries have enhanced growth, productivity, and fitness and are also selected for participation in cultural activities like camel wrestling (Çakirlar and Berthon 2014). However, second-generation (F2) hybrids tend to display behavioral instability and reduced fitness (Faye and Konuspayeva 2012).

More critically, the uncontrolled hybridization between Bactrian and wild camels is now a major conservation concern. Uncontrolled interbreeding can lead to genetic introgression that may erode the adaptive immune diversity of wild populations. At the same time, hybrid individuals may exhibit novel immunogenetic profiles that influence susceptibility or resistance to pathogens. In Mongolia and China, the wild camel populations are increasingly surrounded by large numbers of domestic camels and known fertile hybrids. This proximity facilitates not only disease spillover (Walzer et al. 2012) but also genetic introgression, as documented through mitochondrial, Y-chromosomal, and nuclear DNA markers (Felkel et al. 2019; Silbermayr and Burger 2012; Silbermayr et al. 2010). Some hybridization events are initiated intentionally to enhance fitness of domesticated flocks, further complicating the genetic integrity of wild populations (Yadamsuren et al. 2012).

Demographic reconstructions based on genome-wide data suggest that Old World camel species experienced species-specific bottlenecks. All three diverged over the last 100,000 years, each undergoing population declines during the Last Glacial Period (~100,000–20,000 years ago), with further bottlenecks coinciding with domestication, human-driven habitat loss and hunting pressure (~4000–5000 years ago). While Bactrian camels recovered demographically after the Last Glacial Maximum (~20,000 years ago), dromedary populations remained at low effective population sizes. Notably, *C. dromedarius* has consistently exhibited a lower effective population size (N_e) (a genetic measure reflecting the number of individuals contributing to reproduction and genetic diversity) than *C. bactrianus* over the last 4000 years. At the same time, wild camels suffered prolonged isolation and contraction, contributing to their critically endangered status today (Burger 2016). Despite these constraints, Old World camels have maintained relatively high genetic variation, compared to other domesticated species. This may be due to the absence of breed formation bottlenecks, long-distance cross-border movements for trade and warfare, and the limited application of systematic selection schemes (Abri and Faye 2019; Fitak et al. 2020).

Interestingly, Yuan et al. (2024) emphasize that our current understanding of Old World camel genomics remains relatively incomplete, largely because existing datasets rely exclusively on modern species. They incorporated paleogenomic and stable isotope data from the extinct *Camelus knoblochi* to uncover mismatching mitochondrial and nuclear phylogenies among *C. knoblochi*, *C. ferus*, and *C. bactrianus*, suggesting historical gene flow among all three species. This suggests that hybridization events occurred repeatedly throughout the evolutionary history of the *Camelus* lineage and not only in contemporary populations.

4.3 Camel Immunogenetics and Zoonotic Interfaces

Camels exhibit distinctive influences on genes associated with the immune response due to evolutionary divergence, domestication history, and exposure to pathogens across heterogeneous environments. Multiple genomic studies have identified regions under positive selection, mainly associated with immune regulation, metabolic adaptations, and stress resilience (Al Abri et al. 2023; Almathen 2024; Bahbahani et al. 2019, 2024, 2025; Fitak et al. 2020; Khalkhali-Evrigh et al. 2022).

Genome-wide association studies have further revealed loci contributing to body morphology and growth, coat color, hematological traits, biomechanics, and behavioral attributes in camels (Bitaraf Sani et al. 2022; Guo et al. 2020; Iglesias Pastrana et al. 2024a, b; Sabahat et al. 2022). The functional co-localization of immune genes with those involved in energy homeostasis and thermoregulation highlights the dual pressures of ecological adaptation and disease resilience. Interestingly, a recent study has provided experimental support for this integration by identifying a camel-specific exon in the *NRAP* (nebulin-related-anchoring protein) gene, which is involved in cardiac function and environmental stress responses. By using a mouse knock-in model expressing this camel-specific exon, the authors showed altered thermoregulatory responses under cold stress and changes in the expression of inflammatory cytokine genes. This further supports the idea of a direct link between environmental adaptation mechanisms and immune-related pathways (Lee et al. 2025).

Comparative genomics reveal marked interspecies differences in immune-related gene diversity. For instance, annotations from InterProScan22 (Quevillon et al. 2005) indicate that immunoglobulin-like domains are among the most prevalent functional motifs in camel genomes. This highlights that immune diversity is a key evolutionary asset in camelid biology (Al-Swailem et al. 2010).

Recent advances in sequencing and assembly, including the CamDro3, CamBac2, and CamFer2 genome references (for dromedaries, Bactrian camels, and wild camels, respectively), have enabled high-resolution mapping of critical immune loci, such as major histocompatibility complex (MHC) class I/II, Toll-like receptors (TLRs), and interferon-regulated genes (Lado et al. 2020; Plasil et al. 2016). These tools provide a robust foundation for investigating associations between genotype and phenotype in immune response characteristics.

The following subsections explore in greater detail the structure and function of innate immunity, MHC organization, and host genetic determinants of disease susceptibility in camels, particularly for zoonotic viruses and bacteria of major concern.

4.3.1 Innate Immunity and Antimicrobial Adaptations

Camels inhabit some of the most extreme environments on earth, characterized by high temperatures, limited water availability, nutritional stress, and constant exposure to environmental pathogens which place exceptional demands on their physiology and immune system. To persist under these harsh conditions, camels have evolved immune strategies that emphasize robustness, flexibility, and long-term effectiveness rather than sheer complexity. These adaptations allow camels to maintain immune protection even under heat stress, dehydration, and metabolic constraints, making their immune system particularly well-suited to desert ecosystems.

One of the most distinctive features of the camel immune system is the production of unique heavy-chain antibodies. These antibodies play a crucial role in protecting camels against pathogens in challenging environments. In simple terms, they are smaller, more stable, and more resilient than conventional antibodies, enabling them to remain functional under extreme physiological conditions. These same properties have also attracted considerable interest in human medicine, as camel antibodies offer promising applications in diagnostics, therapeutics, and nanobiotechnology (Arbabi-Ghahroudi 2017; Muyldermans 2013).

Camels produce heavy-chain-only antibodies (HCAbs), also known as nanobodies, which lack light chains and differ structurally from conventional antibodies found in most mammals (Hamers-Casterman et al. 1993). Their compact size and enhanced stability allow them to access cryptic epitopes on pathogens (molecular regions that are often inaccessible to larger antibodies) thereby expanding the range of targets that can be recognized by the immune system.

Due to the placental structure of camels, maternal immunoglobulins are not transferred to the fetus during gestation. Instead, passive immunity is acquired exclusively through colostrum during the first hours of life, until the camel calf's immune system matures. Among the immunoglobulin classes present in camel colostrum, IgG has received particular attention because of its high concentration and its unique structural characteristics compared with immunoglobulins in other species (Hussen and Schuberth 2021). Insufficient colostrum intake during the first 24 h after birth can result in impaired immune protection, substantially increasing the risk of infectious disease and mortality in camel calves (Kamber et al. 2001).

The discovery of the IgG2 and IgG3 isotypes revealed that approximately 70% of camel serum antibodies consist of heavy-chain-only antibodies. Unlike conventional antibodies, camelid antibodies possess an extended complementarity-determining region 3 (CDR3) in the heavy chain instead of the light chain. This elongated CDR3 increases the antigen-contact surface, enabling broader and more flexible antigen recognition (Bannas et al. 2015; Kulkarni and Falzarano 2021; Transue et al. 1998). The variable domain of the heavy chain (VHH) is the key part of the antigen-binding region and is valuable in medicine and biotechnology because it is small, specific, stable and reliable in experiments (De Genst et al. 2006). Importantly, the diversity and functionality of these antibodies are under genetic control, meaning that variation among camel populations may influence the breadth and effectiveness of their immune responses. Such genetic variation in antibody repertoires may contribute to enhanced antiviral and antibacterial defense in camels, with potential implications for their role as reservoirs of zoonotic pathogens. The genetic diversity of heavy-chain antibodies indicates a link between camel immunogenetics and zoonotic risk, although their exact contribution to zoonotic disease dynamics remains the subject of active research.

Beyond antibody-mediated immunity, camels also exhibit distinctive features in their cellular immune responses. Natural killer (NK) cells participate in both innate and adaptive immunity, and their functional diversity is characterized by the NK cell receptors. A recent genomic analysis of camel NK receptor gene families, focusing on the natural killer complex (NKC) and the leukocyte receptor complex (LRC), revealed a low level of polymorphism in killer cell lectin-like receptor (KLR) genes. This pattern is comparable to that observed in domestic pigs and is in contrast to cattle, which have a greater expansion and diversity of NK receptor genes (Futas et al. 2019). Functionally, this limited receptor diversity in camels may reflect adaptation of their NK cell repertoire to their typical microbial environment, rather than an indication of impaired immune function.

Several studies have also examined variations in camel Toll-like receptor (TLR) genes, which are involved in the immune system's recognition of conserved microbial components such as bacterial lipoproteins and toxins. Polymorphisms identified in camel TLR genes may affect the response to infections like brucellosis and tuberculosis (Bishnoi et al. 2024; Choudhary et al. 2021; Dahiya et al. 2014; Gahlot et al. 2023; Suthar et al. 2019).

Additional evidence of species-specific immune adaptation comes from studies of macrophage responses in camels infected with *Mycobacterium avium* subsp. *paratuberculosis* (MAP), which causes Johne's disease in cattle and ruminants. Infected camels show a high expression of the chemokine receptor CX3CR1 in their intestinal macrophages. CX3CR1 promotes immune surveillance at mucosal sites by facilitating macrophage recruitment and close interaction with infected epithelial cells. Its elevated expression suggests enhanced retention of macrophages in the intestinal mucosa, potentially strengthening local immune control of chronic infection. Notably, this feature has been described in camels but not in other livestock species, highlighting a distinctive immune adaptation that may influence host-pathogen interactions in camel populations (Al Ramadan et al. 2021; Regoli et al. 2017).

4.3.2 Major Histocompatibility Complex (MHC) and Immune Diversity

The major histocompatibility complex (MHC) plays a central role in immune defense by enabling the recognition of pathogens and the activation of immune responses. MHC molecules allow immune cells to distinguish between the body's own cells and foreign agents such as viruses or bacteria. Both MHC class I and II genes have been identified and characterized in camels, although they show lower allelic diversity than those in other domesticated mammals such as cattle or pigs (Plasil et al. 2016). This reduced variation is frequently attributed to historical demographic constraints associated with domestication and population bottlenecks, and it may influence the range of pathogens that can be effectively recognized by the immune system. In Old World camelids, the MHC region is located on the long arm of chromosome 20 and retains the conserved mammalian organization of class I, II, and III regions (Plasil et al. 2016, 2019).

Despite their relatively limited allelic diversity, camel MHC molecules exhibit peptide-binding promiscuity, meaning that individual MHC variants can bind and present a wide range of pathogen-derived peptides. This functional flexibility may help maintain effective immune recognition by compensating for reduced genetic variation (Plasil et al. 2016).

Comparative genomic studies indicate that domestic Bactrian camels exhibit the highest average nucleotide diversity across MHC class I and II genes and a broader set of innate and adaptive immune response genes. This elevated diversity likely reflects long-term exposure to heterogeneous pathogen environments, particularly in mixed herding systems where camels coexist with other domestic species (Silbermayr et al. 2010; Ming et al. 2022), as well as a comparatively weaker loss of diversity during domestication. In contrast, the critically endangered wild camel shows consistently low genetic diversity across these immune-related gene categories, in keeping with its prolonged demographic contraction and genetic isolation (Ji et al. 2009; Yadamsuren et al. 2012). Dromedaries display intermediate levels of immune gene diversity, although their genetic profiles tend to resemble those of wild camels more closely than those of Bactrian counterparts, particularly in innate and adaptive immune regions (Plasil et al. 2016). Together, these interspecies patterns illustrate how a variety of factors shape immunogenetic diversity.

Low allelic diversity at some immune loci may also reflect limited variability in specific antigen-binding regions of T-cell receptors, particularly the CDR1 and CDR2 domains, which are highly conserved across mammalian species (Antonacci et al. 2011). Nevertheless, other immune system mechanisms contribute to threat recognition. For example, the TRG (T-cell receptor gamma) and TRD (T-cell receptor delta) genes encode receptors involved in the recognition of nonclassical antigens, thereby expanding immune surveillance capacity (Ciccarese et al. 2019; Kulkarni and Falzarano 2021). Notably, γδ T-cell receptors in dromedaries exhibit an unusually high degree of somatic hypermutation, a mechanism that introduces additional variation after gene rearrangement, a potentially unique way of enhancing immune system flexibility (Ciccarese et al. 2019).

Further distinctive features of camel immunity have been identified at the cellular and tissue levels that align more closely with monogastric mammals than with ruminants. According to recent immunological studies, cellular markers in camels have distinct expression features, particularly in relation to mucosal immunity. For example, both LPAM-1 (lymphocyte Peyer's patch adhesion molecule-1) and MAdCAM-1 (mucosal vascular addressin cell adhesion molecule-1) are expressed in camel mammary gland tissues, similar to the pattern observed in mucosal tissues of monogastric species but largely absent in ruminants (Al Ramadan et al. 2021; Regoli et al. 2017). These features have practical implications for disease epidemiology and herd management, as camels may respond to mucosal pathogens similarly to monogastric livestock. This is particularly relevant in mixed-production systems, where camels are kept near monogastric species, for understanding disease transmission dynamics and designing disease control strategies that account for species-specific immune organization and exposure pathways.

Aside from unusual antigen recognition genes, camelids also possess a distinctive collection of cytotoxic effector genes involved in the elimination of infected or abnormal cells. These include genes encoding perforin (PRF1), granulysin (GNLY), and several granzymes (GZMA, GZMB, GZMH, GZMK, GZMM, GZMO). Comparative analyses show that camelid granzymes share high sequence similarity with those of other ungulates (a group of hoofed mammals that includes camels, cattle, sheep, pigs, and other large herbivores), where they are central to cell-mediated immunity. However, certain loci exhibit species-specific diversification. Genetic polymorphism at these loci is generally low in dromedaries but higher in Bactrian camels, particularly in GNLY and chymotrypsin-like granzymes (Futas et al. 2021).

As mentioned above, genome-wide analyses using improved reference assemblies for *C. dromedarius*, *C. bactrianus*, and *C. ferus* (CamDro3, CamBac2, and CamFer2, respectively) have confirmed marked differences in nucleotide diversity among innate, adaptive, and MHC gene categories (Lado et al. 2020). Consistent with earlier findings, Bactrian camels exhibit the highest genome-wide and immune gene diversity, while wild camels show the lowest levels. These patterns, shaped by demographic history, carry implications for disease susceptibility and conservation efforts.

4.3.3 Host Genetic Determinants in MERS-CoV Infection

The emergence of MERS-CoV has brought significant attention to camel immunogenetics. Dromedaries are the natural reservoir for MERS-CoV, maintaining viral circulation with minimal clinical signs (Sabir et al. 2016). Infected camels may exhibit transient nasal discharge, mild rhinitis, and low fever, although most infections remain subclinical (Adney et al. 2014). Dromedaries aged 2 years and older showed higher seropositivity to the infection due to increased exposure with age (Alagaili et al. 2014; Lado et al. 2021). Host tolerance is hypothesized to be facilitated by specific genetic configurations at loci involved in viral entry, replication, and immune evasion. For instance, DPP4 (dipeptidyl peptidase 4), the receptor for MERS-CoV, shows differential expression across species, which may modulate host range and pathogenicity (Widagdo et al. 2016).

In addition, the virus exhibits multiple lineages and frequent recombination events, particularly at the junction between *ORF1b* (open reading frame) and the spike (S) gene, which are genomic regions associated with viral entry and pathogenicity. Recombination refers to the process by which genetic material from different viral variants is exchanged when they co-infect the same host cell, generating new genetic combinations. This high genetic variability and the presence of numerous quasi-species within camel populations suggest a non-static viral reservoir capable of generating novel variants (Yusof et al. 2017).

Several genes are associated with MERS-CoV infection. For example, the *MAGOHG* gene is involved in RNA splicing and innate immune activation. In simple terms, RNA splicing is a process by which genetic instructions are edited before being translated into functional proteins, allowing cells to produce the correct immune components. Variants of this gene are more frequent in noninfected camels, suggesting a higher resistance to viral infection in these animals (Lado et al. 2021).

Host mutations in MERS-CoV protein spike gene, only found in dromedaries, are related to viral cell entry due to interaction with the DPP4 receptor. Some other mutations in the receptor-binding domain (e.g., Ser32Pro) are found only in camel strains, while others are exclusively in humans. However, there is a genetic variation in one receptor-binding domain (Ile568Thr) present in both hosts, demonstrating a possible cross-species link for this virus (Azhar et al. 2023).

The immune selective pressure determined by differences in geographical distribution also produces genetic variations. Research comparing MERS-CoV isolated from camels in Saudi Arabia (as compared to Africa) detected differences in key genes such as *ORF4b* and *ORF3*, both of which have an impact in the host immune response.

The abovementioned studies suggest potential changes in virulence (i.e., how aggressively the virus damages host tissues) and transmissibility via several mechanisms. For instance, certain mutations might enhance viral evasion of the interferon response, allowing higher replication rates in host cells, thus increasing viral load and shedding. So variants that are more competent in replicating and resisting immune defenses might spread more readily among camels, influencing the zoonotic potential of this infection (El-Kafrawy et al. 2019). Notably, MERS-CoV continues to circulate extensively within the Arabian Peninsula, suggesting that regionally structured viral genotypes may differ in their capacity to cross the species barrier and contribute to human outbreaks (Dudas et al. 2018).

4.3.4 Genetic Susceptibility to Infections

Camelpox is one of the most contagious orthopoxviral infections affecting dromedaries, often causing significant outbreaks in herds. Defense against viruses is mediated by cytokines, particularly type I interferons. In camels, the interferon-alfa (INF alfa) has demonstrated antiviral activity in vitro against camelpox in camel kidney fibroblast cells (Premraj et al. 2020).

Parasitic infections such as trypanosomiasis (*Trypanosoma evansi*) and sarcocystosis (*Sarcosystis* spp.) show differential distribution across populations. These patterns may reflect both environmental exposure and underlying genetic variation, particularly in immune-related loci. Higher immune gene diversity in domestic Bactrian camels suggests active selection in high-pathogen environments, while reduced variation in *C. ferus* raises concerns for long-term disease resilience (Ming et al. 2020a; Mohandesan et al. 2017). Dromedaries generally display intermediate levels of immune gene diversity due to their broader geographic distribution and exposure to a large range of pathogens, compared to Bactrian and wild camels (Futas et al. 2021; Plasil et al. 2016).

Infection with *Trypanosoma evansi* is associated with impaired immune function in camels, facilitating parasite persistence and repeated infections. The high genetic and antigenic variability of this parasite enables it to evade host immune responses and complicates vaccine development. Different *T. evansi* strains express distinct antigenic profiles, particularly through variation in their surface glycoproteins, meaning that a single vaccine is unlikely to provide broad protection across regions. Instead, vaccines targeting a limited set of antigens may only be effective against specific local strains. Similarly, some camels may be resistant to particular strains while remaining susceptible to others. The risk of infection is therefore closely linked to geographic location, due to the limited genetic exchange among *T. evansi* populations across regions (Salim et al. 2011).

Transmissible spongiform encephalopathies (TSEs) have been described in camels in the form of camel prion disease (CPD). Genetic characterization of the prion protein gene (PRNP) indicates that camelids differ from other mammals belonging to the orders Cetartiodactyla and Perissodactyla. Cetartiodactyla includes even-toed hoofed mammals such as cattle, sheep, pigs, and deer, while Perissodactyla comprises odd-toed ungulates such as horses and rhinoceroses. Although these genetic differences highlight the distinct evolutionary position of camelids among large herbivorous mammals, it remains unclear whether they confer resistance or susceptibility to CPD (Wright et al. 2024).

There are some limitations detecting *Mycobacterium avium* subsp. *paratuberculosis* (MAP) in serology tests because camelids have heavy-chain antibodies (HCAbs), affecting the sensitivity of testing. As described in the chapter devoted to HCAbs (Chap. 6), this distinctive antibody structure can alter antigen recognition and influence the sensitivity of standard serology assays, which are typically optimized for conventional IgG responses. However, it is known that in female camels and purebreds there is higher resistance to infection with MAP. The purebred effect refers to genetically homogeneous dromedary lines, usually maintained under breeding systems, leading to lower infection rates due to their reduced genetic makeup (Al Naeem et al. 2024; Ghosh et al. 2012).

Also of relevance for camel-to-human infections, the camel is an important reservoir of *Coxiella burnetii* (Q fever). Its prevalence is associated with sex and age, being more prevalent in camels in females and adults, a pattern that is also seen in humans (Hussein et al. 2015). Additionally, multidrug-resistant gram-negative bacteria (*Escherichia coli*) have been found in camels. Given the importance of this infection in humans, it will be crucial to further study the relationship between human and animal strains (Carvalho et al. 2020).

Molecular characterization of *Cryptosporidium* spp., *Giardia duodenalis*, and *Enterocytozoon bieneusi* in dromedary camels has revealed a wide diversity of species and genotypes. Importantly, several of the detected genotypes (such as *C. parvum*, *G. duodenalis* assemblages A/B, and specific *E. bieneusi* genotypes) are well-known causes of gastrointestinal disease in humans, including diarrhea, abdominal cramps, dehydration, and weight loss. Their presence in camels therefore highlights clear zoonotic potential, indicating that camels can act as reservoirs or sources of infection for humans in close contact with them, particularly pastoralists and farm workers. In camels, these infections are often subclinical or mild, although young or stressed animals may develop diarrhea, reduced appetite, and dehydration. This contrast between mild infection in camels and clinical disease in humans underscores the public health relevance of these findings. Further studies are needed to investigate host genetic susceptibility and transmission dynamics at the human–camel interface (Elmahallawy et al. 2023).

4.3.5 Genetic Modulation of Reservoir Potential

Not all infected camels become effective reservoirs of zoonotic pathogens. Genetic differences among individuals and populations can affect infection characteristics such as viral replication, shedding duration, and transmission potential. In the case of MERS-CoV, younger animals have been shown to shed more virus and may express DPP4 differently (Widagdo et al. 2016).

Region-specific variation in innate immune pathways and receptor expression (e.g., TLRs, ISGs) can further modulate the outcome of infection. Populations with higher heterozygosity may resist infection or limit pathogen load, while genetically eroded herds, especially intensive systems, may act as superspreaders. The extremely low genetic diversity in wild camels (*Camelus ferus*) potentially heightens its susceptibility to emerging infections (Ali et al. 2019), while the high animal density (in addition to reduced genetic diversity) associated with managed herds can exacerbate this problem, through a heightened risk of pathogen amplification and cross-species transmission.

A comparative genome analysis of MAP isolates from camels against a standard cattle isolate shows important differences in single nucleotide polymorphisms (SNPs) and insertions or deletions, likely driven by host-specific selective pressure. In simple terms, SNPs are single-letter changes in the genetic code, while insertions involve the addition of short DNA segments (or loss, in the case of deletion) that can alter how genes function. Several of the affected genes in the camel MAP isolates are involved in host cell invasion, intracellular persistence, and immune modulation. This genetic divergence in camel-adapted MAP strains may influence pathogenicity or modify host immune responses if they are transmitted to humans (Ghosh et al. 2012).

Nucleotide sequence variants in some genes between healthy and trypanosomiasis-affected camels (corresponding with metabolic, immunological, and antioxidant functions) may play a large role in transmitting this infection. Genetic selection, taking into account gene polymorphism, could therefore reduce the incidence and spread of trypanosomiasis (Elsayed et al. 2024).

These insights emphasize the value of integrating host genomic data into One Health surveillance programs. Genomic risk profiling, combined with pathogen sequencing and ecological monitoring, could help identify high-risk individuals, predict outbreak hotspots, and guide tailored interventions.

4.4 Behavioral and Environmental Modulators of Camel Disease Dynamics

The behavioral ecology and environmental resilience of camels are tightly interwoven with their genetic makeup, directly influencing their exposure, susceptibility, and response to zoonotic pathogens. In this section, we explore how ecological plasticity, husbandry practices, social behavior, and environmental stressors intersect with immunogenetic variability to shape disease dynamics in camel populations.

4.4.1 Ecological Plasticity and Pastoralist Systems

Old World camels exhibit extraordinary ecological plasticity. Their resilience is due to the close integration of genetic variation and physiological function, which shape adaptive responses to environmental stressors and infectious challenges. Thus, they can thrive in harsh environments where other livestock species cannot (Alim et al. 2019; Bagiyal et al. 2024; Kandeel et al. 2022; Ramadan and Inoue-Murayama 2017; Wu et al. 2014).

This resilience is closely tied to traditional management systems that involve seasonal migrations, communal grazing lands, and access to shared water sources (Decker et al. 2025; Habte et al. 2022; Wako et al. 2017). Migration patterns may preserve immunogenetic diversity and enhance resistance to locally circulating pathogens by maintaining gene flow across populations (Berthouly et al. 2009). Yet they also increase exposure to diverse environmental conditions and contact with other domestic and wild species. The resulting epidemiological mosaic can enhance viral recombination, bacterial spillover, and the amplification of parasitic vectors (Vicente et al. 2021).

From a sociocultural viewpoint, another risk factor in this context is the level of awareness of pastoralist communities about camel diseases and healthcare (Kuria et al. 2021). Moreover, pastoralist camel production is often associated with limited access to veterinary infrastructure and fewer opportunities for systematic disease surveillance and control (Zhu et al. 2019). Despite these constraints, many pastoralist communities possess sophisticated ethno-veterinary knowledge (Antoine-Moussiaux et al. 2005). Gossner et al. (2016) reported a high degree of awareness regarding seasonality, symptomatology, and transmission of camel diseases, particularly in calves, among Somali camel pastoralists. Milking the camels in a predefined order to diminish the spread of disease in the herd was also reported as a common practice among Kenyan pastoralists (Seligsohn et al. 2025). This suggests that the long tradition of camel herding has resulted in observational learning that has influenced local knowledge of camel diseases. This traditional expertise plays a critical role in early disease recognition and response, especially where diagnostic and therapeutic facilities for zoonoses are lacking. Nonetheless, this knowledge, while valuable, is not a substitute for systematic veterinary diagnosis and genomic surveillance, and its limitations must be acknowledged when designing comprehensive One Health strategies.

4.4.2 Camel Behavior as a Driver of Disease Interface Complexity

Camel behavior plays a critical role in modulating exposure to pathogens and vectors. Close social bonding and communal grazing increase the likelihood of within-herd transmission of respiratory or fecal-oral pathogens (Mirkena et al. 2018). In extremely hot environments, watering points tend to be limited and low in volume, leading to high animal density and elevated frequency of camel-to-camel contact at these sites (Higgins 1986). Mating behavior also involves close physical interaction and exchange of fluids, which may further elevate disease transmission risk, particularly in the absence of assisted-reproduction technologies (Zema et al. 2022).

These natural camel behaviors combine with the host's genetic background to influence disease risk. For example, herds with high heterozygosity in loci for innate immunity may exhibit improved pathogen clearance, even under high-contact scenarios (Minias and Vinkler 2022). However, reduced genomic diversity is not necessarily deleterious, as several studies have identified increased frequency of homozygous genomic segments (e.g., runs of homozygosity or ROH) that harbor favorable genes (Liu et al. 2022). In camels, gene set enrichment analyses of ROH regions have revealed the presence of advantageous immune-related genes including CX3CR1, IL6R, CCR8, CXCL9, CXCL10, and CXCL11 (Ibrahim et al. 2025; Khalkhali-Evrigh and Hedayat 2023; Tao et al. 2023), suggesting adaptative refinement to pathogen-rich environments.

Conversely, human-induced changes in camel behavior and management, such as sedentary rearing, overstocking, and confinement in peri-urban and intensive systems, can induce chronic stress and immunosuppression. This condition impairs host immune defenses and exacerbates disease susceptibility (Inbaraj et al. 2022). In such systems, the selective breeding of a limited number of high-performing individuals may result in higher levels of inbreeding and reduced overall fitness. Despite the potential enrichment of favorable alleles within ROH islands, genomic monitoring remains essential to prevent inbreeding depression and to preserve immune diversity by avoiding mating among related individuals.

4.4.3 Environmental Stressors and Immune Vulnerability

Environmental pressures such as prolonged drought, habitat degradation, and land-use change exert chronic physiological stress on camel populations. These factors can impair immune function by disrupting energy homeostasis, elevating glucocorticoid levels, and compromising nutrient absorption (Carroll et al. 2012; Inbaraj et al. 2022).

Crucially, the impact of these stressors is not uniform across individuals or herds, and some stress adaptations may even have a heritable basis (Tando and Matsui 2023). Genetic background plays an role in modulating physiological and immunological responses, meaning that populations with greater genetic diversity are more likely to buffer these adverse effects and maintain functional immunity, particularly in genes linked to stress and immune regulation (Erwan et al. 2015).

Ecological disruptions also increase pathogen exposure. For example, drought-induced overgrazing and resource scarcity can push herds into novel habitats where they come into contact with previously unencountered pathogen reservoirs or vectors (du Toit et al. 2017).

Vector ecology is also shifting. Climate-driven changes are altering the population dynamics of key vectors such as *Glossina* and *Stomoxys* spp., both implicated in the transmission of camel-relevant zoonoses like trypanosomiasis and piroplasmosis. Extinction risk modelling for tsetse flies (*Glossina* spp.) predicts range constraints under global warming scenarios (Terblanche et al. 2008), although real-world populations may persist in microhabitats with milder thermal profiles (Are and Hargrove 2020). In contrast, biting flies of the genus *Stomoxys* are expected to increase in density, particularly in urban areas, due to higher maximum temperatures (Dafka et al. 2025).

Environmental stressors may amplify vulnerability to infectious disease and facilitate localized outbreaks for camel populations that are geographically isolated and especially intensively selected. These interaction zones represent critical frontiers for disease emergence and underscore the need for geographically focused genetic monitoring.

4.5 Human–Camel Interfaces and Host–Pathogen Coevolution

The camel–human interface is defined by sustained and multifaceted contact. Camels play a central role in providing food security, mobility, and socio-economic stability for traditional pastoralist and seminomadic groups across regions such as the Sahel, Horn of Africa, Middle East, and Central Asia (Faye 2015). Microbial exchange is facilitated through practices including raw milk consumption, cohabitation of living spaces, use in religious rituals and livestock markets, and routine handling during transport or breeding (Amenu et al. 2019; Zhu et al. 2019). Additionally, mass gatherings such as camel races, wrestling events, or beauty contests attract large numbers of spectators, many of whom do not have regular contact with camels and may therefore lack prior immunological exposure. This could increase their susceptibility to potential infections or infestations acquired through close or incidental contact. Moreover, strict screening regulations should be implemented for camels and humans participating in mass gathering events (Petersen et al. 2020; Zumla et al. 2016).

In recent decades, the socio-economic interest in camel production has expanded significantly on a global scale. This trend is transforming the nature of human–camel interactions through the emergence of new production systems and functional niches. Beyond conventional roles in milk, meat, and labor, camels are increasingly involved in recreational activities (e.g., camelback leisure riding), therapeutic programs, and education-related services (Faye 2020; Iglesias Pastrana et al. 2024a, b; Khalafalla et al. 2021). These emerging uses introduce new contexts of close human–camel contact (often in peri-urban or nontraditional environments) which have not yet been systematically integrated into zoonotic risk assessments.

This evolving interface fosters dynamic pathogen exchange. Unlike isolated spillover events, these interactions are often bidirectional and recurrent, providing fertile ground for a wide range of pathogens to adapt to multiple hosts. For instance, in the case of MERS-CoV, Letko et al. (2018) showed that the virus can swiftly adjust to host-specific variations in the DPP4 receptor-binding domain through multiple mutational routes. While this illustrates the adaptive potential of coronaviruses, similar evolutionary mechanisms may operate in other pathogens exposed to sustained interspecies circulation (Mackenzie et al. 2007), underscoring the broader relevance of co-adaptation at the camel–human interface.

At the same time, persistent exposure to shared pathogens may drive immunogenetic selection in both camels and humans, especially in loci related to innate and adaptive immunity (e.g., MHC, ISGs, TLRs). This may result in localized immune adaptation, influencing not only susceptibility and symptomatology but also pathogen shedding duration, which is a key factor in reservoir potential and zoonotic risk (Mackenzie et al. 2007).

These socio-ecological settings have limited access to veterinary services, lack of professional specialization in camel health, infrequent genomic monitoring, and fragmented public health interventions (Mackenzie et al. 2007; Mohamed et al. 2024). Such conditions both amplify hosts' vulnerability to circulating pathogens and delay the detection of potentially more virulent genetic shifts in those pathogens.

Host–pathogen dynamics across this interface are an active evolutionary arena, shaped by the interplay of genetic variability, contact intensity, and microbial adaptability. One Health strategies must therefore adopt a longitudinal, integrative approach that includes real-time genomic surveillance of both hosts and pathogens. They should also account for regional differences in contact patterns, production systems, and management practices. Camelid One Health frameworks must recognize the full spectrum of camel–human interactions, including emerging recreational and therapeutic contexts, for accurate risk assessment and targeted intervention design.

4.6 Genomic Applications for Camel Conservation and One Health

Genomic research has potential for applications in camel conservation and disease mitigation, especially under the One Health paradigm. By linking genetic diversity to pathogen susceptibility, adaptive capacity, and breeding performance, genomics offers powerful tools to preserve camelid biodiversity while managing zoonotic risk in a changing world.

4.6.1 Conservation-Oriented Genomic Strategies

Preserving the genetic diversity of camel populations is not only vital for evolutionary and ecological reasons, but also for safeguarding animal and human health. As outlined in earlier sections, the erosion of immunogenetic variability—whether through domestication bottlenecks, intensive selection, or unmanaged hybridization—can reduce immune responsiveness and increase vulnerability to emerging infectious diseases.

To address these challenges, conservation and livestock development programs are increasingly turning to genomic approaches. Recent advances in long-read and scaffolding technologies have significantly improved the quality of camel genome assemblies. Early drafts of camelid genomes (e.g., Wu et al. 2014; Jirimutu et al. 2013) relied primarily on short-read sequencing and yielded fragmented assemblies with limited resolution in repetitive or immune-relevant regions. In contrast, newer genome assemblies achieve near chromosome-level resolution, allowing a much more accurate reconstruction of the camel genome (Elbers et al. 2019; Ming et al. 2020a). These improvements have enabled detailed characterization of complex immune gene regions, such as immunoglobulin, T-cell receptor, and MHC gene clusters, which were previously difficult to analyze. Such advances are critical for immunogenomic surveillance, as they improve our ability to detect genetic variation in key immune genes involved in pathogen recognition and disease resistance. Overall, modern genomic technologies are providing unprecedented insight into camel immunogenetic architecture and its relevance for health and conservation.

Transcriptomic datasets that reveal gene expression profiles across tissues, physiological states, and environmental conditions complement these efforts. Notably, over 4500 potentially novel or rapidly evolving genes have been identified (Al-Swailem et al. 2010), offering a reservoir of functional diversity for disease resistance and stress tolerance.

Recent resequencing initiatives—covering more than 400 genomes—aimed to map genome-wide variation across the camelid distribution range. These projects have enabled the development of tools like the Illumina® CamelHD BeadChip, a high-density SNP genotyping platform tailored for camel trait selection and diversity monitoring (Bruno et al. 2023; Burger et al. 2019).

Such genomic infrastructure will support:

- The detection of genetic introgression, particularly between *C. ferus* and domestic camels
- Marker-assisted selection for immune-relevant loci in breeding schemes
- Maintenance of diversity in herds managed under traditional systems
- Integration of genomic surveillance into disease early warning systems

In summary, safeguarding camel genomic integrity is foundational to ensure long-term resilience—ecological, immunological, and economic.

4.6.2 Genomics in Disease Monitoring and Biomedical Innovation

Genomic surveillance is becoming a vital component of disease monitoring and risk mitigation in camelid systems. By characterizing variation in immune-related genes and linking it to resistance phenotypes, genomic tools support early detection, resilience-oriented breeding, and targeted intervention which are core pillars of One Health strategies.

Technologies such as whole-genome sequencing (WGS) and genome-wide association studies (GWAS) enable the identification of alleles associated with immune robustness, stress tolerance, and pathogen clearance. These approaches help detect emerging signals of selection across populations, flag genetic erosion before it impacts herd health, and uncover variants that can be leveraged in breeding programs to improve resistance traits, not only to specific pathogens but across diverse infectious threats. In addition, environmental DNA (eDNA) sequencing can provide complementary insights by monitoring pathogen presence and diversity in shared environments, thereby linking host genetic susceptibility with real-world exposure and transmission risk.

Integrating such host genomic data with ecological and management information enables the identification of high-risk herds, genotypes under pressure, and geographic hotspots of reduced genetic variability. This multidimensional perspective is crucial not only for anticipating disease outbreaks but also for designing evidence-based interventions aimed at preserving adaptive potential in camel populations. For instance, genomic surveillance in regions where infectious disease pressures are recurrent can help pinpoint vulnerable or epidemiologically significant lineages. Screening for resilient alleles in key immune loci—such as MHC or TLRs—supports the selection of individuals with enhanced immunocompetence. These data-driven strategies contribute to building herd-level robustness and long-term biosecurity, particularly in the face of intensifying environmental stress and increasingly complex production systems.

Complementary to this, metagenomic sequencing captures interactions between host genetics, microbiota, and pathogens (Miller et al. 2013). This systems-level view can provide insight into how microbial communities modulate camel immune responses and pathogen susceptibility—information that can support more precise vaccination strategies, targeted probiotic use, and better antimicrobial stewardship.

Mitochondrial DNA (mtDNA) is a valuable marker for camel breeding and conservation, offering insights into maternal lineages, gene flow, and introgression—particularly in efforts to preserve *C. ferus*. Specific mtDNA haplotypes have been linked to traits like stamina and thermotolerance, making it a useful tool for selecting resilient phenotypes under environmental and health pressures (Manee et al. 2024).

To operationalize these tools, genetic monitoring must be institutionalized across veterinary and public health infrastructures. This includes:

- Integrating genetic diversity metrics into herd health evaluations
- Establishing local capacity for genomic diagnostics and training
- Promoting cross-sector data-sharing systems

In sum, as genomic resources and technologies become more accessible, they must be deployed not only to understand past domestication and adaptation events but to guide present-day decisions. By embedding immunogenetic and functional diversity criteria into selection frameworks, camel breeding and health management can shift from reactive to proactive—anchored in science, adapted to context, and aligned with One Health principles.

4.7 Conclusions and Future Directions

Understanding the genetic foundations of disease resistance in camels is no longer a scientific curiosity, it is a necessity. As camel populations expand their ecological and socio-economic roles, particularly under shifting environmental and production paradigms, the need to integrate immunogenetic diversity into health and conservation strategies becomes urgent.

This chapter demonstrates that genetic variation, shaped by evolutionary history, domestication, and hybridization, modulates susceptibility to zoonotic pathogens. Immune gene diversity, maintained through extensive management and population mobility, acts as a buffer against emerging infectious threats. However, this natural resilience is increasingly threatened by demographic bottlenecks, inbreeding, and selective pressures from intensifying production systems.

At the same time, the camel–human interface is evolving, creating novel zones of high-contact and microbial exchange—from traditional pastoralist systems to emerging niches like tourism, therapy, and peri-urban farming. These contexts challenge existing models of zoonotic risk and call for surveillance systems that reflect the spatial, behavioral, and genetic complexity of camelid systems.

Modern genomics offers the tools to address these challenges. Whole-genome and transcriptomic resources, immune gene mapping, and field-deployable sequencing can now inform breeding, conservation, and pathogen monitoring. However, translating this capacity into operational One Health strategies requires institutional investment in diagnostics, data infrastructure, and interdisciplinary training, especially in regions where camels are most vital and least studied.

Future research should prioritize:

- Longitudinal studies linking genotypes to immunity and infection outcomes
- Functional validation of candidate immune variants
- Spatial integration of genomic, ecological, and behavioral datasets
- Conservation genomics for critically endangered populations like *C. ferus*
- Inclusive, locally grounded approaches to capacity building and technology transfer

In summary, genomic diversity in camels is both a biological resource and a public health asset. Safeguarding and leveraging this diversity is essential to navigating the dual pressures of zoonotic emergence and sustainable livestock development in camel-dependent regions.

Acknowledgements **This work was carried out during the tenure of a "Juan de la Cierva" postdoctoral contract (2023 Call), funded by the Spanish State Agency for Research.**

References

Abri MAA, Faye B (2019) Genetic improvement in dromedary camels: challenges and opportunities. Front Genet 10:167

Adney DR, van Doremalen N, Brown VR, Bushmaker T, Scott D, de Wit E, Bowen RA, Munster VJ (2014) Replication and shedding of MERS-CoV in upper respiratory tract of inoculated dromedary camels. Emerg Infect Dis 20(12):1999

Al Abri M, Alfoudari A, Mohammad Z, Almathen F, Al-Marzooqi W, Al-Hajri S, Al-Amri M, Bahbahani H (2023) Assessing genetic diversity and defining signatures of positive selection on the genome of dromedary camels from the southeast of the Arabian Peninsula. Front Vet Sci 10:1296610

Al Naeem A, Salem M, Housawi F, Al-Mohammed Salem K, Hussen J, Fayez M, Zaghawa A, Kostoulas P (2024) Epidemiological insights into paratuberculosis in camels in Saudi Arabia: Bayesian estimation of true prevalence and identification of risk factors. PLoS One 19(3):e0299881

Al Ramadan SY, Salem KTA-M, Alshubaith IH, Al-Ali AM, Abohelaika S, Moqbel MS, Alluwaimi AM (2021) Selected aspects of camel immune system and immune responses. Open J Vet Med 11(6):177–211

Alagaili AN, Briese T, Mishra N, Kapoor V, Sameroff SC, de Wit E, Munster VJ, Hensley LE, Zalmout IS, Kapoor A (2014) Middle East respiratory syndrome coronavirus infection in dromedary camels in Saudi Arabia. MBio 5(2):e00884-14. https://doi.org/10.1128/mbio.00884-00814

Ali A, Baby B, Vijayan R (2019) From desert to medicine: a review of camel genomics and therapeutic products. Front Genet 10:17

Alim FZD, Romanova EV, Tay Y-L, Rahman AYBA, Chan K-G, Hong K-W, Rogers M, Southey BR, Greenwood MP, Mecawi AS (2019) Seasonal adaptations of the hypothalamo-neurohypophyseal system of the dromedary camel. PLoS One 14(6):e0216679

Almathen F (2024) Genomic signatures of positive selection in Awarik dromedary camels from southwestern of Saudi Arabia. Front Vet Sci 11:1443748

Almathen F, Charruau P, Mohandesan E, Mwacharo JM, Orozco-terWengel P, Pitt D, Abdussamad AM, Uerpmann M, Uerpmann H-P, De Cupere B (2016) Ancient and modern DNA reveal dynamics of domestication and cross-continental dispersal of the dromedary. Proc Natl Acad Sci 113(24):6707–6712

Al-Swailem AM, Shehata MM, Abu-Duhier FM, Al-Yamani EJ, Al-Busadah KA, Al-Arawi MS, Al-Khider AY, Al-Muhaimeed AN, Al-Qahtani FH, Manee MM (2010) Sequencing, analysis, and annotation of expressed sequence tags for *Camelus dromedarius*. PLoS One 5(5):e10720

Amenu K, Wieland B, Szonyi B, Grace D (2019) Milk handling practices and consumption behavior among Borana pastoralists in Southern Ethiopia. J Health Popul Nutr 38:1–12

Antoine-Moussiaux N, Faye B, Vias GF (2005) Tuareg ethnodiagnostic skill of camel diseases in Agadez area (Niger). J Camel Pract Res 12(2):85

Antonacci R, Mineccia M, Lefranc M-P, Ashmaoui HM, Lanave C, Piccinni B, Pesole G, Hassanane MS, Massari S, Ciccarese S (2011) Expression and genomic analyses of *Camelus dromedarius* T cell receptor delta (TRD) genes reveal a variable domain repertoire enlargement due to CDR3 diversification and somatic mutation. Mol Immunol 48(12–13):1384–1396

Arbabi-Ghahroudi M (2017) Camelid single-domain antibodies: historical perspective and future outlook. Front Immunol 8:1589

Are EB, Hargrove JW (2020) Extinction probabilities as a function of temperature for populations of tsetse (*Glossina* spp.). PLoS Negl Trop Dis 14(5):e0007769

Azhar EI, Velavan TP, Rungsung I, Traore T, Hui DS, McCloskey B, El-Kafrawy SA, Zumla A (2023) Middle East respiratory syndrome coronavirus—a 10-year (2012-2022) global analysis of human and camel infections, genomic sequences, lineages, and geographical origins. Int J Infect Dis 131:87–94

Bagiyal M, Parsad R, Ahlawat S, Gera R, Chhabra P, Sharma U, Arora R, Sharma R (2024) Review on camel genetic diversity: ecological and economic perspectives. Mamm Genome 35(4):1–12

Bahbahani H, Musa HH, Wragg D, Shuiep ES, Almathen F, Hanotte O (2019) Genome diversity and signatures of selection for production and performance traits in dromedary camels. Front Genet 10:893

Bahbahani H, Alfoudari A, Al-Ateeqi A, Al Abri M, Almathen F (2024) Positive selection footprints and haplotype distribution in the genome of dromedary camels. Animal 18(3):101098

Bahbahani H, Mohammad Z, Alfoudari A, Al Abri M (2025) Genomic insights into racing camels: inbreeding levels and positive selection linked to athletic traits. Animal 19(4):101467

Balard A, Heitlinger E (2022) Shifting focus from resistance to disease tolerance: a review on hybrid house mice. Ecol Evol 12(5):e8889

Bannas P, Lenz A, Kunick V, Well L, Fumey W, Rissiek B, Haag F, Schmid J, Schütze K, Eichhoff A (2015) Molecular imaging of tumors with nanobodies and antibodies: timing and dosage are crucial factors for improved in vivo detection. Contrast Media Mol Imaging 10(5):367–378

Berthouly C, Do Ngoc D, Thevenon S, Bouchel D, Nhu Van T, Danes C, Grosbois V, Hoang Thanh H, Vu Chi C, Maillard JC (2009) How does farmer connectivity influence livestock genetic structure? A case-study in a Vietnamese goat population. Mol Ecol 18(19):3980–3991

Bishnoi S, Jyotsana B, Kumar V, Prakash V, Ranjan R, Narnaware SD, Pannu U (2024) Toll like receptor 4 (TLR4) gene polymorphism and its association with somatic cell score and milk production traits in Indian dromedary camels. Anim Biotechnol 35(1):2331642

Bitaraf Sani M, Zare Harofte J, Banabazi MH, Faraz A, Esmaeilkhanian S, Naderi AS, Salim N, Teimoori A, Bitaraf A, Zadehrahmani M (2022) Identification of candidate genes for pigmentation in camels using genotyping-by-sequencing. Animals 12(9):1095

Bruno S, Senczuk G, Landi V, Di Civita M, Brooks SA, Almathen F, Faye B, Gaouar SSB, Piro M, Kim KS, Dadi H, Iglesias Pastrana C, Al-Haddad H, Al-Abri M, David X, Eggen A, Burger P, Ciani E (2023) The illumina® agricultural greater good initiative: development of a medium-Density SNP chip for camels. In: The 6th conference of the International Society of Camelid Research and Development (ISOCARD), Al Ahsa, Saudi Arabia

Burger PA (2016) The history of Old World camelids in the light of molecular genetics. Trop Anim Health Prod 48(5):905–913

Burger PA, Ciani E, Faye B (2019) Old World camels in a modern world–a balancing act between conservation and genetic improvement. Anim Genet 50(6):598–612

Çakirlar C, Berthon R (2014) Caravans, camel wrestling and cowrie shells: towards a social zooarchaeology of camel hybridization in Anatolia and adjacent regions. Anthropozoologica 49(2):237–252

Carroll J, Burdick N, Chase C Jr, Coleman S, Spiers D (2012) Influence of environmental temperature on the physiological, endocrine, and immune responses in livestock exposed to a provocative immune challenge. Domest Anim Endocrinol 43(2):146–153

Carvalho I, Tejedor-Junco MT, González-Martín M, Corbera JA, Silva V, Igrejas G, Torres C, Poeta P (2020) Escherichia coli producing extended-spectrum β-lactamases (ESBL) from domestic camels in the Canary Islands: a one health approach. Animals 10(8):1295

Choudhary J, Purva M, Milind M, Gahlot K, Bhati T, Maherchandani S, Pandey A (2021) Phylogenetic and protein domain architecture analysis of toll-like receptor 4-10 genes in camel (*Camelus dromedarius*). Int J Curr Microbiol App Sci 10(01):2887–2895

Ciccarese S, Burger PA, Ciani E, Castelli V, Linguiti G, Plasil M, Massari S, Horin P, Antonacci R (2019) The camel adaptive immune receptors repertoire as a singular example of structural and functional genomics. Front Genet 10:997

Dafka S, Osman AM, Koua HK, Vieira RF, Rocklöv J (2025) Impact of climate change and variability on the occurrence and distribution of Trypanosoma vectors in the Gambia. Parasitol Res 124(3):1–18

Dahiya SS, Nagarajan G, Bharti VK, Swami SK, Mehta SC, Tuteja FC, Narnaware SD, Patil N (2014) Sequence analysis of the Toll-like receptor 2 gene of old world camels. J Adv Res 5(6):695–704

De Genst E, Saerens D, Muyldermans S, Conrath K (2006) Antibody repertoire development in camelids. Dev Comp Immunol 30(1–2):187–198

Decker E, Flueckiger RM, Frumkin M, Jeudin RH, Celestin C, Essa AKO, Yilmaz EE (2025) Camel leasing as a resilience-building practice: insights from Somali pastoralist households and dairy farms. World Dev Perspect 37:100668

du Toit JT, Cross PC, Valeix M (2017) Managing the livestock–wildlife interface on rangelands. In: Rangeland systems: processes, management and challenges. Springer, pp 395–425

Dudas G, Carvalho LM, Rambaut A, Bedford T (2018) MERS-CoV spillover at the camel-human interface. elife 7:e31257

Elbers JP, Rogers MF, Perelman PL, Proskuryakova AA, Serdyukova NA, Johnson WE, Horin P, Corander J, Murphy D, Burger PA (2019) Improving Illumina assemblies with Hi-C and long reads: an example with the North African dromedary. Mol Ecol Resour 19(4):1015–1026

El-Kafrawy SA, Corman VM, Tolah AM, Al Masaudi SB, Hassan AM, Müller MA, Bleicker T, Harakeh SM, Alzahrani AA, Alsaaidi GA (2019) Enzootic patterns of Middle East respiratory syndrome coronavirus in imported African and local Arabian dromedary camels: a prospective genomic study. Lancet Planet Health 3(12):e521–e528

Elmahallawy EK, Köster PC, Dashti A, Alghamdi SQ, Saleh A, Gareh A, Alrashdi BM, Hernández-Castro C, Bailo B, Lokman MS (2023) Molecular detection and characterization of *Cryptosporidium* spp., *Giardia duodenalis*, and *Enterocytozoon bieneusi* infections in dromedary camels (*Camelus dromedaries*) in Egypt. Front Vet Sci 10:1139388

Elsayed AA, Barghash SM, El-Kattan AM, Wassif IM, Osman WA, Ateya AI (2024) Analysis of potential genes, immunological and antioxidant profiles associated with trypanosomiasis susceptibility in dromedary camels. Vet Parasitol 331:110264

Erwan Q, Galan M, Cosson J-F, Klein F (2015) Immunogenetic heterogeneity in a widespread ungulate. Mol Ecol 24:3873–3887

Faye B (2015) Role, distribution and perspective of camel breeding in the third millennium economies. Emir J Food Agric 27(4):318–327

Faye B (2019) TADs in the dromedary. In: Moustafa Kardjadj AD, Lancelot R (eds) Transboundary animal diseases in Sahelian Africa and connected regions. Springer, pp 91–103

Faye B (2020) How many large camelids in the world? A synthetic analysis of the world camel demographic changes. Pastoralism 10(1):25

Faye B, Konuspayeva G (2012) The encounter between Bactrian and dromedary camels in Central Asia, vol 451. Verlag der Österreichischen Akademie der Wissenschaften

Felkel S, Wallner B, Chuluunbat B, Yadamsuren A, Faye B, Brem G, Walzer C, Burger PA (2019) A first Y-chromosomal haplotype network to investigate male-driven population dynamics in domestic and wild Bactrian camels. Front Genet 10:423

Fitak RR, Mohandesan E, Corander J, Yadamsuren A, Chuluunbat B, Abdelhadi O, Raziq A, Nagy P, Walzer C, Faye B (2020) Genomic signatures of domestication in Old World camels. Commun Biol 3(1):316

Futas J, Oppelt J, Jelinek A, Elbers JP, Wijacki J, Knoll A, Burger PA, Horin P (2019) Natural killer cell receptor genes in camels: another mammalian model. Front Genet 10:620

Futas J, Oppelt J, Burger PA, Horin P (2021) A deadly cargo: gene repertoire of cytotoxic effector proteins in the camelidae. Gene 12(2):304

Gahlot RK, Jyotsana B, Prakash V, Narnaware SD, Rathore N (2023) Molecular characterization of toll-like receptor 2 gene in Bikaneri camels (*Camelus dromedarius*). Indian J Anim Sci 93(7):706–710

Gentry A, Clutton-Brock J, Groves CP (2004) The naming of wild animal species and their domestic derivatives. J Archaeol Sci 31(5):645–651

Ghosh P, Hsu C, Alyamani EJ, Shehata MM, Al-Dubaib MA, Al-Naeem A, Hashad M, Mahmoud OM, Alharbi KB, Al-Busadah K (2012) Genome-wide analysis of the emerging infection with *Mycobacterium avium* subspecies paratuberculosis in the Arabian camels (*Camelus dromedarius*). PLoS One 7(2):e31947

Gossner C, Danielson N, Gervelmeyer A, Berthe F, Faye B, Kaasik Aaslav K, Adlhoch C, Zeller H, Penttinen P, Coulombier D (2016) Human–dromedary camel interactions and the risk of

acquiring zoonotic Middle East respiratory syndrome coronavirus infection. Zoonoses Public Health 63(1):1–9
Guo F, Ming L, Si R, Yi L, He J, Ji R (2020) A genome-wide association study identifies quantitative trait loci affecting hematological traits in *Camelus bactrianus*. Animals 10(1):96
Habte M, Eshetu M, Maryo M, Andualem D, Legesse A (2022) Effects of climate variability on livestock productivity and pastoralists perception: the case of drought resilience in Southeastern Ethiopia. Vet Anim Sci 16:100240
Hamers-Casterman C, Atarhouch T, Muyldermans SA, Robinson G, Hammers C, Songa EB, Bendahman N, Hammers R (1993) Naturally occurring antibodies devoid of light chains. Nature 363(6428):446–448
Higgins A (1986) The camel in health and disease. Baillière Tindall London
Hussein MF, Alshaikh MA, Al-Jumaah RS, GarelNabi A, Al-Khalifa I, Mohammed OB (2015) The Arabian camel (*Camelus dromedarius*) as a major reservoir of Q fever in Saudi Arabia. Comp Clin Pathol 24(4):887–892
Hussen J, Schuberth H-J (2021) Recent advances in camel immunology. Front Immunol 11:614150
Ibrahim MA, Tolone M, Barbato M, Alsubaie FM, Alrefaei AF, Almutairi M (2025) Geographical distribution, genetic diversity, and environmental adaptations of dromedary camel breeds in Saudi Arabia. Front Vet Sci 11:1490186
Iglesias Pastrana C, Navas González FJ, Delgado Bermejo JV, Baum D, Crenshaw V, Larsson M, Brachmann J, Adams C, Arevalo S, Kaufmann M (2024a) Camel-assisted services (CAS): treatment, education and support programmes. J Camel Pract Res 31(3):257–268
Iglesias Pastrana C, Navas González FJ, Macri M, Martínez Martínez MDA, Ciani E, Delgado Bermejo JV (2024b) Identification of novel genetic loci related to dromedary camel (*Camelus dromedarius*) morphometrics, biomechanics, and behavior by genome-wide association studies. BMC Vet Res 20(1):418
Inbaraj S, Sejian V, Ramasamy S (2022) Role of environmental stressor-host immune system–pathogen interactions in development of infectious disease in farm animals. Biol Rhythm Res 53(5):707–724
Ji R, Cui P, Ding F, Geng J, Gao H, Zhang H, Yu J, Hu S, Meng H (2009) Monophyletic origin of domestic Bactrian camel (*Camelus bactrianus*) and its evolutionary relationship with the extant wild camel (*Camelus bactrianus ferus*). Anim Genet 40(4):377–382
Jirimutu WZ, Ming L (2018) Genome sequencing highlights the origin and evolution of Bactrian camel. In: 3rd international scientific conference—impact of climate change and environment on animal production: advantages, constraints and perspectives for the Dromedar, Faculté des Sciences Ben M'Sik, Casablanca, Morocco
Jirimutu WZ, Ding G, Chen G, Sun Y, Sun Z, Zhang H (2013) The Bactrian camels genome sequencing and analysis consortium. Correction: corrigendum: genome sequences of wild and domestic Bactrian camels. Nat Commun 4:2089
Kamber R, Farah Z, Rüsch P, Hässig M (2001) Studies on the supply of immunoglobulin G to newborn camel calves (*Camelus dromedarius*). J Dairy Res 68(1):1–7
Kandeel M, Al-Taher A, Venugopala KN, Marzok M, Morsy M, Nagaraja S (2022) Camel proteins and enzymes: a growing resource for functional evolution and environmental adaptation. Front Vet Sci 9:911511
Kena D (2022) Review on camel production and marketing status in Ethiopia. Pastoralism 12(1):38
Khalafalla AI (2023) Zoonotic diseases transmitted from the camels. Front Vet Sci 10:1244833
Khalafalla AI, Hussein MF, Bornstein S (2021) Evolution, distribution, and economic importance of the camels. In: Infectious diseases of dromedary camels: a concise guide. Springer, pp 1–19
Khalkhali-Evrigh R, Hedayat N (2023) Identification of genomic runs of homozygosity and investigation of related genes in the dromedary camels using whole-genome sequencing data. Iran J Anim Sci 54(2):105–116
Khalkhali-Evrigh R, Hedayat N, Ming L, Jirimutu. (2022) Identification of selection signatures in Iranian dromedary and Bactrian camels using whole genome sequencing data. Sci Rep 12(1):9653

Khomeiri M (2015) Introduction to camel origin, history, raising, characteristics, and wool, hair and skin, a review. EPH-Int J Agric Environ Res 1(1):1–12

Kulkarni SS, Falzarano D (2021) Unique aspects of adaptive immunity in camelids and their applications. Mol Immunol 134:102–108

Kuria S, Agwanda B, Ogada S, Panyako P, Mungube E, Masika M, Ommeh S (2021) Risk factors predisposing dromedary camels and pastoral communities to zoonotic diseases in Kenya. J Agric Sci Technol 20(2):jagst.v20i2

Lado S, Elbers J, Doskocil A, Ciani E, Burger P (2018) Genome-wide diversity and demographic history in the global dromedary population. In: 5th conference of ISOCARD 'Recent advances in camelids: biology, health and production'

Lado S, Elbers JP, Rogers MF, Melo-Ferreira J, Yadamsuren A, Corander J, Horin P, Burger PA (2020) Nucleotide diversity of functionally different groups of immune response genes in Old World camels based on newly annotated and reference-guided assemblies. BMC Genomics 21:1–17

Lado S, Elbers JP, Plasil M, Loney T, Weidinger P, Camp JV, Kolodziejek J, Futas J, Kannan DA, Orozco-terWengel P (2021) Innate and adaptive immune genes associated with MERS-CoV infection in dromedaries. Cells 10(6):1291

Lee SY, Lee BY, Lim B, Uzzaman R, Jang G, Kim KS (2025) Exploring the importance of predicted camel NRAP exon 4 for environmental adaptation using a mouse model. Anim Genet 56(1):e13490

Lei Y, Hare J, Guoying Y, Yun C (2012) The status of the wild camel in China, vol 451. Verlag der Österreichischen Akademie der Wissenschaften

Letko M, Miazgowicz K, McMinn R, Seifert SN, Sola I, Enjuanes L, Carmody A, van Doremalen N, Munster V (2018) Adaptive evolution of MERS-CoV to species variation in DPP4. Cell Rep 24(7):1730–1737

Lin P, Adem A, Murphy D, Greenwood M (2025) How does the camel survive in the desert without drinking? In: Evolutionary and comparative neuroendocrinology, pp 389–423. https://doi.org/10.1007/978-3-031-80209-6_13

Liu Y, Zhao G, Lin X, Zhang J, Hou G, Zhang L, Liu D, Li Y, Li J, Xu L (2022) Genomic inbreeding and runs of homozygosity analysis of indigenous cattle populations in southern China. PLoS One 17(8):e0271718

Mackenzie J, Childs J, Richt J (2007) The biology, circumstances and consequences of cross-species transmission. Springer

Magee P (2015) When was the dromedary domesticated in the ancient Near East? Zeitschrift für Orient-Archaeologie 8:252–277

Manee MM, Al-Shomrani BM, Alqahtani FH (2024) Mitochondrial DNA of the Arabian Camel *Camelus dromedarius*. Animals 14(17):2460

Miller RR, Montoya V, Gardy JL, Patrick DM, Tang P (2013) Metagenomics for pathogen detection in public health. Genome Med 5:1–14

Ming L, Yi L, Sa R, Wang Z, Wang Z, Ji R (2017) Genetic diversity and phylogeographic structure of Bactrian camels shown by mitochondrial sequence variations. Anim Genet 48(2):217–220

Ming L, Wang Z, Yi L, Batmunkh M, Liu T, Siren D, He J, Juramt N, Jambl T, Li Y (2020a) Chromosome-level assembly of wild Bactrian camel genome reveals organization of immune gene loci. Mol Ecol Resour 20(3):770–780

Ming L, Yuan L, Yi L, Ding G, Hasi S, Chen G, Jambl T, Hedayat-Evright N, Batmunkh M, Badmaevna GK (2020b) Whole-genome sequencing of 128 camels across Asia reveals origin and migration of domestic Bactrian camels. Commun Biol 3(1):1

Ming L, Siren D, Hasi S, Jambl T, Ji R (2022) Review of genetic diversity in Bactrian camel (*Camelus bactrianus*). Anim Front 12(4):20–29

Minias P, Vinkler M (2022) Selection balancing at innate immune genes: adaptive polymorphism maintenance in toll-like receptors. Mol Biol Evol 39(5):msac102

Mirkena T, Walelign E, Tewolde N, Gari G, Abebe G, Newman S (2018) Camel production systems in Ethiopia: a review of literature with notes on MERS-CoV risk factors. Pastoralism 8(1):30

Mirzaei F (2000) Production and trade of camel products in some Middle East countries. Population 10000(16603147):3781158

Mohamed M-YI, Lakshmi GB, Sodagari H, Habib I (2024) A one health perspective on camel meat hygiene and zoonoses: insights from a decade of research in the Middle East. Vet Sci 11(8):344

Mohandesan E, Fitak RR, Corander J, Yadamsuren A, Chuluunbat B, Abdelhadi O, Raziq A, Nagy P, Stalder G, Walzer C (2017) Mitogenome sequencing in the genus Camelus reveals evidence for purifying selection and long-term divergence between wild and domestic Bactrian camels. Sci Rep 7(1):9970

Müller MA (2014) MERS-CoV. BMC Infect Dis 14(Suppl 2):S22

Muyldermans S (2013) Nanobodies: natural single-domain antibodies. Annu Rev Biochem 82(1):775–797

Petersen E, Memish ZA, Zumla A, Al Maani A (2020) Transmission of respiratory tract infections at mass gathering events. Curr Opin Pulm Med 26(3):197–202

Plasil M, Mohandesan E, Fitak RR, Musilova P, Kubickova S, Burger PA, Horin P (2016) The major histocompatibility complex in Old World camelids and low polymorphism of its class II genes. BMC Genomics 17:1–17

Plasil M, Wijkmark S, Elbers JP, Oppelt J, Burger PA, Horin P (2019) The major histocompatibility complex of Old World camels—a synopsis. Cells 8(10):1200

Premraj A, Aleyas AG, Nautiyal B, Rasool TJ (2020) Camelid type I interferons: identification and functional characterization of interferon alpha from the dromedary camel (*Camelus dromedarius*). Mol Immunol 119:132–143

Quevillon E, Silventoinen V, Pillai S, Harte N, Mulder N, Apweiler R, Lopez R (2005) InterProScan: protein domains identifier. Nucleic Acids Res 33(suppl_2):W116–W120

Ramadan S, Inoue-Murayama M (2017) Advances in camel genomics and their applications: a review. J Anim Genet 45(2):49–58

Regoli M, Bertelli E, Gulisano M, Nicoletti C (2017) The multifaceted personality of intestinal CX3CR1+ macrophages. Trends Immunol 38(12):879–887

Sabahat S, Nadeem A, Brauning R, Thomson PC, Khatkar MS (2022) Genome wide association study for growth in Pakistani dromedary camels using genotyping-by-sequencing. Anim Biosci 36(7):1010

Sabir JS, Lam TT-Y, Ahmed MM, Li L, Shen Y, Abo-Aba SEM, Qureshi MI, Abu-Zeid M, Zhang Y, Khiyami MA (2016) Co-circulation of three camel coronavirus species and recombination of MERS-CoVs in Saudi Arabia. Science 351(6268):81–84

Salim B, De Meeûs T, Bakheit MA, Kamau J, Nakamura I, Sugimoto C (2011) Population genetics of *Trypanosoma evansi* from camel in The Sudan. PLoS Negl Trop Dis 5(6):e1196

Seligsohn D, Abey KA, Chenais E (2025) Camel pastoralists' perceptions of udder health: results from a qualitative interview study in Northern Kenya. BMC Vet Res 21(1):270

Sequencing, T. B. C. G, Consortium, A (2013) Corrigendum: genome sequences of wild and domestic bactrian camels. Nat Commun 4:2089

Silbermayr K, Burger P (2012) Hybridization: a threat to the genetic distinctiveness of the last wild old world camel species, vol 451. Verlag der Österreichischen Akademie der Wissenschaften

Silbermayr K, Orozco-terWengel P, Charruau P, Enkhbileg D, Walzer C, Vogl C, Schwarzenberger F, Kaczensky P, Burger P (2010) High mitochondrial differentiation levels between wild and domestic Bactrian camels: a basis for rapid detection of maternal hybridization. Anim Genet 41(3):315–318

Suthar D, Purva M, Gahlot K, Deora A, Maherchandani S (2019) Sequence analysis of toll like receptor 1 gene of camel (*Camelus dromedarius*). J Camel Pract Res 26(1):41–47

Tando Y, Matsui Y (2023) Inheritance of environment-induced phenotypic changes through epigenetic mechanisms. Environ Epigenetics 9(1):dvad008

Tao W, Aniwar L, ZuliPicar A, Tulafu H, Zhang R, Liu B, Wu W, Huang J (2023) Analysis of genetic diversity and population structure of Tarim and Junggar Bactrian camels based on simplified GBS genome sequencing. Animals 13(14):2349

Terblanche JS, Clusella-Trullas S, Deere JA, Chown SL (2008) Thermal tolerance in a south-east African population of the tsetse fly *Glossina pallidipes* (Diptera, Glossinidae): implications for forecasting climate change impacts. J Insect Physiol 54(1):114–127

Transue TR, De Genst E, Ghahroudi MA, Wyns L, Muyldermans S (1998) Camel single-domain antibody inhibits enzyme by mimicking carbohydrate substrate. Proteins Struct Funct Bioinf 32(4):515–522

Trinks A, Burger P, Benecke N, Burger J (2012) Ancient DNA reveals domestication process: the case of the two-humped camel, vol 451. Verlag der Österreichischen Akademie der Wissenschaften

Uerpmann H-P, Uerpmann M (2002) The appearance of the domestic camel in south-east Arabia. J Oman Stud 12:235–260

Vicente J, Vercauteren KC, Gortázar C (2021) Diseases at the wildlife-livestock interface. Springer

Volpato G (2025) Renegotiating identities around camels and the badiya. Berghahn Books

Vyas S, Sharma N, Sheikh F, Singh S, Sena D, Bissa U (2015) Reproductive status of *Camelus bactrianus* during early breeding season in India. Asian Pac J Reprod 4(1):61–64

Wako G, Tadesse M, Angassa A (2017) Camel management as an adaptive strategy to climate change by pastoralists in Southern Ethiopia. Ecol Process 6(1):1–12

Walzer C, Kaczensky P, Dulamtseren E, Yadamsuren A (2012) Working in a freezer: capturing and collaring wild Bactrian camels, vol 451. Verlag der Österreichischen Akademie der Wissenschaften

Widagdo W, Raj VS, Schipper D, Kolijn K, Van Leenders GJ, Bosch BJ, Bensaid A, Segalés J, Baumgärtner W, Osterhaus AD (2016) Differential expression of the Middle East respiratory syndrome coronavirus receptor in the upper respiratory tracts of humans and dromedary camels. J Virol 90(9):4838–4842

Wright EA, Reddock MB, Roberts EK, Legesse YW, Perry G, Bradley RD (2024) Genetic characterization of the prion protein gene in camels (Camelus) with comments on the evolutionary history of prion disease in Cetartiodactyla. PeerJ 12:e17552

Wu H, Guang X, Al-Fageeh MB, Cao J, Pan S, Zhou H, Zhang L, Abutarboush MH, Xing Y, Xie Z (2014) Camelid genomes reveal evolution and adaptation to desert environments. Nat Commun 5(1):5188

Yadamsuren A, Dulamtseren E, Reading RP (2012) The conservation status and management of wild camels in Mongolia, vol 451. Verlag der Österreichischen Akademie der Wissenschaften

Yi L, Ai Y, Ming L, Hai L, He J, Guo F-C, Qiao X-Y, Ji R (2017) Molecular diversity and phylogenetic analysis of domestic and wild Bactrian camel populations based on the mitochondrial ATP8 and ATP6 genes. Livest Sci 199:95–100

Yuan J, Hu J, Liu W, Chen S, Zhang F, Wang S, Zhang Z, Wang L, Xiao B, Li F (2024) *Camelus knoblochi* genome reveals the complex evolutionary history of Old World camels. Curr Biol 34(11):2502–2508.e2505

Yusof MF, Queen K, Eltahir YM, Paden CR, Al Hammadi ZMAH, Tao Y, Li Y, Khalafalla AI, Shi M, Zhang J (2017) Diversity of Middle East respiratory syndrome coronaviruses in 109 dromedary camels based on full-genome sequencing, Abu Dhabi, United Arab Emirates. Emerg Microbes Infect 6(1):1–10

Zema E, Monti S, Biondi V, Faraz A, Pugliese M, Marino G, Passantino A (2022) European regulations on camel germplasm movement within the European Union: a current framework based on safety. Animals 12(17):2255

Zhu S, Zimmerman D, Deem SL (2019) A review of zoonotic pathogens of dromedary camels. EcoHealth 16:356–377

Zumla A, Alagaili AN, Cotten M, Azhar EI (2016) Infectious diseases epidemic threats and mass gatherings: refocusing global attention on the continuing spread of the Middle East respiratory syndrome coronavirus (MERS-CoV). BMC Med 14:1–4

Camel-Omics: Technologies for Camel Research

5

Silvia Bruno, Rashid Mehmood, and Elena Ciani

Abstract

This chapter examines how omics technologies can strengthen One Health approaches by enabling integrated analysis of the links among animal, human, and environmental health, using camels as a model system. It reviews the contributions of genomics, transcriptomics, epigenomics, proteomics, metagenomics, and in silico biology to understanding camel adaptation, health, productivity, and zoonotic disease dynamics. The chapter highlights how these approaches reveal the molecular basis of camel resilience to harsh environments, support disease surveillance and antimicrobial resistance monitoring, and offer new opportunities for sustainable livestock management. It also discusses current limitations in camel omics research and emphasizes the importance of multi-omics integration for advancing One Health strategies in arid and pastoral ecosystems.

Keywords

Camel adaptation · Antimicrobial resistance · Disease surveillance · Metagenomics

S. Bruno · E. Ciani
Department of Biosciences, Biotechnologies and Environment, University of Bari, Bari, Italy
e-mail: silvia.bruno@uniba.it; elena.ciani@uniba.it

R. Mehmood (✉)
Department of Life Sciences, College of Science and General Studies, Alfaisal University, Riyadh, Saudi Arabia
e-mail: rmehmood@alfaisal.edu

S. A. Gordine et al. (eds.), *Camels and One Health*,
https://doi.org/10.1007/978-981-92-2246-9_5

5.1 Introduction

The One Health concept emphasizes the fundamental interconnectedness of human, animal, and environmental health and employs cohesive, interdisciplinary strategies to address both emerging and recurring health challenges (Pitt and Gunn 2024). One Health has long been recognized as a useful model, but its implementation has been limited by the incomplete characterization of health and disease states. Recent advances in omics technologies have fundamentally transformed the diagnostic possibilities. They enable multilayered, system-wide analyses of organisms and their interactions with the environment, bringing us closer to the holistic, integrative model at the core of the One Health vision (Ukpoju-Ebonyi et al. 2025).

Omics is defined as comprehensive, large-scale analyses of biological molecules, including genomics and transcriptomics and, more recently, proteomics, metabolomics, epigenomics, and metagenomics (Andreu-Perez et al. 2015; Calcino et al. 2024; Dai and Shen 2022; Farschtschi et al. 2022; Franco-Martínez et al. 2020; Heyer et al. 2025; Suravajhala et al. 2016). Genomics has emerged as a cornerstone of modern One Health research through unprecedented insights into disease dynamics across species and ecosystems (Scarpa and Casu 2024). In particular, next-generation sequencing (NGS) enables rapid, high-resolution characterization of genomes from human, animal, and environmental sources. Researchers can use NGS to identify strain-specific features and evolution, trace transmission pathways, and detect reservoirs of various pathogens. Such capabilities are critical for the management of zoonotic diseases and have dramatically improved our ability to mount timely responses to disease outbreaks (Sundermann et al. 2026) and to understand cross-species transmission mechanisms (Carrascosa-Sàez et al. 2024; Du et al. 2025).

Genomic studies of pathogens also play a pivotal role in combating antimicrobial resistance (AMR), a global health priority (Liu and Pandey 2024). Genomic methods can detect AMR genes and elucidate their mechanisms of action and propagation, thus providing essential information to coordinate mitigation strategies (Chigozie et al. 2025). Further, genomic studies can uncover genes that influence disease susceptibility and resistance. This information enables predictive assessments of risk and personalized interventions, including targeted vaccination and monitoring in both humans and animals. In turn, genomics are used in vaccine development, where detailed analyses of virulence and antigenic possibilities guide the design of strain-specific immunization strategies (Din et al. 2025).

Genomic technologies are also transforming animal breeding and health management by helping select individual animals with genetic variants for improved disease resistance, productivity, and environmental adaptation (Jinya 2024; Kasimanickam et al. 2025; Meuwissen et al. 2016; Van Eenennaam 2025). In the future, gene-editing approaches hold the potential to increase disease resistance, reduce zoonotic spillover risks, and contribute to food security—all of which align closely with One Health priorities (Gao et al. 2023; Islam et al. 2020; Kim et al. 2025; Liu et al. 2022; Yuan et al. 2024).

Our understanding of organisms' interactions with their environment can also be enhanced with other omics disciplines. Transcriptomics captures dynamic gene expression as a response to various stressors; proteomics reveals which proteins mediate these responses, while metabolomics characterizes active biochemical processes in the organism; and microbiome-focused approaches measure the interactions between hosts and their microbial communities. By integrating these datasets, we can get a holistic representation of the organism's biological function and fitness (Molla and Bitew 2024; Morabito et al. 2025; Wang et al. 2024). Such multi-omics studies are increasingly used as a foundation for predictive, preventive, and species-specific health strategies.

Domesticated camels (*Camelus dromedarius* and *Camelus bactrianus*) are interesting model organisms for omics studies. This is due to their unique physiological adaptations to arid and harsh environments, including resilience to heat, dehydration, and nutritional stress (Ali et al. 2023; Alkhatabi et al. 2025; Guo et al. 2025; Hoter et al. 2019; Ibrahim and Almundarij 2023; Tibary and El Allali 2020; Wu et al. 2014) and shaped by millennia of unique evolutionary history (Lado et al. 2020). Domesticated camels serve as sources of milk, meat, fiber, transport, and cultural value across large parts of Africa, the Middle East, and Central Asia (Faye 2016, 2022). They are one of the few livestock species capable of thriving in arid and semi-arid areas. From a One Health perspective, their relevance extends far beyond their socioeconomic importance, as they are implicated in zoonotic disease ecology (Conzade et al. 2018; Eckstein et al. 2022; Rooney et al. 2024; Zhu et al. 2019). Camels may act as an early warning signal for climate-driven shifts in disease dynamics, due to their high mobility across rapidly changing environments. Their involvement in zoonotic disease spread highlights the need to deepen our molecular understanding of the interactions among camels, pathogens, and their environment. Moreover, intensified camel farming operations (Faye et al. 2023), coupled with increasing human–camel interactions, further underscore the need for integrated health surveillance and sustainable management strategies.

Although omics-based research in camels is at an early stage, essential genomic resources are available to advance the field quickly. These resources include chromosome-level reference genomes and genome-wide SNP arrays (Burger and Ciani 2022). They have made it possible to study the genetic diversity and population structure of camel populations worldwide (Lado et al. 2020) and to identify markers associated with economically important traits. So far, the traits include morphometry, athletic performance in racing camels, properties of the wool, and milk and meat production (Alsubaie et al. 2024; Bahbahani et al. 2025; Bitaraf Sani et al. 2021; Iglesias Pastrana et al. 2024; Yao et al. 2024). Emerging omics studies are shedding light on the molecular and physiological foundations of camel adaptations. In addition to genomic analyses, transcriptomic and multiomic studies have documented gene expression responses to water restriction (Alvira-Iraizoz et al. 2021; Wu et al. 2014). A recent metabolomic study highlighted shifts in metabolite levels in follicular fluid, indicating metabolic plasticity possibly linked with reproductive adaptation to environmental conditions (Abdoon et al. 2025). An integrated proteomic and metabolomic study of muscle tissue from Bactrian camels highlighted distinguishing pathways among different muscles related to meat quality (Si et al. 2025). Meta-omics explorations of the camel microbiome are rapidly emerging (Gao et al. 2024; Gharechahi and Salekdeh 2018; He et al. 2019). They can contribute to further highlighting the role of host–microbe–environment interactions in antimicrobial resistance, gut function, and disease ecology. From a One Health perspective, these findings carry significant implications. Insights into camel adaptations can inform sustainable livestock management under climate change, while better understanding of camel-associated pathogens improves zoonotic threat preparedness. Moreover, the integration of omics data across biological layers provides a model for system-level investigations of health and disease that extend far beyond camel biology.

5.2 Omics in One Health

Organisms feature increasing levels of complexity, from the molecular, through the cellular and organ system layers, up to the whole body, with interactions between them controlling processes from the cellular level up to systemic environmental responses. Studying the different levels in isolation provides only partial understanding of the organism, while integrated analysis can show the regulation of molecular networks and the emergence of system-level properties and in response to stimuli. By combining insights across these layers, we gain a comprehensive view of the organism, linking molecular interactions to whole-organism outcomes and ecological interactions. In the following sections, we examine how the various omics disciplines, together with in silico approaches, further our understanding of camel biology and reinforce One Health strategies at the human–animal–environment interface.

5.3 Genomics

Genomics is the comprehensive study of the genetic material (genome) of an organism. It includes the sequencing, analysis, and interpretation of DNA sequences in order to understand emergent characteristics, such as body structure, organ function, and the species' evolution. The genome represents the organism's blueprint and contains all the hereditary information that governs its development, physiology, and adaptation to its environment. In a One Health context, genomics can reveal the shared genetic mechanisms underlying adaptation, disease susceptibility, and host–pathogen interactions.

Genomics has evolved into specialized subdisciplines that focus on different areas:

- *Structural genomics* deals with genome sequencing, assembly and annotation.
- *Functional genomics* aims to characterize the genes' purpose and how they are regulated.

- *Comparative genomics* analyzes genomic similarities and differences across species to identify conserved elements and infer evolutionary relationships between them.
- *Population genomics* examines genome-wide patterns of genetic variation within and among populations.

These subdisciplines have significantly improved our understanding of gene structure, expression patterns, regulatory processes, and the genetic basis of diseases. Additionally, advances in these subdisciplines have paved the way for the development of personalized medicine and innovative biotechnological applications. In this way, they support applications that are central to the One Health framework, such as conservation of genetic diversity and mitigation of zoonotic disease risks.

5.3.1 Methods

Genomics is composed of a range of advanced methods and technologies that enable the exploration and manipulation of genetic material (Metzker 2010).

Sanger Sequencing laid the foundation of genomic research, allowing the determination of nucleotide sequence of small DNA fragments with high precision. Slower than more modern genetic sequencing technologies, Sanger sequencing is very accurate and was instrumental in early genomic research. It played a central role in the Human Genome Project (HGP), a landmark international effort to sequence the entire human genome, initiated in 1990. Using Sanger-based automated sequencing instruments, researchers were able to map the complete human genome, made up of approximately 3 billion base pairs (Lander et al. 2001). Because the Sanger method is so accurate and reliable, it continues to be the gold standard for sequence validation and small-scale studies, even though newer, high-throughput methods have surpassed it in speed and scalability.

Next-Generation Sequencing (NGS) While Sanger sequencing laid the foundations of genomics research, NGS revolutionized the field by enabling high-throughput, cost-effective sequencing of entire genomes. Sanger sequencing analyzes one DNA sequence at a time in a slow, precise manner, while NGS is much faster, examining millions of DNA sequences simultaneously.

NGS can be used to perform whole-genome sequencing (WGS) to read the complete DNA sequence of an organism and to identify various genetic features. These features include single-nucleotide polymorphisms (SNPs) (small variations in which a nucleotide differs among individuals within a population) as well as insertions, deletions, copy number variations, and large structural variants across the genome. WGS has significantly improved our ability to detect disease-associated mutations and characterize genetic diversity across individuals and evolutionarily related species. In contrast to WGS, whole exome sequencing (WES) reads only the parts of DNA that code for proteins, which is where most disease-related mutations occur, making it faster and cheaper to find the genetic differences most likely to affect the organism.

Gene Editing Gene editing tools have enabled precise, targeted modifications of DNA sequences within living organisms. Among these, CRISPR-Cas9 has become the most versatile and widely used method due to its efficiency and adaptability (Cong et al. 2013). The CRISPR-Cas9 is an adaptive bacterial immune mechanism that has been repurposed for precise gene editing in eukaryotes. Its core components are the Cas9 nuclease, which acts as molecular scissors, and a single-guide RNA (sgRNA), which directs Cas9 to a specific DNA sequence via complementary base pairing. Cas9 recognizes the target site and creates a double-stranded break in the DNA, allowing researchers to modify the DNA sequence by knocking out genes, inserting new sequences, or correcting mutations. CRISPR-Cas9 and subsequent advancements, such as single base editing, have opened new avenues for functional genomics by allowing precise modifications of specific genomic loci to study gene function and develop therapeutic strategies for genetic diseases (Cong et al. 2013; Komor et al. 2016).

Together, genomic methods provide a comprehensive toolkit to understand the structure, function, and evolution of genomes and to translate genomic discoveries into biomedical and biotechnological innovations.

5.3.2 Studies in Camels and One Health

Although various livestock species have been investigated through genomic methods, camel genomics has lagged behind. The first comprehensive camel genomics study was conducted in 2012, when the genomes of domesticated (*Camelus bactrianus*) and wild (*Camelus ferus*) Bactrian camels were sequenced and published, a major milestone in understanding the species' genetic makeup and adaptive traits (Bactrian Camels Genome Sequencing and Analysis Consortium et al. 2012). Using whole genome sequencing and comparative genomics, the study produced the first camel reference genome, containing approximately 20,821 protein-coding genes. Aside from describing basic genetic traits (gene annotation, catalogues of repeats, and heterozygosity estimates), the study identified rapidly evolving genes enriched in metabolic pathways and candidate genes that may be responsible for camels' remarkable adaptations (e.g., salt tolerance and immune system uniqueness). Later studies corroborated these findings and expanded the scope to other camel species, including dromedary camels (*Camelus dromedarius*). The studies also revealed gene families and pathways that may have been evolutionarily selected for fat and water metabolism, as well as stress responses to heat, aridity, intense ultraviolet radiation, and dust exposure (Ming et al. 2020; Wu et al. 2014).

Beyond evolutionary adaptations, recent studies have begun to link genomic variation in camels with productivity, morphology, behavior, and environmentally driven traits. For instance, Yao et al. (2024) identified candidate genes and SNPs in camels associated with milk yield and protein and fat content. At the same time, a genome-wide association study (GWAS) in camels has begun to uncover genetic loci related to morphometrics, biomechanics, and behavior that are important for racing or performance (Iglesias Pastrana et al. 2024).

Genomics also holds promise in understanding zoonotic disease transmission from camels. For example, sequencing studies of MERS-CoV isolates from dromedary camels and humans have revealed identical viral genomes, confirming camels as a significant reservoir and shedding light on the molecular pathways of interspecies transmission (Azhar et al. 2014). Moreover, genomic analysis of *Brucella* and *Coxiella burnetii* isolates from camels has helped trace infection sources and identify shared strains between camels, livestock, and humans.

While camel genomics is steadily advancing, its application in large-scale genomic selection and precision breeding remains in its infancy. Unlike the well-established genomic selection for cattle, sheep, or poultry, the resources for camels, such as an extensive reference population, high-density SNP arrays, and comprehensive genomic databases, are still lacking. Furthermore, the absence of standardized phenotype recording methods and well-documented pedigrees create additional bottlenecks, severely limiting the power and accuracy of genome-wide association studies, which are essential for identifying markers linked to desirable traits. However, the growing availability of whole-genome data coupled with improved bioinformatics tools provides the potential to integrate genomic information into breeding programs. Hence, transformation of camel breeding from traditional selection to large-scale genomic selection and precision breeding remains an emerging frontier.

5.4 Transcriptomics

Transcriptomics is the study of the RNA molecules, collectively known as the transcriptome, that are transcribed from the genome in a cell, tissue, or organism. Unlike the genome, which does not change over time, the transcriptome changes significantly in response to developmental stages, environmental stimuli, and disease states. It encompasses several classes of RNA, including messenger RNA (mRNA), noncoding RNA (ncRNA), microRNA (miRNA), and long noncoding RNA (lncRNA), each playing a distinct role in gene regulation and cellular function. The transcriptome links the genome to the proteome, enabling genetic information to be translated into functional proteins. Through transcriptomic analysis, we can explore the regulation of gene expression, as well as alternative splicing and RNA processing events. By understanding the transcriptome, we can better measure at the molecular level how cells and organisms respond to stressors, such as environmental pressures, pathogens, and physiological challenges.

5.4.1 Methods

Transcriptomic methods have evolved dramatically, transforming the field into a high-resolution, quantitative science. Some of these methods are described below.

cDNA Libraries A complementary DNA (cDNA) library is a collection of cloned cDNA fragments synthesized from the mRNA of a specific tissue or organism. These libraries reflect the actively expressed genes at the time of RNA extraction and have historically been used for gene identification, expression studies, and functional screening. Before automated sequencing technologies emerged, cDNA libraries provided fundamental information on gene structure and expression, even though they are labor-intensive to produce.

DNA Microarrays DNA microarrays represented a major advance by enabling the simultaneous detection of thousands of expressed genes. In this method, fluorescently labeled cDNA is hybridized to probes fixed on a solid surface. The resulting fluorescence intensity indicates relative expression levels. Although microarrays require prior knowledge of the gene sequences and have limited dynamic range, they were instrumental in large-scale gene expression profiling and disease classification studies during the early 2000s.

RNA Sequencing (RNA-Seq) RNA-seq, based on next-generation sequencing (NGS), has become the gold standard in transcriptomics. It enables comprehensive, quantitative, and unbiased analysis of all transcripts in a sample. Unlike microarrays, RNA-seq does not rely on predefined probes, so it can provide a high-resolution overview of the transcriptome, detecting previously uncharacterized genes, various forms of RNA, and measurements of the relative expression levels of genes.

Single-Cell Transcriptomics (scRNA-Seq) Single-cell RNA-seq extends traditional RNA sequencing analysis to the level of individual cells, allowing it to resolve cell-to-cell variability. It enables identification of rare cell populations, differential developmental trajectories, and transcriptional heterogeneity within tissues.

5.4.2 Studies in Camels and One Health

Transcriptomic studies elucidate how camels maintain health and productivity in harsh environments while serving as potential reservoirs for zoonotic diseases (Ma et al. 2025; Wu et al. 2014).

The first transcriptomic analysis of Bactrian camels highlighted unique mechanisms for osmoregulation, osmoprotection, and water conservation through elevated blood glucose levels (Wu et al. 2014). Additional studies of Bactrian camels under salt and water-deprivation stress identified differential expression of several genes through mRNAs, miRNAs, and lncRNAs. These genes are associated with a mechanism to restrict sodium influx into cells and to conserve water under extreme conditions. Other works have shed light on molecular mechanisms related to reproduction, organ development, and milk yield and composition (Liu et al. 2025; Yao et al. 2023).

Other studies have measured camels' immune response to pathogens such as *MERS-CoV, Brucella*, and *Trypanosoma evansi*, providing insights into disease resistance and transmission risks between animals and humans. Transcriptomic analyses of differential gene expression, pathway enrichment, and network modeling have suggested candidate genes for further validation. These works highlight the power of transcriptomics not only to advance fundamental knowledge of adaptation and evolution, but also for applied objectives related to One Health–oriented management of camel populations.

5.5 Epigenomics

Epigenomics is the study of heritable changes in gene function that occur without modifying the underlying DNA sequence. Epigenomic mechanisms regulate gene expression in a dynamic and context-dependent manner and include DNA methylation, histone modifications, and noncoding RNAs. Like the transcriptome, the epigenome is highly responsive to transient conditions in the cell's environment. Chemical tags on DNA and on histones (proteins that DNA wraps around) act like switches that enable and disable expression of specific genes. For instance, DNA methylation is generally associated with repression of transcription, while histone acetylation facilitates gene activation through an open chromatin configuration. Combinatorial patterns of these modifications constitute an "epigenetic code" that governs cell identity, differentiation, and adaptive responses to environmental change. Thus, epigenomics provides critical insights into how organisms translate genetic information into context-specific phenotypes and how environmental factors can influence biological function across the organisms' lifetime.

From a One Health perspective, epigenomics is particularly relevant because it captures the molecular consequences of environmental exposures (such as climate stress, nutritional variation, pollutants, and pathogens) on gene regulation. Epigenetic modifications can influence disease susceptibility, immune responses, metabolic regulation, and developmental outcomes and in some cases may be transmitted across generations, thereby linking environmental conditions to long-term health effects at the population level.

5.5.1 Methods

Several methods have been developed for determining the epigenetic profile of genes.

DNA Methylation DNA methylation can be investigated using several techniques.

Bisulfite Sequencing Bisulfite treatment converts unmethylated cytosines to uracil but leaves methylated ones unchanged, helping researchers see exactly where methylation has occurred. This can be done on part of the genome or the whole genome.

Methylated DNA Immunoprecipitation Sequencing (MeDIP-Seq) Antibodies that recognize methylated DNA are used to enrich these regions, which are then sequenced to reveal methylation patterns across the genome.

5.5.1.1 Histone Modifications

Chromatin Immunoprecipitation Sequencing (ChIP-Seq) This method uses antibodies that can recognize specific chemical tags on histones. These antibodies pull down the DNA wrapped around these tagged histones, and then this DNA is sequenced to see where the tags are located across the whole genome.

5.5.1.2 Chromatin Conformation and 3D Architecture

Hi-C Gene regulation involves both short-range (local promoter–enhancer contact) and long-range interactions. Distant regulatory elements (enhancers, silencers) loop around to physically contact gene promoters, forming 3-dimensional chromatin structures to precisely control specific genes' expression for some period. The Hi-C method captures genome-wide chromatin interactions and is the primary tool for mapping the 3D folding of chromosomes, including compartments and topologically associating domains (TADs). These interactions are captured by crosslinking chromatin, digesting it, and ligating nearby DNA fragments to reveal the physical architecture underlying long-range gene regulation.

5.5.2 Studies in Camels and One Health

Epigenetic mechanisms allow organisms to adapt to changing environments by providing a critical molecular interface between environmental conditions and gene expression. Camels inhabit some of the most extreme ecosystems on Earth and maintain close interactions with humans and livestock. Their ecological niche suggests that epigenomic regulation plays an important role in shaping camels' physiological resilience, as they are exposed to chronic heat stress, dehydration, nutritional scarcity, and pathogens. Notably, to date, there is a dearth of epigenomic studies in camels.

Analysis of CpG islands in the camel genome suggests the presence of regulatory regions that may be subject to epigenetic control (Barazandeh et al. 2016). Emerging evidence also highlights the role of epigenetics in camel reproduction and development. For example, studies have shown that assisted reproductive technologies, such as oocyte cryopreservation followed by in vitro fertilization, can disrupt normal epigenetic landscapes during early embryogenesis, underscoring the importance of epigenetic regulation in the reproductive and developmental processes (Moulavi et al. 2021).

Overall, epigenomic research in camels may reveal mechanisms of environmental adaptation, inform disease risk assessment, and support sustainable management strategies, in alignment with the One Health framework.

5.6 Proteomics

Proteomics is the study of the proteome, the set of proteins expressed by a cell, tissue, or organism at a given time and under specific physiological conditions. As proteins are the primary functional molecules executing most cellular processes, proteomics represents a critical extension of genomics, transcriptomics, and epigenomics. While genomics describes the genetic blueprint, transcriptomics reveals which genes are actively transcribed, and epigenomics explains how gene activity is regulated, and proteomics captures the proteins that are produced and active within the cell as the final output of these regulatory layers.

Beyond protein abundance, proteomics investigates protein function through the proteins' structure, subcellular localization, interactions, and posttranslational modifications (PTMs). Proteomics provides direct insights into signaling pathways, metabolic processes, immune responses, and cellular adaptation to environmental stimuli. In a One Health context, proteomics enables the functional characterization of biological responses that link animal physiology, environmental stressors, and human health outcomes.

5.6.1 Methods

Mass Spectrometry (MS) MS is one of the main tools used to identify and characterize proteins. It turns proteins into charged particles in order to measure their weight, helping to determine which proteins are present and whether they carry any chemical modifications. This allows researchers to accurately profile very complex protein mixtures.

Two-Dimensional Gel Electrophoresis (2D-GE) 2D-GE separates proteins by isoelectric point and molecular weight.

Liquid Chromatography (LC) Coupled with MS (LC-MS) LC-MS integrates the separation capabilities of liquid chromatography with the precise mass detection and analysis provided by mass spectrometry.

5.6.2 Studies in Camels and One Health

Proteomics studies in camels provide valuable insights into the molecular basis of their unique physiology, nutrition, and environmental adaptations, with direct implications for One Health. Comprehensive analyses of camel milk, including its whey, milk fat globule membranes, and exosomes, have identified immune and bioactive proteins that are beneficial not only for human nutrition and antimicrobial activity, but also for therapeutic and industrial applications (Alhassani 2024; Bakry et al. 2021; Ming et al. 2025). Milk protein composition was shown to be affected by heat treatment, the camel's lactation stage, and the environmental conditions (Benabdelkamel et al. 2017; Xiao et al. 2022).

Proteomic studies of camel urine and organs including liver, heart, brain, kidney, and hump fat have demonstrated the species' exceptional adaptations to dehydration and heat stress (Alhaider et al. 2012; Warda et al. 2014). Such studies are important for sustainable livestock operations in desert ecosystems by helping farmers to understand how camels maintain homeostasis in extreme environments.

Finally, camel proteomics has emerging relevance in zoonotic disease surveillance and food safety. Proteomic tools can help characterize host–pathogen interactions in infections such as *Brucella* or MERS-CoV, providing early disease indicators. Collectively, these studies link animal welfare with human health outcomes and sustainability. Thus, integrating proteome data into One Health strategies enhances our ability to safeguard both camel populations and the communities that depend on them.

5.7 Metagenomics

Metagenomics seeks to provide a complete picture of the microorganisms present in a biological sample through genetic sequencing. In contrast, traditional diagnostic methods are limited by their focus on detecting a single known pathogen in a sample. These limitations are most apparent when infections are caused by multiple species, the pathogens are unknown, or animals show no clinical signs (Zhao et al. 2024). Metagenomic approaches are especially valuable for animal disease surveillance, as they can detect pathogens even in clinically healthy animals (von Fricken et al. 2023). Metagenomics also supports zoonotic risk assessments by identifying microorganisms circulating in livestock or wildlife that may cross species barriers and, potentially, infect humans. This capability has become increasingly important in the wake of the COVID-19 pandemic, which demonstrated the speed with which viral pathogens can evolve and spread globally (Ko et al. 2022). In this context, viral metagenomic next-generation sequencing (vmNGS) enables the detection of zoonotic viruses without prior knowledge of their genetic identity, thus allowing global surveillance efforts to recognize viral threats earlier and to respond more selectively (Russell et al. 2025). Metagenomics can also be used to study microbiome function, revealing how microbial communities in host animals support digestion, metabolism, and immune responses. In many livestock species, the microbiome plays a crucial role in nutrient acquisition and overall fitness, and it changes in response to diet, climate, and management practices (Wei et al. 2025). Since these changes influence animal health and productivity, metagenomic data can also be used to monitor ecological and environmental health. Surveillance of antimicrobial resistance (AMR) is another essential application. Resistance genes can move between microbes, hosts, and environments as mobile genetic elements, meaning resistance can spread even without direct antibiotic exposure. AMR is leading to a global health crisis and is projected to cause more than ten million human deaths per year by 2050 (O'Neill 2016). It also threatens livestock production and food security (Jonas 2017), especially in regions where veterinary antibiotics are widely used and sanitation infrastructure is limited. This risk for camel operations is increasing, as species-specific therapeutics are now more widely used, and the close interactions among animals, people, and shared environments can facilitate the emergence and spread of resistant microbes. Metagenomics allows researchers to detect and track AMR genes circulating in livestock, wildlife, soils, and water systems, providing insight into how resistance spreads and how it might be contained (Klümper et al. 2025). Taken together with a One Health perspective, metagenomics provides both diagnostic and ecological benefits, revealing hidden infections and highlighting microbial networks that support or threaten cross-species health.

5.7.1 Methods

Metagenomic studies rely on sequencing-based approaches that differ in resolution and applicability.

Metabarcoding The amplicon-based (or metabarcoding) approach focuses on the selective PCR amplification of "marker genes," also called "barcode genes" (e.g., 16S rRNA for prokaryotes, ITS for fungi or small eukaryotes, COI for eukaryotes). A barcode gene suitable for metagenomic studies must have several characteristics. It must:

(i) Be ubiquitous or shared in the taxonomic group of interest.
(ii) Be variable enough in order to discriminate between different microbial species.
(iii) Have a size compatible with available NGS platforms.
(iv) Contain variable regions flanked by conserved sites, to allow both species discrimination (variable regions) and PCR primer binding and amplification (conserved regions).

Following PCR amplification of the target regions, the resulting amplicon libraries are sequenced by NGS. Metabarcoding is generally the most cost-effective strategy in terms of both sequencing and computational analysis.

Shotgun In the shotgun metagenomics approach all the extracted nucleic acids from a sample are sequenced without targeting specific genomic regions. Sequencing typically generates millions of reads that represent the genomic content of the entire microbial community. After that, the reads are analyzed throughout dedicated bioinformatic pipelines cataloguing and annotating all microbial genomes constituting a microbiome. Despite the recent improvement in computational analysis, shotgun sequencing is still more expensive and more complex than metabarcoding but offers a much broader and more detailed overview of the microbiome.

5.7.2 Studies in Camels and One Health

Metagenomic studies of dromedary camels are relatively recent compared to other livestock species, but the research in this area is growing rapidly. These studies have explored a range of biological materials, including blood, feces, respiratory secretions, rumen contents, milk, and even associated arthropod vectors, to better understand both the microbial diversity of the camel and its role within the broader One Health ecosystem. One of these studies focused on blood-associated microbial communities, revealing the presence of tick-borne bacteria and protozoa such as *Anaplasma*, *Theileria*, and *Trypanosoma evansi*, even in animals without clinical symptoms (Mohamed et al. 2021). This demonstrates that dromedaries can function as silent reservoirs, maintaining transmission cycles that involve wildlife, livestock, and vector populations. Such infections might have gone undetected in traditional, culture-based diagnostics, demonstrating the value of metagenomics for animal disease surveillance and for understanding subclinical pathogen circulation. Metagenomic studies have also characterized the camel virome, identifying a surprising diversity of viruses in fecal (Woo et al. 2014) and respiratory (Hassan et al. 2025; Li et al. 2017) samples collected from both healthy and ill camels. Overall, circoviruses, picobirnaviruses, and several respiratory pathogens have been detected, highlighting the camel's role in the ecology of viral diseases, including zoonotic agents such as MERS-CoV. The upper respiratory tract appears to be a dynamic interface for viral circulation and diversification, likely due to high viral receptor expression (Alnaeem et al. 2020), intense viral replication, and frequent close contact among camels and between camels and humans (Adney et al. 2014). Thus, the upper respiratory tract is a key site for monitoring emerging infectious risks.

A study of camel milk has shown that its microbiota varies with hygiene practices, environmental conditions, and animal health (Rahmeh et al. 2022). Since camel milk is traditionally consumed raw in many regions, understanding these microbial profiles has direct implications for food safety and public health, highlighting the value of metagenomics for monitoring both nutritional and zoonotic dimensions of camel-derived food. Another major focus has been the rumen (Chaiebeddraa 2025; Gharechahi et al. 2022; Hinsu et al. 2021; Kholousi Adab et al. 2024; Rajabi et al. 2022; Tulsani et al. 2021) and gut (Chaiebeddraa 2025) microbiome, because they play an essential role in the camel's ability to thrive on fibrous desert vegetation (Hinsu et al. 2021). Metagenomic data reveal microbial communities filled with carbohydrate-activated enzymes capable of breaking down tough lignocellulosic plant material. Several enzymes isolated from camel rumen metagenomes, such as novel cellulases and xylanases, have shown promise for industrial bioprocessing and biofuel production. Furthermore, studies comparing the gastrointestinal microbiomes between wild and captive populations of Bactrian camels (*Camelus ferus*) (Yan et al. 2022) indicate that diet and environmental exposure strongly affect microbiome composition and metabolic function.

Finally, a recent metagenomic investigation (Gao et al. 2024) has begun to map antimicrobial resistance (AMR) within camel-associated microbiomes, identifying Bacillota and Bacteroidota as the main reservoirs of AMR genes, with vancomycin resistance genes being the most prevalent. Similarly, in 2023, Akunda et al. characterized antimicrobial resistance in *Escherichia coli* from camel fecal samples and found a high prevalence of β-lactamase-producing and multidrug-resistant strains, indicating that even environments with apparently low exposure to antibiotics can harbor clinically relevant AMR genes. Notably, almost half (48%) of the *E. coli* isolates in this study exhibited antimicrobial resistance, including resistance to second-, third-, and fourth-generation cephalosporins, a marked increase compared to previous reports from the same region (Nüesch-Inderbinen et al. 2020). A complementary study highlighted the presence of clinically relevant AMR bacteria in *Hyalomma dromedarii* ticks: notably, all gram-negative isolates were resistant to cefazolin, while all gram-positive isolates were resistant to fusidic acid, with occasional resistance to other antibiotic molecules (Aljasham et al. 2025). These studies highlighted that AMR genes can be detected even in settings with limited antibiotic exposure. The observed rise in resistance (Akunda et al. 2023) may result from:

(i) The environmental persistence of antibiotics
(ii) Their nonregulated use in livestock
(iii) Indirect transmission through shared water sources
(iv) Wildlife–livestock interactions

Taken together, these factors highlight the complex ecological variables of AMR circulation in camel populations. This positions the camel as a sentinel species for monitoring the evolution of AMR in pastoral and desert ecosystems.

5.8 In Silico Biology

In silico biology is an umbrella term for biological analyses conducted entirely through computational methods, without the need for in vivo or in vitro experimentation. These methods have grown in relevance as modern omics technologies generate enormous volumes of data. In silico biology encompasses a broad spectrum of approaches:

- Sequence analysis and comparative genomics
- Structural modeling and molecular dynamics
- Reconstruction of metabolic, regulatory, and host–pathogen interaction networks
- Ecological and epidemiological simulations

Machine learning and artificial intelligence methods are featured more prominently in such analyses. Applications of in silico biology are wide-ranging. Genomic surveillance enables the rapid detection of emerging pathogens and variants through automated sequence comparison and phylogenetic reconstruction.[1] Host–microbiome studies allow researchers to move from microbiome composition to functional predictions, identifying metabolic capabilities or immune-modulating effects

[1] The scientific process of inferring how different organisms are related to each other through common ancestors by creating a branching diagram called a phylogenetic tree.

encoded in microbial genomes. In veterinary epidemiology and One Health modeling, in silico tools can simulate disease spread across animal and human populations, assisting in risk assessment and intervention planning. Structural bioinformatics and molecular docking approaches support the identification of vaccine targets or drug candidates, while metabolic modeling can predict how organisms respond to nutritional changes or environmental stressors. In this context, in silico biology acts as a core interpretation layer to connect molecular data with biological meaning and health decision-making across humans, animals, and environments.

5.8.1 Methods

In silico biology includes a wide range of computational approaches that can be categorized into several methodological domains.

Sequence Analysis and Comparative Genomics These sequencing-based approaches rely on algorithms for genome assembly (e.g., de Bruijn graph–based and long-read assembly approaches), annotation pipelines, variant calling tools, and phylogenetic inference methods. These analyses can identify single-nucleotide polymorphisms (SNPs), structural variants, and selective sweeps (i.e., the process through which a new beneficial mutation's frequency increases and becomes fixed in a population or breed), as well as to perform phylogenetic reconstruction. The latter is a crucial tool in genomic surveillance, where the rapid comparison of sequences helps track the emergence and spread of new viral or bacterial strains.

Structural Bioinformatics and Molecular Modeling These approaches use computational modeling to predict macromolecular structure, functions, and interactions. Techniques such as homology modeling refer to constructing an atomic-resolution model of the "target" protein from its amino acid sequence and an experimental three-dimensional structure of a related homologous protein. In addition, molecular docking algorithms estimate the binding affinities between ligand and receptor, and molecular dynamics simulations assess the conformational stability of protein complexes under different physiological conditions.

Network Reconstruction and Systems Biology It provides a framework for studying how genes, proteins and metabolites work together in a coordinated way. By reconstructing metabolic pathways or gene-regulatory networks, in silico approaches can predict how organisms respond to changes in diet, infection, or environmental stress. As an example, constraint-based modeling (CBM) is a powerful computational framework that uses biochemical network maps and physical/physiological limits (constraints) to predict cell behavior, like metabolism, from genotype, by assuming steady states and defining reaction ranges. It helps to understand cellular function, optimize industrial processes, and link genes to traits. These models provide mechanistic insights into physiological adaptation, host–pathogen interaction, and the metabolic consequences of environmental stress.

Machine Learning and Artificial Intelligence These methods are particularly powerful when dealing with the large, multidimensional datasets generated by omics technologies. AI tools can detect subtle patterns that would be difficult to uncover manually, classify disease states, predict phenotypic traits from genetic variants, or infer transmission routes during outbreaks. In livestock research, machine learning also supports precision farming by analyzing sensor-derived data to detect early signs of stress or disease.

Epidemiological and Ecological Modeling It uses computational simulations to explore how diseases spread across populations and how environmental or management changes might influence transmission. These models help evaluate potential interventions, anticipate risks and inform One Health decision-making by integrating data from humans, animals, and the environment.

5.8.2 Studies in Camels and One Health

In silico biology is increasingly being applied to the study of camel genetics, physiology, and adaptation. A key step forward has been the publication of high-quality and chromosome-level reference genomes for both the dromedary and the Bactrian camel, together with technological advances in whole-genome sequencing (WGS) and genotyping approaches such as genotyping-by-sequencing (GBS) and SNP array platforms. These resources allow researchers to perform genome-wide analyses aimed at understanding the biological mechanisms underpinning camel adaptation to extreme environments. For example, genomic studies have investigated the genetic basis of heat stress tolerance (Wu et al. 2014) and have identified selection signatures linked to adaptive and productive traits, including immune response, lipid metabolism, energy homeostasis, adaptation to arid environments, and milk characteristics (Al Abri et al. 2023; Bahbahani et al. 2019). Genome-wide association studies (GWASs) have also been used to explore traits associated with growth, morphometrics, pigmentation, and behavior (Bitaraf Sani et al. 2021, 2022a, b, 2023; Iglesias Pastrana et al. 2024; Sabahat et al. 2023). In parallel, transcriptomic analyses have begun to elucidate functional gene expression changes in camels under various stress conditions. For example, a recent RNA-sequencing study of Bactrian camels' mammary gland tissue revealed differentially expressed genes enriched in immune-related pathways, suggesting that specific gene families contribute to the camel's resilience to infectious challenges, such as subclinical mastitis (Ma et al. 2025). These insights may support the development of diagnostic and preventive strategies, benefiting both the camel milk industry and public health in the future. In addition, structural modeling approaches have also focused on camelid single-domain antibodies (VHHs), which are uniquely small and stable, thus making them valuable therapeutic and diagnostic tools in both human (Harmsen and De Haard 2007) and veterinary medicine (Liu et al. 2021; Mir et al. 2020).

Beyond molecular investigations, in silico biology is now increasingly valuable in livestock management and animal health surveillance. Computational modeling and digital herd simulation platforms are used to optimize breeding strategies, predict productivity under different feeding regimes, and assess how environmental conditions influence animal welfare and performance (Džermeikaitė et al. 2025; Shoyombo et al. 2024). In dairy systems, metabolic modeling allows prediction of nutrient utilization efficiency and methane emissions, supporting climate-smart feeding and breeding strategies (Sahraei et al. 2025). Likewise, machine learning tools are increasingly applied to precision livestock farming data (such as infrared thermography, accelerometer-based behavior monitoring, and automated milking sensors) to detect early signs of heat stress (Ciliberti et al. 2025), mastitis (Fadul-Pacheco et al. 2021; Kalkan et al. 2025) or metabolic disorders (Michelena et al. 2024) before clinical symptoms appear. Although these applications are currently more developed in other livestock species, they represent transferable frameworks that can be adapted to camel systems as datasets and digital infrastructures expand. Taken together, these advances illustrate how in silico biology provides the computational foundation for interpreting molecular and production data, linking them to practical decision-making in animal health and management, and ultimately advancing sustainable and One Health–oriented livestock systems.

Increasingly, research in both livestock and camel biology is moving from single-layer analyses to multi-omics data integration (Suravajhala et al. 2016). By combining genomic, transcriptomic, proteomic, metabolomic, and metagenomic data, it becomes possible to capture the dynamic regulatory relationships that link genotype to phenotype (Upadhyay et al. 2024; Wang et al. 2025). In practical terms, multi-omics frameworks can reveal how gene expression is modulated by diet, environment, or health status and how these molecular changes translate into differences in productivity, resilience, or disease susceptibility. This shift is particularly relevant in One Health contexts, where animal health, productivity, microbiome composition, pathogen exposure, and environmental conditions are tightly interconnected. In camel research, multi-omics datasets are only beginning to emerge, but conceptual frameworks and analytical strategies have already been shaped in cattle and small ruminants. For example, integrative studies in dairy cows have linked genomic variants with mammary gland transcriptomes, milk proteomic profiles, and microbiome composition to identify molecular signatures of mastitis resistance and lactation efficiency (Cai et al. 2025; Zhang et al. 2024; Zheng et al. 2025). Similar multi-omics pipelines have been used to characterize metabolic adaptation under heat stress by combining genomic markers of thermotolerance, skin and liver transcriptomics, blood metabolomics, and microbiome-derived short-chain fatty acid profiles (Halli et al. 2025; Li et al. 2025; Xue et al. 2020; Yan et al. 2022). These approaches have revealed regulatory networks that determine resilience to climatic stress that are likely to be of particular relevance for camels, given their strong physiological adaptation to arid environments.

5.9 Limitations and Future Directions

Despite significant progress in recent years, omics-based research in camels still faces a number of shared and interrelated limitations that currently constrain its full potential within both livestock science and One Health frameworks. Across genomics, transcriptomics, epigenomics, and proteomics, a recurring challenge is the limited availability of well-annotated reference resources. In genomics, despite the availability of high-quality reference genomes for dromedary and Bactrian camels, the relatively small sample sizes, together with the lack of standardized phenotypic data and the absence of high-density camel-specific SNP arrays, still hamper robust genome-wide association studies. Similar limitations affect transcriptomic analyses, where incomplete genome annotations hinder accurate read mapping, isoform detection, and functional interpretation, particularly for alternative splicing and noncoding RNAs. In addition to reference-related constraints, many omics studies in camels are limited by small and heterogeneous datasets. Transcriptomic investigations often focus on a narrow range of tissues, developmental stages, or environmental conditions, reducing statistical power and limiting the generalizability of findings across breeds, production systems, and ecological contexts. The predominant use of bulk RNA sequencing further masks cell-type-specific expression patterns, especially in complex tissues such as immune organs. Epigenomic research is even more constrained, as genome-wide, tissue-specific epigenomic maps are largely lacking, limiting insights into how DNA methylation, histone modifications, and regulatory elements contribute to adaptation, stress resilience, and disease susceptibility. Proteomics faces additional technical challenges, including the wide dynamic range of protein abundance, limited camel-specific protein reference databases, and difficulties in detecting low-abundance regulatory proteins and posttranslational modifications, all of which complicate reproducibility and cross-study comparisons. Metagenomics and in silico biology, while offering powerful tools to study host–microbe interactions, pathogen surveillance, and antimicrobial resistance, present their own methodological and interpretative challenges. Metabarcoding approaches are constrained by marker choice and reference database completeness, limiting taxonomic resolution and the detection of novel taxa, whereas shotgun metagenomics remains costly and computationally demanding. These issues are amplified in camel research by fragmented microbiome reference databases, heterogeneous sampling strategies, and nonstandardized analytical pipelines. Similarly, in silico approaches are often limited by data scarcity, uneven data quality, and the high computational demands required for large-scale, multilayer analyses.

Importantly, significant gaps also persist in the practical implementation of comprehensive omics studies in domestic camel species. Many camel-rearing regions face infrastructural, economic, and logistical constraints that limit large-scale sampling, long-term monitoring, and the integration of omics data into breeding, health management, or surveillance programs. Addressing these challenges will require coordinated and transdisciplinary collaboration across veterinary sciences, genomics, public health, ecology, and social sciences, including careful consideration of ethical, cultural, and community dimensions that shape camel research and management practices.

Looking forward, future research in camels should prioritize coordinated efforts to generate high-quality and multi-omics data, including standardized phenotypes and documented pedigrees, that will be essential to enable large-scale GWAS, genomic selection, and precision breeding. In parallel, larger and better-designed transcriptomic studies, including single-cell and spatial approaches, will be critical to resolve cellular heterogeneity and immune responses at high resolution. The development of genome-wide epigenomic reference maps across key tissues and stress conditions will further clarify the role of epigenetic plasticity in adaptation to arid environments. Advances in quantitative and targeted proteomics, including posttranslational modification profiling, will be needed to link molecular regulation to physiological function. A major future direction across all omics layers lies in multi-omics integration, combining genomic, transcriptomic, epigenomic, proteomic, and metagenomic data to reconstruct regulatory and metabolic networks that link genotype, environment, and phenotype. While such integrative frameworks have already proven informative in other livestock species, their application to camels remains limited by differences in data types, measurement scales, computational complexity, and the lack of standardized pipelines. Addressing these challenges through harmonized experimental designs, shared databases and scalable analytical workflows will be crucial. Within a One Health perspective, integrating host omics data with microbiome profiles, pathogen genomics, and environmental information will strengthen disease surveillance, improve understanding of zoonotic risks, and support sustainable management of camel-based production systems in arid and pastoral ecosystems. In summary, although substantial challenges remain, omics sciences, by elucidating the molecular networks connecting animals, humans, and their environments, offer a powerful means to overcome long-standing fragmentation in health research, helping to translate the One Health concept into effective and operational practice.

References

Abdoon A, Soliman S, Hussein N, Haggag S, El-Sanea A, Abdel-Hamid A-H (2025) Metabolomic profile of dromedary camel follicular fluid during the breeding and non-breeding seasons. Sci Rep 15:8923. https://doi.org/10.1038/s41598-025-91710-9

Adney DR, van Doremalen N, Brown VR, Bushmaker T, Scott D, de Wit E, Bowen RA, Munster VJ (2014) Replication and shedding of MERS-CoV in upper respiratory tract of inoculated dromedary camels. Emerg Infect Dis 20(12):1999–2005. https://doi.org/10.3201/eid2012.141280

Akunda IK, Kariuki DW, Matulis G, Mwaura P, Maina B, Mohammed H, Paul A et al (2023) Antimicrobial resistance patterns and characterisation of emerging beta-lactamase-producing *Escherichia coli* in camels sampled from Northern Kenya. Vet Med Sci 9(3):1407–1416. https://doi.org/10.1002/vms3.1090

Al Abri M, Alfoudari A, Mohammad Z, Almathen F, Al-Marzooqi W, Al-Hajri S, Al-Amri M, Bahbahani H (2023) Assessing genetic diversity and defining signatures of positive selection on the genome of dromedary camels from the southeast of the Arabian Peninsula. Front Vet Sci 10:1296610. https://doi.org/10.3389/fvets.2023.1296610

Alhaider AA, Bayoumy N, Argo E, Gader AGMA, Stead DA (2012) Survey of the camel urinary proteome by shotgun proteomics using a multiple database search strategy. Proteomics 12(22):3403–3406. https://doi.org/10.1002/pmic.201100631

Alhassani WE (2024) Camel milk: nutritional composition, therapeutic properties, and benefits for human health. Open Vet J 14(12):3164–3180. https://doi.org/10.5455/OVJ.2024.v14.i12.2

Ali MA, Damir HA, Adem MA, Ali OM, Amir N, Shah AAM, Al Muhairi SSM et al (2023) Effects of long-term dehydration on stress markers, blood parameters, and tissue morphology in the dromedary camel (*Camelus dromedarius*). Front Vet Sci 10:1236425. https://doi.org/10.3389/fvets.2023.1236425

Aljasham AT, Raghunandanan S, Farzan R, Abdulwahed AM, Aneesh EM, Alharbi S, Shukri Y, Alshammari M, Alanazi F (2025) *Hyalomma dromedarii* infesting camels in Hail Province, Saudi Arabia, carry antimicrobial resistant bacteria. Front Vet Sci 12:1662637. https://doi.org/10.3389/fvets.2025.1662637

Alkhatabi HA, Alhashmi M, Alkhatabi HA, Altayb HN (2025) In silico analysis of temperature-induced structural, stability, and flexibility modulations in camel cytochrome c. Animals 15(3):381. https://doi.org/10.3390/ani15030381

Alnaeem A, Kasem S, Qasim I, Al-Doweriej A, Refaat M, Al-Shabebi A, Hemida MG (2020) The dipeptidyl peptidase-4 expression in some MERS-CoV naturally infected dromedary camels in Saudi Arabia 2018–2019. VirusDisease 31(2):200–203. https://doi.org/10.1007/s13337-020-00586-y

Alsubaie FM, Ibrahim MA, Wani SY, Awad A, Alrefaei AF, Almutairi MH (2024) Genome-wide association study for high milk yield in Saudi Arabian dromedary using whole genome sequencing, ed. Xinqing Xiao. Adv Agric 2024(1):7452228. https://doi.org/10.1155/2024/7452228

Alvira-Iraizoz F, Gillard BT, Lin P, Paterson A, Pauža AG, Ali MA, Alabsi AH et al (2021) Multiomic analysis of the Arabian camel (*Camelus dromedarius*) kidney reveals a role for cholesterol in water conservation. Commun Biol 4(1):779. https://doi.org/10.1038/s42003-021-02327-3

Andreu-Perez J, Poon CCY, Merrifield RD, Wong STC, Yang G-Z (2015) Big data for health. IEEE J Biomed Health Inform 19(4):1193–1208. https://doi.org/10.1109/JBHI.2015.2450362

Azhar EI, El-Kafrawy SA, Farraj SA, Hassan AM, Al-Saeed MS, Hashem AM, Madani TA (2014) Evidence for camel-to-human transmission of MERS coronavirus. N Engl J Med 370(26):2499–2505. https://doi.org/10.1056/NEJMoa1401505

Bactrian Camels Genome Sequencing and Analysis Consortium, Jirimutu, Wang Z, Ding G, Chen G, Sun Y, Sun Z et al (2012) Genome sequences of wild and domestic Bactrian camels. Nat Commun 3:1202. https://doi.org/10.1038/ncomms2192

Bahbahani H, Musa HH, Wragg D, Shuiep ES, Almathen F, Hanotte O (2019) Genome diversity and signatures of selection for production and performance traits in dromedary camels. Front Genet 10:893. https://doi.org/10.3389/fgene.2019.00893

Bahbahani H, Mohammad Z, Alfoudari A, Al Abri M (2025) Genomic insights into racing camels: inbreeding levels and positive selection linked to athletic traits. Animal 19(4):101467. https://doi.org/10.1016/j.animal.2025.101467

Bakry IA, Yang L, Farag MA, Korma SA, Khalifa I, Cacciotti I, Ziedan NI et al (2021) A comprehensive review of the composition, nutritional value, and functional properties of camel milk fat. Foods 10(9):2158. https://doi.org/10.3390/foods10092158

Barazandeh A, Mohammadabadi MR, Ghaderi-Zefrehei M, Nezamabadi-pour H (2016) Genome-wide analysis of CpG islands in some livestock genomes and their relationship with genomic features. Czeh J Anim Sci 61(11):487–495. https://doi.org/10.17221/78/2015-CJAS

Benabdelkamel H, Masood A, Alanazi IO, Alzahrani DA, Alrabiah DK, AlYahya SA, Alfadda AA (2017) Proteomic profiling comparing the effects of different heat treatments on camel (*Camelus dromedarius*) milk whey proteins. Int J Mol Sci 18(4):721. https://doi.org/10.3390/ijms18040721

Bitaraf Sani M, Harofte JZ, Banabazi MH, Esmaeilkhanian S, Naderi AS, Salim N, Teimoori A et al (2021) Genomic prediction for growth using a low-density SNP panel in dromedary camels. Sci Rep 11(1):7675. https://doi.org/10.1038/s41598-021-87296-7

Bitaraf Sani M, Roudbari Z, Karimi O, Banabazi MH, Esmaeilkhanian S, Asadzadeh N, Harofte JZ, Naderi AS, Burger PA (2022a) Gene-set enrichment analysis for identifying genes and biological activities associated with growth traits in dromedaries. Animals 12(2):184. https://doi.org/10.3390/ani12020184

Bitaraf Sani M, Harofte JZ, Banabazi MH, Faraz A, Esmaeilkhanian S, Naderi AS, Salim N et al (2022b) Identification of candidate genes for pigmentation in camels using genotyping-by-sequencing. Animals 12(9):1095. https://doi.org/10.3390/ani12091095

Bitaraf Sani M, Karimi O, Burger PA, Javanmard A, Roudbari Z, Mohajer M, Asadzadeh N et al (2023) A genome-wide association study of morphometric traits in dromedaries. Vet Med Sci 9(4):1781–1790. https://doi.org/10.1002/vms3.1151

Burger PA, Ciani E (2022) Structural and functional genomics in old world camels—where do we stand and where to go. Anim Front 12(4):30–34. https://doi.org/10.1093/af/vfac047

Cai W, Cole JB, Goddard ME, Li J, Zhang S, Song J (2025) Mammary gland multi-omics data reveals new genetic insights into milk production traits in dairy cattle. PLoS Genet 21(4):e1011675. https://doi.org/10.1371/journal.pgen.1011675

Calcino A, Cooke I, Cowman P, Higgie M, Massault C, Schmitz U, Whittaker M, Field MA (2024) Harnessing genomic technologies for one health solutions in the tropics. Glob Health 20(1):78. https://doi.org/10.1186/s12992-024-01083-3

Carrascosa-Sàez M, Buigues J, Viñals A, Andreu-Moreno I, Martínez-Recio R, Soriano-Tordera C, Monrós JS, Cuevas JM, Sanjuán R (2024) Genetic diversity and cross-species transmissibility of bat-associated picornaviruses from Spain. Virol J 21(1):193. https://doi.org/10.1186/s12985-024-02456-1

Chaiebeddraa A (2025) Metagenomic investigation of the rumen and caecum microbial communities of the Saharan dromedary camel. Turk J Vet Anim Sci 49(2):6

Chigozie V, Aniokete U, Ogbonna P, Iroha R (2025) Transforming antimicrobial resistance mitigation: the genomic revolution in one health and public health. Discov Appl Sci 7:1187. https://doi.org/10.1007/s42452-025-07053-7

Ciliberti MG, Albenzio M, Sevi A (2025) From sensors to sustainability: integrating welfare, management, and climate resilience in small ruminant farm systems. Animals 15(22):3240. https://doi.org/10.3390/ani15223240

Cong L, Ann Ran F, Cox D, Lin S, Barretto R, Habib N, Hsu PD et al (2013) Multiplex genome engineering using CRISPR/Cas systems. Science (New York, NY) 339(6121):819–823. https://doi.org/10.1126/science.1231143

Conzade R, Grant R, Malik MR, Elkholy A, Elhakim M, Samhouri D, Ben Embarek PK, Van Kerkhove MD (2018) Reported direct and indirect contact with dromedary camels among laboratory-confirmed MERS-CoV cases. Viruses 10(8):425. https://doi.org/10.3390/v10080425

Dai X, Shen L (2022) Advances and trends in omics technology development. Front Med 9:911861. https://doi.org/10.3389/fmed.2022.911861

Din MU, Liu X, Jiang H, Ahmad S, Xiangdong L, Wang X (2025) Advancing vaccine development in genomic era: a paradigm shift in vaccine discovery. Prog Biomed Eng 7(2):adb2c8. https://doi.org/10.1088/2516-1091/adb2c8

Du W, Chen S, Jiang R, Zhou H, Li Y, Ouyang D, Gong Y, Yao Z, Ye X (2025) Inferring *Staphylococcus aureus* host species and cross-species transmission from a genome-based model. BMC Genomics 26(1):149. https://doi.org/10.1186/s12864-025-11331-4

Džermeikaitė K, Šidlauskaitė M, Antanaitis R, Anskienė L (2025) Enhancing genomic selection in dairy cattle through artificial intelligence: integrating advanced phenotyping and predictive models to advance health, climate resilience, and sustainability. Dairy 6(5):50. https://doi.org/10.3390/dairy6050050

Eckstein S, Ehmann R, Gritli A, Rhaiem MB, Yahia HB, Diehl M, Wölfel R et al (2022) Viral and bacterial zoonotic agents in dromedary camels from Southern Tunisia: a seroprevalence study. Microorganisms 10(4):727. https://doi.org/10.3390/microorganisms10040727

Fadul-Pacheco L, Delgado H, Cabrera V (2021) Exploring machine learning algorithms for early prediction of clinical mastitis. Int Dairy J 119:105051. https://doi.org/10.1016/j.idairyj.2021.105051

Farschtschi S, Riedmaier-Sprenzel I, Phomvisith O, Gotoh T, Pfaffl MW (2022) The successful use of -omic technologies to achieve the "One Health" concept in meat producing animals. Meat Sci 193:108949. https://doi.org/10.1016/j.meatsci.2022.108949

Faye B (2016) The camel, new challenges for a sustainable development. Trop Anim Health Prod 48(4):689–692. https://doi.org/10.1007/s11250-016-0995-8

Faye B (2022) Is the camel conquering the world? Anim Front 12(4):8–16. https://doi.org/10.1093/af/vfac034

Faye B, Konuspayeva G, Magnan C (2023) Large camel farming: a care-management guide from breeding to camel products. Springer Netherlands, Dordrecht. https://doi.org/10.1007/978-94-024-2237-5

Franco-Martínez L, Martínez-Subiela S, Cerón JJ, Tecles F, Eckersall PD, Oravcova K, Tvarijonaviciute A (2020) Biomarkers of health and welfare: a one health perspective from the laboratory side. Res Vet Sci 128:299–307. https://doi.org/10.1016/j.rvsc.2019.12.012

Gao F, Li P, Yin Y, Xuguang D, Cao G, Sen W, Zhao Y (2023) Molecular breeding of livestock for disease resistance. Virology 587:109862. https://doi.org/10.1016/j.virol.2023.109862

Gao Y, Jiangchao W, Zeng J, Huo X, Lou K (2024) Beyond the desert sands: decoding the relationship between camels, gut microbiota, and antibiotic resistance through metagenomics. Sci One Health 3:100071. https://doi.org/10.1016/j.soh.2024.100071

Gharechahi J, Salekdeh GH (2018) A metagenomic analysis of the camel rumen's microbiome identifies the major microbes responsible for lignocellulose degradation and fermentation. Biotechnol Biofuels 11:216. https://doi.org/10.1186/s13068-018-1214-9

Gharechahi J, Sarikhan S, Han J-L, Ding X-Z, Salekdeh GH (2022) Functional and phylogenetic analyses of camel rumen microbiota associated with different lignocellulosic substrates. NPJ Biofilms Microbi 8(1):46. https://doi.org/10.1038/s41522-022-00309-9

Guo X, Jirimutu LM, Wang Z (2025) Deciphering genetic adaptations of old world camels through comparative genomic analyses across all camelid species. iScience 28(5):112477. https://doi.org/10.1016/j.isci.2025.112477

Halli K, Yin T, Koch C, Krebs S, König S (2025) Heat stress induces specific methylation, transcriptomic and metabolic pattern in dairy cows and their female progeny. Sci Rep 15(1):17021. https://doi.org/10.1038/s41598-025-01082-3

Harmsen MM, De Haard HJ (2007) Properties, production, and applications of camelid single-domain antibody fragments. Appl Microbiol Biotechnol 77(1):13–22. https://doi.org/10.1007/s00253-007-1142-2

Hassan AM, Yasir M, Al-Subhi T, Sohrab SS, Abbas AT, El-Kafrawy SA, Kumosani T, Azhar EI (2025) Analysis of the metagenomic assembled genome of MERS-CoV and alpha-CoV from *Camelus dromedarius* in Saudi Arabia. J Infect Public Health 18(11):102908. https://doi.org/10.1016/j.jiph.2025.102908

He J, Hai L, Orgoldol K, Yi L, Ming L, Guo F, Li G, Ji R (2019) High-throughput sequencing reveals the gut microbiome of the Bactrian camel in different ages. Curr Microbiol 76(7):810–817. https://doi.org/10.1007/s00284-019-01689-6

Heyer R, Wolf M, Benndorf D, Uzzau S, Seifert J, Grenga L, Pabst M et al (2025) Metaproteomics in the one health framework for unraveling microbial effectors in microbiomes. Microbiome 13:134. https://doi.org/10.1186/s40168-025-02119-5

Hinsu A, Nilam T, Panchal K, Ramesh P, Jyotsana B, Dafale N, Patil N et al (2021) Characterizing rumen microbiota and CAZyme profile of Indian dromedary camel (*Camelus dromedarius*) in response to different roughages. Sci Rep 11:9400. https://doi.org/10.1038/s41598-021-88943-9

Hoter A, Rizk S, Naim HY (2019) Cellular and molecular adaptation of Arabian camel to heat stress. Front Genet 10:588. https://doi.org/10.3389/fgene.2019.00588

Ibrahim Z, Almundarij T (2023) Morphology of the dromedary camel stomach with reference to physiological adaptation. Slov Vet Res 60:341–352. https://doi.org/10.26873/SVR-1621-2022

Iglesias Pastrana C, González FJN, Macri M, del Amparo M, Martínez M, Ciani E, Bermejo JVD (2024) Identification of novel genetic loci related to dromedary camel (*Camelus dromedarius*) morphometrics, biomechanics, and behavior by genome-wide association studies. BMC Vet Res 20:418. https://doi.org/10.1186/s12917-024-04263-w
Islam MA, Rony SA, Rahman MB, Cinar MU, Villena J, Uddin MJ, Kitazawa H (2020) Improvement of disease resistance in livestock: application of immunogenomics and CRISPR/Cas9 technology. Animals 10(12):2236. https://doi.org/10.3390/ani10122236
Jinya L (2024) Genomic selection in livestock breeding: advances and applications. Anim Mol Breed 14(3):239–251. https://animalscipublisher.com/index.php/amb/article/view/3813. 25 Feb 2026
Jonas O, World Team (2017) Drug-resistant infections: a threat to our economic future. World Bank, Washington, DC
Kalkan A, Tepeli M, Göde A (2025) Mastitis diagnosis with machine learning algorithms. Neural Comput & Applic 37:12351–12372. https://doi.org/10.1007/s00521-025-11176-8
Kasimanickam R, Ferreira JCP, Kastelic J, Kasimanickam V (2025) Application of genomic selection in beef cattle disease prevention. Animals 15(2):277. https://doi.org/10.3390/ani15020277
Kholousi Adab F, Yaghoobi MM, Gharechahi J (2024) Enhanced crystalline cellulose degradation by a novel metagenome-derived cellulase enzyme. Sci Rep 14(1):8560. https://doi.org/10.1038/s41598-024-59256-4
Kim S, Park Y-G, Choi J-H, Moon S-H (2025) CRISPR technology for livestock improvement: advances and future directions. Mol Biotechnol 68(5):1–16. https://doi.org/10.1007/s12033-025-01532-3
Klümper U, Fang P, Bing Li Y, Xia DF, Hamilton KA, Quon H, Berendonk TU, de la Cruz M, Barron. (2025) Towards the integration of antibiotic resistance gene mobility into environmental surveillance and risk assessment. npj Antimicrob Resist 3(1):81. https://doi.org/10.1038/s44259-025-00154-8
Ko KKK, Chng KR, Nagarajan N (2022) Metagenomics-enabled microbial surveillance. Nat Microbiol 7(4):486–496. https://doi.org/10.1038/s41564-022-01089-w
Komor AC, Kim YB, Packer MS, Zuris JA, Liu DR (2016) Programmable editing of a target base in genomic DNA without double-stranded DNA cleavage. Nature 533(7603):420–424. https://doi.org/10.1038/nature17946
Lado S, Elbers JP, Doskocil A, Scaglione D, Trucchi E, Banabazi MH, Almathen F et al (2020) Genome-wide diversity and global migration patterns in dromedaries follow ancient caravan routes. Commun Biol 3:387. https://doi.org/10.1038/s42003-020-1098-7
Lander ES, Linton LM, Birren B, Nusbaum C, Zody MC, Baldwin J, Devon K et al (2001) Initial sequencing and analysis of the human genome. Nature 409(6822):860–921. https://doi.org/10.1038/35057062
Li Y, Khalafalla AI, Paden CR, Yusof MF, Eltahir YM, Al ZM, Hammadi YT et al (2017) Identification of diverse viruses in upper respiratory samples in dromedary camels from United Arab Emirates. PLoS One 12(9):e0184718. https://doi.org/10.1371/journal.pone.0184718
Li M, Wang Z, Zhu L, Wang Y, Zhang L, Jia H, Yang Z, Karrow NA, Mao Y (2025) Integrative blood transcriptomic and metabolomic profiling reveals biomarkers of natural heat tolerance in Holstein cows. J Dairy Sci 108(11):12824–12837. https://doi.org/10.3168/jds.2025-26997
Liu CSC, Pandey R (2024) Integrative genomics would strengthen AMR understanding through ONE Health approach. Heliyon 10(14):e34719. https://doi.org/10.1016/j.heliyon.2024.e34719
Liu D, Li L, Cao X, Jinyan W, Guoyu D, Shang Y (2021) Selection and identification of single-domain antibody against Peste Des Petits Ruminants virus. J Vet Sci 22(4):e45. https://doi.org/10.4142/jvs.2021.22.e45
Liu Z, Tianwen W, Xiang G, Wang H, Wang B, Feng Z, Yulian M, Li K (2022) Enhancing animal disease resistance, production efficiency, and welfare through precise genome editing. Int J Mol Sci 23(13):7331. https://doi.org/10.3390/ijms23137331
Liu J, Yao X, Meng J, Wang J, Zeng Y, Li L, Ren W (2025) Whole-transcriptome sequencing and differential expression analysis of the epididymis in Junggar Bactrian camels before and after sexual maturity. Biology 14(7):760. https://doi.org/10.3390/biology14070760

Ma W, Yao H, Zhang L, Zhang Y, Wang Y, Wang W, Liu Y et al (2025) Transcriptomics-based study of immune genes associated with subclinical mastitis in Bactrian camels. Vet Sci 12(2):121. https://doi.org/10.3390/vetsci12020121

Metzker ML (2010) Sequencing technologies—the next generation. Nat Rev Genet 11(1):31–46. https://doi.org/10.1038/nrg2626

Meuwissen T, Hayes B, Goddard M (2016) Genomic selection: a paradigm shift in animal breeding. Anim Front 6:6. https://doi.org/10.2527/af.2016-0002

Michelena Á, Fontenla-Romero O, Calvo-Rolle J (2024) A review and future trends of precision livestock over dairy and beef cow cattle with artificial intelligence. Log J IGPL 33:jzag020. https://doi.org/10.1093/jigpal/jzae111

Ming L, Yuan L, Yi L, Ding G, Hasi S, Chen G, Jambl T et al (2020) Whole-genome sequencing of 128 camels across Asia reveals origin and migration of domestic Bactrian camels. Commun Biol 3(1):1. https://doi.org/10.1038/s42003-019-0734-6

Ming L, Yi L, Ai Y, Ji R (2025) A new insight into the exosome protein and lipid composition in camel colostrum and mature milk using comparative proteome and lipidomics analyses. Food Chem X 29:102729. https://doi.org/10.1016/j.fochx.2025.102729

Mir MA, Mehraj U, Sheikh BA, Hamdani SS (2020) Nanobodies: the "magic bullets" in therapeutics, drug delivery and diagnostics. Hum Antibodies 28(1):29–51. https://doi.org/10.3233/HAB-190390

Mohamed WMA, Ali AO, Mahmoud HYAH, Omar MA, Chatanga E, Salim B, Naguib D et al (2021) Exploring prokaryotic and eukaryotic microbiomes helps in detecting tick-borne infectious agents in the blood of camels. Pathogens 10(3):351. https://doi.org/10.3390/pathogens10030351

Molla G, Bitew M (2024) Revolutionizing personalized medicine: synergy with multi-omics data generation, main hurdles, and future perspectives. Biomedicine 12(12):2750. https://doi.org/10.3390/biomedicines12122750

Morabito A, De Simone G, Pastorelli R, Brunelli L, Ferrario M (2025) Algorithms and tools for data-driven omics integration to achieve multilayer biological insights: a narrative review. J Transl Med 23(1):425. https://doi.org/10.1186/s12967-025-06446-x

Moulavi F, Saadeldin IM, Swelum AA, Tasdighi F, Hosseini-Fahraji H, Hosseini SM (2021) Oocyte vitrification induces loss of DNA methylation and histone acetylation in the resulting embryos derived using ICSI in dromedary camel. Zygote 29(5):383–392. https://doi.org/10.1017/S0967199421000150

Nüesch-Inderbinen M, Kindle P, Baschera M, Liljander A, Jores J, Corman VM, Stephan R (2020) Antimicrobial resistant and extended-spectrum ß-lactamase (ESBL) producing *Escherichia coli* isolated from fecal samples of African dromedary camels. Sci Afr 7:e00274. https://doi.org/10.1016/j.sciaf.2020.e00274

O'Neill J (2016) Tackling drug-resistant infections globally: final report and recommendations. Wellcome Trust, London, p 84

Pitt SJ, Gunn A (2024) The one health concept. Br J Biomed Sci 81:12366. https://doi.org/10.3389/bjbs.2024.12366

Rahmeh R, Akbar A, Alomirah H, Kishk M, Alateeqi A, Al-Milhm S, Shajan A et al (2022) Camel Milk microbiota: a culture-independent assessment. Food Res Int 159:111629. https://doi.org/10.1016/j.foodres.2022.111629

Rajabi M, Nourisanami F, Ghadikolaei KK, Changizian M, Noghabi K, Zahiri H (2022) Metagenomic psychrohalophilic xylanase from camel rumen investigated for bioethanol production from wheat bran using Bacillus subtilis AP. Sci Rep 12(1):8152. https://doi.org/10.1038/s41598-022-11412-4

Rooney T, Fèvre EM, Villinger J, Brenn-White M, Cummings CO, Chai D, Kamau J et al (2024) *Coxiella burnetii* serostatus in dromedary camels (*Camelus dromedarius*) is associated with the presence of *C. burnetii* DNA in attached ticks in Laikipia County, Kenya. Zoonoses Public Health 71(5):503–514. https://doi.org/10.1111/zph.13127

Russell T, Formiconi E, Casey M, McElroy M, Mallon PWG, Gautier VW (2025) Viral metagenomic next-generation sequencing for one health discovery and surveillance of (re)emerging viruses: a deep review. Int J Mol Sci 26(19):9831. https://doi.org/10.3390/ijms26199831

Sabahat S, Nadeem A, Brauning R, Thomson PC, Khatkar MS (2023) Genome wide association study for growth in Pakistani dromedary camels using genotyping-by-sequencing. Anim Biosci 36(7):1010–1021. https://doi.org/10.5713/ab.22.0181

Sahraei A, Knob D, Lambertz C, Gattinger A, Breuer L (2025) Modeling enteric methane emission from dairy cows using deep learning approach. Sci Total Environ 984:179713. https://doi.org/10.1016/j.scitotenv.2025.179713

Scarpa F, Casu M (2024) Genomics and bioinformatics in one health: transdisciplinary approaches for health promotion and disease prevention. Int J Environ Res Public Health 21(10):1337. https://doi.org/10.3390/ijerph21101337

Shoyombo A, Popoola M, Kuusu D, Yisah L, Adebayo O, Yakubu H, Ndiomu E, Ake A (2024) Computational approaches in livestock breeding: a review. IEEE. https://doi.org/10.1109/SEB4SDG60871.2024.10630145

Si R, Ming L, Yun X, He J, Yi L, Na Q, Ji R, Dong T (2025) Proteomics integrated with metabolomics: analysis of the internal mechanism underlying changes in meat quality in different muscles from Bactrian camels. Food Chem X 26:102230. https://doi.org/10.1016/j.fochx.2025.102230

Sundermann AJ, Kumar P, Griffith MP, Waggle KD, Srinivasa VR, Raabe N, Mills EG et al (2026) Real-time genomic surveillance for enhanced healthcare outbreak detection and control: clinical and economic impact. Clin Infect Dis 82(1):134–141. https://doi.org/10.1093/cid/ciaf216

Suravajhala P, Kogelman LJA, Kadarmideen HN (2016) Multi-omic data integration and analysis using systems genomics approaches: methods and applications in animal production, health and welfare. Genet Sel Evol 48(1):38. https://doi.org/10.1186/s12711-016-0217-x

Tibary A, El Allali K (2020) Dromedary camel: a model of heat resistant livestock animal. Theriogenology 154:203–211. https://doi.org/10.1016/j.theriogenology.2020.05.046

Tulsani NJ, Mishra P, Jakhesara SJ, Srivastava S, Jyotsana B, Dafale NA, Patil NV, Purohit HJ, Joshi CG (2021) Isolation, purification and characterization of a novel esterase from camel rumen metagenome. Protein Expr Purif 187:105941. https://doi.org/10.1016/j.pep.2021.105941

Ukpoju-Ebonyi O, Ebonyi J-M, Ogunjobi T, Obasi D, Ugwu D, Anyanwu G, Abolade S (2025) Integrating multi-omics and personalized medicine in one health: a systems biology approach to precision healthcare. Medinfo. https://doi.org/10.47852/bonviewMEDIN52025831

Upadhyay VR, Ramesh V, Harshit Kumar YM, Somagond SP, Kuniyal A, Prakash V, Sahoo A (2024) Phenomics in livestock research: bottlenecks and promises of digital phenotyping and other quantification techniques on a global scale. OMICS 28(8):380–393. https://doi.org/10.1089/omi.2024.0109

Van Eenennaam AL (2025) Current and future uses of genetic improvement technologies in livestock breeding programs. Anim Front 15(1):80–90. https://doi.org/10.1093/af/vfae042

von Fricken ME, Melendrez MC, Linton Y-M, Takhampunya R (2023) Editorial: metagenomics for epidemiological surveillance in one health. Front Microbiol 14:1191946. https://doi.org/10.3389/fmicb.2023.1191946

Wang N, Zhang G, Li M (2024) Emerging trends in systems biology: multi-omics integration and beyond. Comput Mol Biol 14(5):211–219. https://bioscipublisher.com/index.php/cmb/article/view/3986. 25 Feb 2026

Wang N-N, Cao F, Zhang L-H, Zheng Y-F, Da X (2025) Multi-omics: a bridge connecting genotype and phenotype for epilepsy? Biomark Res 13(1):86. https://doi.org/10.1186/s40364-025-00798-8

Warda M, Prince A, Kim HK, Khafaga N, Scholkamy T, Linhardt RJ, Jin H (2014) Proteomics of old world camelid (*Camelus dromedarius*): better understanding the interplay between homeostasis and desert environment. J Adv Res 5(2):219–242. https://doi.org/10.1016/j.jare.2013.03.004

Wei J, Wei L, Ullah A, Geng M, Zhang X, Wang C, Khan MZ, Wang C, Zhang Z (2025) Metagenomic applications to herbivore gut microbiomes: a comprehensive review of microbial diversity and host interactions. Animals 15(20):2938. https://doi.org/10.3390/ani15202938

Woo PCY, Lau SKP, Teng JLL, Tsang AKL, Joseph M, Wong EYM, Tang Y et al (2014) Metagenomic analysis of viromes of dromedary camel fecal samples reveals large number and high diversity of circoviruses and picobirnaviruses. Virology 471–473:117–125. https://doi.org/10.1016/j.virol.2014.09.020

Wu H, Guang X, Al-Fageeh MB, Cao J, Pan S, Zhou H, Zhang L et al (2014) Camelid genomes reveal evolution and adaptation to desert environments. Nat Commun 5(1):5188. https://doi.org/10.1038/ncomms6188

Xiao Y, Yi L, Ming L, He J, Ji R (2022) Changes in milk components, amino acids, and fatty acids of Bactrian camels in different lactation periods. Int Dairy J 131:105363. https://doi.org/10.1016/j.idairyj.2022.105363

Xue M-Y, Sun H-Z, Xue-Hui W, Liu J-X, Guan LL (2020) Multi-omics reveals that the rumen microbiome and its metabolome together with the host metabolome contribute to individualized dairy cow performance. Microbiome 8(1):64. https://doi.org/10.1186/s40168-020-00819-8

Yan L, Tang L, Zhou Z, Wei L, Wang B, Sun Z, Jiang X et al (2022) Metagenomics reveals contrasting energy utilization efficiencies of captive and wild camels (*Camelus ferus*). Integr Zool 17(3):333–345. https://doi.org/10.1111/1749-4877.12585

Yao H, Liang X, Dou Z, Zhao Z, Ma W, Hao Z, Yan H et al (2023) Transcriptome analysis to identify candidate genes related to mammary gland development of Bactrian camel (*Camelus bactrianus*). Front Vet Sci 10:1196950. https://doi.org/10.3389/fvets.2023.1196950

Yao H, Pan Z, Ma W, Zhao Z, Zhanqiang S, Yang J (2024) Whole-genome resequencing analysis of the *Camelus bactrianus* (Bactrian camel) genome identifies mutations and genes affecting milk production traits. Int J Mol Sci 25(14):7836. https://doi.org/10.3390/ijms25147836

Yuan Y-G, Liu S-Z, Farhab M, Lv M-Y, Zhang T, Cao S-X (2024) Genome editing: an insight into disease resistance, production efficiency, and biomedical applications in livestock. Funct Integr Genomics 24(3):81. https://doi.org/10.1007/s10142-024-01364-5

Zhang C, Liu H, Jiang X, Zhang Z, Hou X, Wang Y, Wang D et al (2024) An integrated microbiome- and metabolome-genome-wide association study reveals the role of heritable ruminal microbial carbohydrate metabolism in lactation performance in Holstein dairy cows. Microbiome 12(1):232. https://doi.org/10.1186/s40168-024-01937-3

Zhao Y, Zhang W, Zhang X (2024) Application of metagenomic next-generation sequencing in the diagnosis of infectious diseases. Front Cell Infect Microbiol 14:1458316. https://doi.org/10.3389/fcimb.2024.1458316

Zheng W, Zhang J, Huang J, Han B, Liu Y, Sun D (2025) A pilot multi-omics study reveals genetic mechanisms regulating milk component traits in dairy cattle. Commun Biol 8(1):1150. https://doi.org/10.1038/s42003-025-08615-6

Zhu S, Zimmerman D, Deem SL (2019) A review of zoonotic pathogens of dromedary camels. EcoHealth 16(2):356–377. https://doi.org/10.1007/s10393-019-01413-7

Camels for Health: Development of Nanobody Research

6

Mehdi Arbabi Ghahroudi

Abstract

Scientific discoveries and advancements have historically come from improving our understanding of the natural world and its biological processes. The most notable example in the biological sciences is the discovery of penicillin by Sir Alexander Fleming in 1928 and his famous remark, "Nature makes penicillin; I just found it." Penicillin, which is produced by the mould *Penicillium notatum* and inhibits the growth of *Staphylococcus* bacteria, was the first effective antibiotic against bacterial infections. This discovery not only transformed medical practice but also created the field of targeted bacterial inhibition and set the cornerstone of modern antimicrobial therapy.

The discovery of unique antibodies in dromedaries (*Camelus dromedarius*) by Professor Raymond Hamers and his research team at Vrije Universiteit Brussel, Belgium, in the late 1980s followed a similar path. Conventional immunoglobulins or antibodies are composed of two heavy and two homodimeric light chains. They are produced by almost all mammals, including South American camelids and camels as well as humans, as a second line of defence mechanism against invading pathogens and have a central role in the health and well-being of living organisms. However, the Camelidae family has been shown to possess a unique, smaller immunoglobulin subclass that contains the heavy chain, but lacks the light chain (referred to as heavy chain only antibodies or HCAbs). These are naturally occurring, distinct from other genetically engineered antibody fragments.

M. Arbabi Ghahroudi (✉)
Human Health Therapeutics Research Centre, Life Sciences Division, National Research Council Canada, Ottawa, Ontario, Canada

Department of Biochemistry, Microbiology and Immunology, Faculty of Medicine, University of Ottawa, Ottawa, Ontario, Canada

Department of Biology, Carleton University, Ottawa, Ontario, Canada
e-mail: Mehdi.ArbabiGhahroudi@nrc-cnrc.gc.ca

S. A. Gordine et al. (eds.), *Camels and One Health*,
https://doi.org/10.1007/978-981-92-2246-9_6

In recent years, HCAbs and specifically nanobodies have gained particular attention in human health research. Nanobodies have unique biophysicochemical characteristics which can serve as building blocks for the development of therapeutics against cancer, inflammation, neurodegenerative diseases, and infectious diseases. The first nanobody-based therapeutic, caplacizumab, was approved by the FDA in 2019 as a drug for human application to treat acquired thrombotic thrombocytopenic purpura disorder. In this chapter, we provide a historical perspective on the discovery of camelid antibodies and their applications in human health and well-being.

Keywords

Single-domain antibodies · Immunotherapy · Cancer research · Molecular recognition · Biomedical innovation · Next-generation therapeutics

6.1 Introduction

Over millions of years, living organisms have developed interesting functional features which are imprinted in their genomes. The development of these features was driven by mutations and other evolutionary processes and phenotypic expression of biomolecules. The resultant features give unique advantages and survival capabilities in extreme conditions such as changes in temperature or exposure to virulent factors.

During a student-led project at Professor Raymond Hamers' laboratory at VUB in Brussels, Belgium, it was serendipitously discovered that the serum of camels has novel smaller-sized antibodies besides the conventional antibodies. These novel antibodies were later termed heavy chain only antibodies (HCAbs). At first, this chance observation was not considered highly important because these types of HCAbs had already been reported in humans and mice. In humans and mice, an immunological disease caused abnormal random mutations which led to the production of these heavy chain antibodies (Fleischman et al. 1962). Yet, as Professor Hamers was always attentive regarding the occurrence of biological phenomena, he followed this observation with great enthusiasm and requested additional experiments to be performed. As a next step, the roles of these novel antibodies in the camel immunoglobulin repertoire were investigated by fractionation of the serum and purification of the antibodies from a dromedary immunized against the *Trypanosoma* parasite. The main question was whether these smaller-sized antibodies were functional. Hamers' research team demonstrated that HCAbs can bind specifically to the trypanosome's lysate, which they published in *Nature* in 1993 (Hamers-Casterman et al. 1993). This specific binding could have neutralization properties, demonstrating that the HCAbs were indeed biologically functional. Following this, the identity and

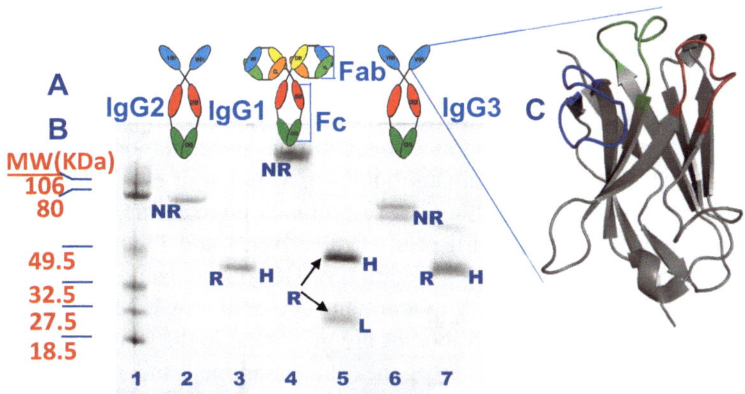

Fig. 6.1 Schematic representation of IgGs from the serum of dromedary camel obtained by fractionation on protein A and protein G columns. (**a**) The IgG isotype from each fraction has been shown on top of lane 2, 4, and 6. (**b**) Dromedary immunoglobulin serum was fractionated on protein G and A and ran on reducing and non-reducing SDS-PAGE. Lane 1, MWM; lane 2, IgG2 (protein A) (non-reduced: NR), lane 3, IgG2 (reduced: R); lane 4, IgG1 (protein A and G) (NR); lane 5, IgG1 (R); lane 6, IgG3 (protein A and G) (NR); lane 7, IgG3 (R). (**c**) The VHH folding structure containing two β-sheets with five and four β-strands is shown on the right with CDR loops shown in dark green (CDR1), red (CDR2), and blue (CDR3)

nature of these antibodies were investigated by molecular cloning techniques, and a molecular and genetic map was established (Vu et al. 1997). It was shown that the camel heavy chains not only lost their light chain partner but also have a shorter heavy chain, lacking the well-conserved CH1 domain, which is involved with light chain pairing. The conventional structural pairing of light chain (VL and CL domains) and heavy chain (VH and CH1 domains) provides structural integrity to the antibody molecule and forms the antigen-binding fragment (Fab). Additionally, the interaction of variable domains (VH and VL) as well as the first two constant domains (CH1 and CL) conceals large hydrophobic interfaces, ensuring a proper folding, stability, secretion, and specific binding to antigens (Fig. 6.1). HCAbs, on the other hand, can bind antigen with high affinity with only a single antigen-binding variable domain (VHH) on each heavy chain.

Conventional IgG is composed of two heavy and two light chains (Fig. 6.1). Heavy chain has a variable domain (VH) connected to the CH1 domain by a flexible linker. The CH1 domain is connected to the CH2 and CH3 constant domains by the hinge region. In contrast, the VHH domain in the heavy chain–only antibodies is directly connected to the CH2–CH3 by the hinge region. The VHH domain could be expressed as a single domain by antibody engineering techniques.

6.2 Molecular Mechanisms Behind the Generation of HCAbs

The 1993 *Nature* publication revolutionized the accepted view of conventional two-chain antibodies. It was demonstrated that camel heavy chain–only antibodies can perform the function of four-chain antibodies in a similar or even superior manner. This opened a new horizon in the field of antibody engineering, which was in its infancy phase in the 1990s (Bird et al. 1988; Skerra and Pluckthun 1988). The molecular mechanism behind the generation of HCAbs in dromedaries was further investigated by Professor Serge Muyldermans (Vu et al. 1997; Muyldermans et al. 1994). It was found that these antibodies are generated in the same way as conventional antibodies in B cells, where genes are assembled and transcribed in the nucleus and transported to the cytoplasm for translation into proteins.

Detailed sequencing data showed that the genes which encode the variable domains of heavy chain only antibodies (known as VHH [also named camelid single-domain antibodies {sdAbs} or nanobodies {Nbs}] where "VH" stands for the variable domain of the heavy chain and "H" stands for being derived from heavy chain–only antibodies) harbour four site-specific mutations (V42F/Y, G49E, L50R, and W52G/F) which otherwise would interact with the former VL domain. These mutations turn the hydrophobic VHH surface, which interacts with the VL domain, into a hydrophilic surface. Therefore, the VHH domains become quite soluble in the absence of its VL binding partner, as compared to a conventional human VH domain. Deletion of the CH1 domain resulted from a genetic point mutation of G to A in the gene encoding heavy-chain antibodies. This leads to a direct connection of the VHH domain to the hinge region, followed by the CH2 and CH3 constant domains, which eventually translated into functional heavy-chain antibodies (Vu et al. 1997; Muyldermans et al. 1994; Nguyen et al. 2000) (Fig. 6.1).

During the 1990s, scientists in the field of immunology, molecular biology, and biotechnology around the globe were developing ways to clone and express antibody genes in a simple bacterial system and to demonstrate their in vivo functional binding (McCafferty et al. 1990; Griffiths 1993). At the same time and due to the complexity of the mammalian immune system alongside difficulties in isolation and scale-up selection of antibodies from hybridoma cells (Kohler and Milstein 1975), scientists in the United States and UK were able to develop a protein fusion strategy. Here, the genes encoding the peptides and antibody fragments of interest were attached to the genes encoding the coat protein of filamentous bacteriophages resulting in the peptide/antibody-coat protein fusion. Upon expression and assembly of the viral particles in bacteria, the viral coat protein fusion is displayed on the phage surface (Jespers et al. 1994; Hoogenboom 2005). The process of fusing the genes of interest followed by the expression of the fusion protein is called genotype–phenotype coupling. This imitates a similar process happening in vivo in the B lymphocyte, where a single B cell harbours an assembled antibody gene and the antibody is displayed on the cell surface. The circulating B cells then bind to

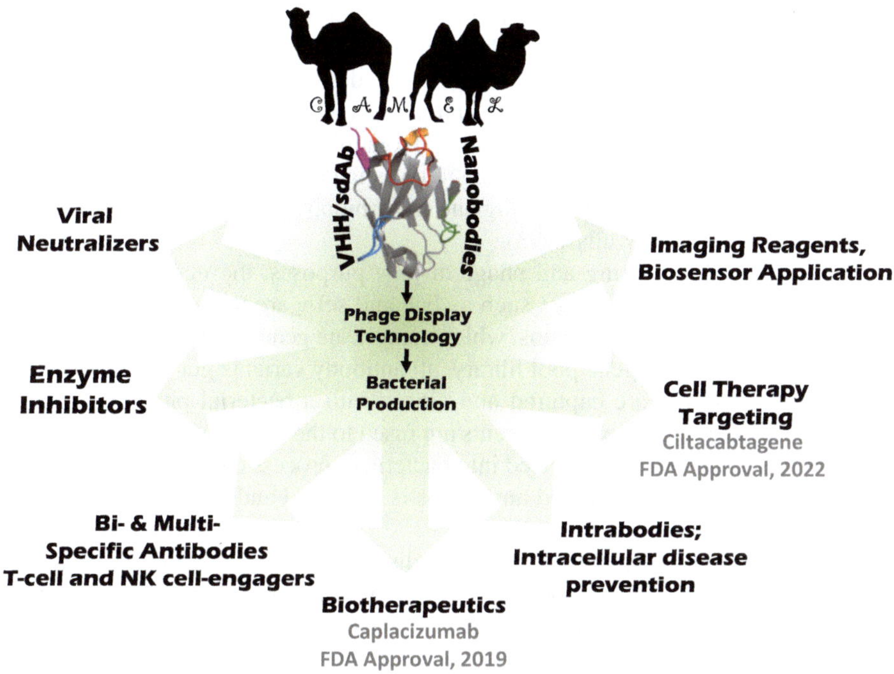

Fig. 6.2 There are multiple applications of camel single domain antibodies (nanobodies) in various field of research such as diagnostic reagents and biosensor application, viral neutralizers, or enzyme inhibitors. Nanobodies could also be used in bivalent or multivalent formats for disease treatments or as bi-and multi-specific formats to target cancer cells and /or engage the immune effector cells. Unlike conventional antibodies, nanobodies can be produced and remain functional inside living cells. Their small size and ability to fold correctly in the cytoplasm make them particularly useful for targeting disease-causing proteins located inside cells

intruding foreign molecules and trap them for neutralization and destruction by engaging other arms of the immune system (Janeway Jr et al. 2016; Winter et al. 1994). By mimicking this principle, the research teams of the Nobel prize laureates Drs. George P. Smith and Sir Gregory P. Winter were able to display a repertoire of random peptides and immune antibody fragments on the surface of filamentous bacteriophages and use them as tools for screening against a target of interest. This technique is now known as phage display biopanning (Barderas and Benito-Pena 2019) (Fig. 6.2).

6.3 Antibody Engineering Around the World

The pioneering work of phage display has been the cornerstone of all target-specific peptides and antibody fragments isolated by engineering since the 1990s and is still the preferred method practiced in all antibody engineering labs around the world.

Antibody engineering uses the in vivo process of antibody generation and transfers the antibodies into a test tube in a highly skilled laboratory. For various applications such as cancer immunotherapy, there is a need to design an antibody molecule with specific features such as optimum size, unique specificity, and the ability to attach a drug. Moreover, this approach is not limited to antibody and peptides display. It is also a useful tool for protein engineering to investigate the effects of multiple amino acid mutations to improve the binding and other biophysico-chemical properties of a protein (Lowman and Wells 1993).

For antibody engineering and phage display purposes, the main drawbacks of conventional antibody fragments such as Fab and scFv are the gene assembly and expression of two or more domains, which complicate gene cloning and expression. In building an antibody gene pool library, all antibody variants generated in an animal by immunization are captured and cloned into a bacterial phage expression vector. In this vector antibody fragments are fused to the coat protein of filamentous phages. This pool is then transferred into bacteria (a process called transformation), in which the repertoire of cloned antibodies is produced and displayed on the surface of phage particles.

Camel single-domain antibodies were discovered at the right time because researchers in the field of antibody engineering were looking for ways to reduce the size of the antibody fragments for efficient phage display and protein production (Ward et al. 1989). Reducing the size of antibody fragments also has important implications for cancer therapy and detection, because smaller antibody fragments can more deeply penetrate heterogenous solid tumour tissues. They are also advantageous as their rapid clearance from the body via the kidneys reduces nonspecific binding to other targets.

In 1997, the first generation of a camelid-based library was reported introduced the camel single-domain antibody as an alternative to the existing conventional antibody fragments (scFv, Fab) with many obvious advantages, including reduced domain complexity, ease of cloning and library construction, and efficient expression and stability in bacterial system selection of VHH domains against virtually any target antigen. Additionally, the VHH antibodies isolated in the report showed inhibitory and neutralization behaviour toward the target proteins, including enzymes and toxins. This property is rarely reported for conventional antibodies.

6.4 Camel Single-Domain Antibodies as Novel Therapeutics

The publication of HCAbs in dromedaries by Hamers-Casterman and colleagues in 1993 re-ignited the idea of single-domain antibodies, which was originally captured by Ward and colleagues in the 1989 (Ward et al. 1989). Yet the first movement toward using these domain antibodies as potential therapeutics only started about a decade later with the founding of Ablynx as a biopharmaceutical/VIB (Vlaames Instituute voor Biotechnologie)-funded spin-off company in December 2001. There were several major scientific developments in the first decade (1993–2003) following the discovery of camel heavy-chain antibodies including (a) the identification of the genes and genetic mechanisms involved in the generation of HCAbs and their close similarity to the most common human variable domain (Nguyen et al. 1998, 2002; Harmsen et al. 2000), (b) the establishment of antibody gene pool cloning and applying phage display technology to select target-specific VHHs (Arbabi Ghahroudi et al. 1997; Desmyter et al. 2002), (c) structural studies of the VHHs and comparisons with conventional VH domain antibodies (Decanniere et al. 1999; Spinelli et al. 2004; Desmyter et al. 2001; Perez et al. 2001; Dumoulin et al. 2002), (d) cloning and expression of target-specific VHHs in bacteria and yeast systems as cost-effective alternative expression systems to mammalian cells (Liu and Huang 2018; de Marco 2020), and (e) the application of VHHs as reagents in immunoaffinity purification (Verheesen et al. 2003).

During the second decade (2003–2013) of VHH development, exponential growth of publications and patent applications occurred. The most significant advances of this decade were the launch of preclinical and clinical investigations of several nanobodies, exploring their use as therapeutics and imaging reagents (Saerens et al. 2008; Wesolowski et al. 2009; Steeland et al. 2016), including the development of VHH targeting: (a) blood glycoprotein vWF to control platelet aggregation and clot formation (Peyvandi et al. 2016), (b) respiratory syncytial virus (RSV) (Schepens et al. 2011), (c) venom toxins (Hmila et al. 2010), (d) IL6-R for treatment of rheumatoid arthritis (Van Roy et al. 2015), and (e) the use of radiolabelled nanobodies for cancer imaging (Berland et al. 2021). Moreover, technological advancements were made to produce VHH antibodies and their multimer formats in various living organisms to increase the yield, stability, and efficacy of the recombinant products.

During the third decade (2013–2023), more VHHs entered clinical trials or advanced closer to market. Sanofi acquired Ablynx in 2018 and expanded its collaborations with large biopharma players such as Merck and Boehringer Ingelheim, with more than 20 pre-clinical and clinical programs. In 2018, the European Medicines Agency (EMA) approved the first nanobody-based drug (caplacizumab), a bivalent anti-vWF nanobody for treating rare blood clotting disorders. The FDA (Food and Drug Administration in the United States) approved caplicizumab in February 2019, and it finally entered the market (Peyvandi et al. 2016; Morrison 2019). Caplacizumab is the eleventh antibody fragment approved by the FDA after several Fabs and scFvs, which have entered the clinic application in the past two decades (Jin et al. 2022) (Table 6.1).

With the diminishing of IP limitations on the composition of VHHs, more biotechnology companies have shown interest in the commercialization of these domain antibodies as therapeutics, diagnostics, and research reagents (Steeland et al. 2016). Taisho Pharmaceutical in Japan has used an anti-TNF (tumour necrosis factor) VHH generated by Ablynx and developed a humanized tri-specific nanobody. This nanobody has three VHHs linked together, targeting three different antigenic determinants (epitopes): two on the TNFα molecule and one on human serum albumin. The resulting therapeutic drug, ozoralizumab (Nanozora), has been approved by Japan's Pharmaceuticals and Medical Devices Agency (PMDA) (Ishiwatari-Ogata et al. 2022). As of 2025 more than two dozen companies are

Table 6.1 VHH-based therapies approved by regulatory agencies for various disease indications as of March 2026

Name/brand name (approval year)	Sponsor	Antibody format/ production or expression host	Target	Indication(s)
Caplacizumab/ Cablivi (EMA 2018/FDA 2019)	Sanofi (Ablynx)	Humanized VHH/*E. coli*	vWF	Acquired thrombotic thrombocytopenic purpura (aTTP)
Envafolimab/ Enweida (China 2021)	Alphamab Oncology, China	VHH-Fc	PD-L1	Solid tumours
Ciltacabtagene autoleucel/Carvykti (FDA/EMA 2022)	Janssen, Legend Biotech Corp	CAR-T-(VHH)2	BCMA	Relapsed or refractory multiple myeloma
Ozoralizumab/ Nanozora (Japan 2022)	Taisho Pharmaceutical Co, Ltd., Japan	Trimeric VHH	TNFα	Rheumatoid arthritis (RA)

Adapted from reference Fridy et al. (2024)

collaborating or directly working with Ablynx (Sanofi) on different areas of camel VHH development, ranging from services in isolation and building libraries to using VHHs for human diagnostics and therapeutics (Table 6.1). At the same time, more than 40 therapeutic nanobodies are currently in clinical trials, and many others will enter clinical trials soon (Alexander and Leong 2024).

The first concept of single-domain antibodies was introduced by Saly Ward in 1989. Research on camelid heavy chain–only antibodies was initiated in the late 1980s and resulted in the first publication in Nature in 1993.The first structural features of a camelid VHH were reported in 1996, followed by a second phage display report in 1997. Multiple patents on camelid VHH were filed between 1996 and 2001 and Ablynx was established in 2001. Research and development continued from 2001 to 2017, and the first VHH-based therapeutic got EMA approval in 2018 and FDA approval in 2019. Since then, three more VHH-based drugs, namely, Enweida, Carvikti, and Nanozora, were approved in the United States, Japan, and China (see Table 6.1).

6.5 Camelid Nanobodies: From Isolation to Applications in Human Health and Well-Being

Camelid nanobodies also have several unique and inherent features that are ideal for many applications, such as cancer therapy and imaging. Despite their single-domain nature (~15 kDa), they have comparable affinity and specificity to that of monoclonal antibodies (mAbs) and fragments thereof. A mAb is an antibody that is produced to target a specific antigen. Compared to mAbs, the small size of nanobodies (2.5 nm in diameter and about 4 nm in length) gives them a unique advantage for rapidly crossing from blood vessels into tissues, allowing rapid clearance from the blood and deep tumour penetration (Debie et al. 2020). The short half-life of VHHs in circulation is well suited to certain applications such as tumour imaging or delivery of toxins or radioisotopes to diseased tissues, where rapid clearance is required (Jin et al. 2022). The most notable example is the development of a VHH-based imaging reagent to detect brain metastasis in breast cancer patients using PET/CT scans, which is currently in a phase II clinical trial (NCT03331601) (Prantner et al. 2018; Yang and Shah 2020).

Due to high sequence homology with the human VH3 family, VHHs are expected to have minimal immunogenicity, i.e. the ability to induce an immune response, during therapeutic use in humans (Rossotti et al. 2022; Ackaert et al. 2021). However, genetic engineering approaches have been applied to humanize and further reduce the immunogenicity of camelid VHH domains (Rossotti et al. 2022; Vincke et al. 2009; Sulea 2022). Additional notable features of VHHs include their high thermostability (high refolding capacity) and relative resistance to harsh environments such as the gastrointestinal (GI) tract and chemical denaturation, non-physiological pH, and high pressure. These features make them ideal reagents for oral and aerosol therapy, antibody drug conjugates, and immobilization on microchip biosensors (Aksu et al. 2024; Medina Perez et al. 2024; Ta et al. 2016; Riazi et al. 2013; Hussack et al. 2014).

Today, there are many reports about the isolation and characterization of the VHHs against numerous targets relevant to oncology, in vivo imaging, haematology, infectious diseases, and neurological and inflammatory disorders (Wesolowski et al. 2009; Jin et al. 2022). One area of particular interest for VHHs is therapeutic targets with antigenic epitopes. Antigenic epitopes are small, highly conserved fragments of a pathogen that can be bound by an antibody as part of the specific immune response to this pathogen. VHHs have the size advantage to target sites that are difficult to access by large molecules such as conventional mAbs. Examples include intracellular targets (McGonigal et al. 2009) or epitopes concealed from mAbs in protein structures (Stijlemans et al. 2004), G-protein-coupled receptors (Bradley et al. 2015), and ion channels (Martina et al. 2024; Hartung et al. 2020).

In camelid VHH antibodies, the part that forms the antigen binding site, known as the complementarity-determining region 3 (CDR3), is longer than average. This longer CDR3 helps VHHs form a complete antigen-binding site without needing a light chain and allows them to reach epitopes that are otherwise difficult to access. This is important because the CDR3 is the major determinant of antigen specificity in antibodies and T-cell receptors due to being the most conserved region in terms of its sequence diversity and length. These features combined with their small size and target specificity make VHHs an ideal fit for the above-mentioned applications as therapeutic targets, while bypassing the drawbacks of small-molecule drugs which have lower specificity and off-target toxicity (Baker 2005). Small-molecule drugs are typically generated by chemical methodologies and are cost-effective and more accessible to the public; however, they often interact with multiple targets, leading to many side effects.

Another potential application of VHHs is the treatment of neurodegenerative diseases. It is well known that the delivery of protein therapeutics to the brain has proven to be quite difficult and challenging, because only a small fraction of injected mAbs (0.1–0.2%) reaches the brain due to the tight blood–brain barrier (BBB) (Soleimanizadeh et al. 2021; Yu et al. 2011). VHH-based therapeutics with the ability to transmigrate the BBB may open a new window of opportunity for brain research and treatment. A first report was published in 2002 (Muruganandam et al. 2002) which described a VHH transmigrating into human brain endothelial cell layers through a receptor-mediated process with the ability to deliver attached cargo (e.g. therapeutics) into the central nervous system in rodents. Several VHHs isolated from non-immunized or immunized libraries (e.g. FC5, FC44, and anti-IGF1R5, and anti-VCAM VHHs) improved the delivery of target peptides/cargo up to 10–30-fold when compared to the control with no VHH (Muruganandam et al. 2002; Farrington et al. 2014; Alata et al. 2022; Nachit et al. 2023). In a recent study, a type of nanobody has been developed that effectively crosses the blood–brain barrier which could have benefits for the treatment of conditions such as schizophrenia. In mouse models of schizophrenia, a single injection of this nanobody restored behavioural function and improved cognition for up to 7 days (Oosterlaken et al. 2025). Further work is still needed to address potential side effects and to determine optimal clinical formats and routes of administration. These and other published studies have demonstrated the proof of concept of nanobody-based immunotherapeutics for the treatment of brain disorders. However, substantial work is still needed to identify optimal target proteins implicated in neurodegenerative diseases and to improve the delivery of antibody-based therapeutics into the brain.

Due to their small size, high solubility and stability, and ability to fold in the cytoplasm, VHHs are also ideally suited for intracellular expression. These features, combined with the preferred recognition of conformational epitopes (where the binding site is made up of amino acids that are far apart in sequence, but close together when folded), make VHHs promising tools to visualize and monitor the conformational state of proteins inside the cells. As potential therapeutics, VHHs could also be used to disrupt cellular pathways, target certain disease targets inside the cells, or combat intracellular pathogens. Soetens and colleagues (Soetens et al. 2020; Kirchhofer et al. 2010) have reviewed numerous examples of VHHs as blocking reagents when expressed intracellularly. Examples range from inhibiting cellular pathways to preventing the development of diseases, such as cancer and neuroinflammation, to blocking viral and bacterial multiplication cycles (e.g. HIV, HCV, HBV, RSV, *Salmonella enterica*) (Soetens et al. 2020; Rossey et al. 2017; Huen et al. 2019).

The modular nature of VHHs makes them ideal building blocks for generating bivalent, bispecific, or more complex antibody molecules which harbour multiple VHH domains targeting different antigens on tumour cells. The single-domain nature of a VHH and its smaller size relative to other antibody fragments bypass issues such as their propensity to aggregate together and immunogenicity (where the antibody treatment causes an unintended immune response) (Holt et al. 2003). VHHs with unique specificities could be linked genetically to interact with multiple tumour targets on cancer cells to increase therapeutic efficacy. They could also bridge multiple surface receptors or ligands to block or to activate synergistic signal pathways on a cell, or to trigger contacts between cancer cells and immune cells such as T lymphocytes (T cells) and natural killer cells (NK cell) (Mazor et al. 2017; Mullard 2020; Zhao et al. 2020; Wang et al. 2021). This is known as "immune cell engagers" which can attract immune cells, including T cells or NK cells, to the site of a tumour (Safarzadeh Kozani et al. 2022; Tapia-Galisteo et al. 2023). VHHs can also be used to make hybrid T cell lymphocytes (CAR-T), which specifically target tumour cells with their VHH arm (Steeland et al. 2016; Safarzadeh Kozani et al. 2022). Additionally, the modular nature of VHHs makes it possible to link multiple domains with different specificities while still preserving their stability and high production yields (Henry et al. 2015; van Faassen et al. 2020). This includes incorporating an anti-human serum albumin (HSA) VHH to extend the nanobody's half-life. Several therapeutic nanobodies, in bi-/multi-valent or bi-/multi-specific formats, have advanced into pre-clinical and clinical use (Steeland et al. 2016; Jin et al. 2023). The first VHH-based CAR-T (ciltacabtagene) with two anti-BCMA VHHs in its binding domain was approved by the FDA in March 2022 for the treatment of multiple myeloma (Mullard 2020).

VHHs also offer distinct advantages in the study of infectious diseases and emerging pathogens. As noted earlier, structural analyses of VHH–antigen complexes have shown that VHHs often recognize structural features on proteins that are more rigid and evolutionarily conserved. This has important implications in VHH therapeutics that are cross-reactive (bind more than one target) against infectious disease targets, which constantly mutate to escape the host immune system. Currently, several VHHs developed against viral and other pathogens are in different phases of clinical trials (Wang et al. 2021). Many VHHs with the ability to block or neutralize SARS-CoV-1 and SARS-CoV-2 have been isolated from naïve, synthetic/semi-synthetic and immune llama libraries using phage or ribosome display technologies (Xu et al. 2021). As of July 2021, the Coronavirus Antibody Database (www.opig.stats.ox.ac.uk) lists 664 single domain antibody entries submitted by various research groups worldwide, including 65 patent applications to protect SARS-CoV-2 nanobodies. Infections such as SARV-CoV possess mutations that allow them to evade the immune response of the host. It was shown that some of the reported VHHs, when used in tandem, could tolerate such immune-escape mutations found in almost all SARS-CoV-2 lineages (Guttler et al. 2021). Interestingly, this group also reports some VHHs with the ability to assist the folding of the receptor-binding domain (RBD) of SARS-CoV-2 in *E. coli*. This finding opens the possibility of producing otherwise difficult-to-express proteins in a simple, low-cost bacterial system for future vaccine development and may be broadly applicable to other challenging antigens.

6.6 Limitations of Nanobodies or VHHs as Therapeutic and Imaging Agents

Despite numerous unique applications and advantages of VHHs as therapeutics, imaging agents, and versatile research tools, several limitations have to be considered. For the reason that each VHH domain is a highly complex molecule in which individual amino acids, depending on their position, can directly or indirectly influence stability and structural integrity, there are inherent constraints on how much manipulation and engineering antigen-specific VHHs can tolerate (Arbabi-Ghahroudi 2022).

Another limitation of VHHs is their relatively low propensity to bind small molecules, which is thought to stem from their unique surface topology relative to conventional antibody fragments (e.g. scFv, Fab).

A rather weak heavy chain–only immune response has been observed against several small molecules such as haptens and carbohydrate antigens in several llama immunization trials (unpublished data). However, it has been shown that repeated immunization of camelid species with small molecules that are attached to larger proteins has led to the successful isolation of VHHs that possess weak to medium affinities against their target (Ladenson et al. 2006; Spinelli et al. 2000; David et al. 2014). The biophysico-chemical properties of VHHs suggest that they would be well-suited for many immunodiagnostic platforms to detect small molecules and environmental chemicals. However, the isolation of VHHs that bind with high affinity to their targets, which is more suitable for such diagnostic applications, is still proving to be a challenge.

At last, a key drawback of isolating nanobodies for human therapeutic use is the challenge of immunizing large, genetically diverse animals, which requires specialized facilities. This becomes especially relevant when alternative immunization approaches such as DNA immunization are used. DNA immunization has shown limited success in camelids and other large animals, with reproducibility remaining a major concern. Nevertheless, VHHs with favourable biophysico-chemical properties have been successfully isolated against therapeutically important targets (Maussang et al. 2013). To overcome this limitation, researchers have used transgenic and gene-manipulation technologies and generated genetically altered (transgenic) mice, bearing either a rearranged dromedary heavy chain or hybrid llama/human antibody genes, which produce a modified dromedary or human heavy chain–only antibodies (Nguyen et al. 2003; Zou et al. 2005). Interestingly, Xu and his colleagues (2021) have developed genetically altered mice harbouring alpaca, dromedary, and Bactrian camel VHH encoding genes. Upon immunization to SARS-CoV-2, VHHs are produced that have a strong target effect against that specific immunized strain, as well as other strains of that virus.

More than three decades after their discovery, immunization of dromedaries, camels, llamas, or alpacas remains the most effective method for isolating high-affinity VHH single-domain antibodies. Nonetheless, there have been reports of generating naïve or synthetic VHH libraries as a more universal approach to obtain VHHs against virtually any target antigen with medium to high affinities. This could offer a time advantage in the event of a future disease outbreak (Yan et al. 2015; Goldman et al. 2006; Monegal et al. 2009). Additionally, several studies also utilize a universally well-characterized VHH scaffold with randomized CDR regions to isolate VHHs with therapeutically important target antigens (Moutel et al. 2016; Liu and Yang 2022; Murakami et al. 2022). It is expected that in the future, further technological approaches will speed up the process of VHH antibody isolation and potentially bypass the need to use animals for their production.

6.7 Concluding Remarks

More than three decades have passed since the first observation of camel HCAbs by Hamers and colleagues. This finding was a significant milestone in the field of antibody engineering and opened many new opportunities and applications. Yet it took several years of intensive efforts to demonstrate that these antibodies are naturally derived and not a disease by-product. Molecular cloning, cDNA sequencing, and genomic and structural data helped investigators to piece the puzzle together which is now known as the field of VHHs, camelid sdAbs, or nanobodies. The unique and extraordinary features of HCAbs and VHHs have attracted many researchers and commercial entities to the field of antibody engineering. These novel antibodies are either in clinical trials or close to approval as pharmaceutical drugs to fight a wide range of diseases, including cancer, inflammation, haematology, and respiratory diseases. Additionally, 40 VHH-based drugs are currently in various stages of preclinical and clinical development (Jin et al. 2022, 2023). VHHs have also been shown to be effective as anti-infectious reagents, particularly as robust reagents in passive immunotherapy, as well as for diagnostic and imaging applications. With the approval of the first VHH-based therapeutic (caplacizumab) by the EMA and the FDA (Peyvandi et al. 2016), VHHs are gradually finding their niche in the biologics market despite earlier challenges and setbacks. It is predicted that in the next decade, many more lower-cost VHH-based therapeutic and diagnostic reagents will enter the market which will be either superior or complementary to classical mAbs and fragments thereof (Alexander and Leong 2024).

In conclusion, camelid VHH antibodies have major implications for human health and they can complement conventional antibodies or provide a better alternative for the following three reasons: (a) their high specificity and stability makes them ideal for use in diagnostic tools such as imaging reagents for cancer detection and in biosensors for the detection of infectious disease like COVID-19 and influenza; (b) their small size make them ideal reagents for cancer therapy, neurological and infectious diseases; and (c) they are more heat-and pH-stable than conventional antibodies, permitting alternative delivery routes such as inhalable nanobody sprays or oral formulations for gut-related diseases.

Conflict of Interest Statement The author has no conflict of interest to declare.

Authors and Contributors MAG conceived and wrote the manuscript.

References

Ackaert C, Smiejkowska N, Xavier C, Sterckx YGJ, Denies S, Stijlemans B, Elkrim Y, Devoogdt N, Caveliers V, Lahoutte T et al (2021) Immunogenicity risk profile of nanobodies. Front Immunol 12:632687

Aksu M, Kumar P, Guttler T, Taxer W, Gregor K, Mussil B, Rymarenko O, Stegmann KM, Dickmanns A, Gerber S et al (2024) Nanobodies to multiple spike variants and inhalation of nanobody-containing aerosols neutralize SARS-CoV-2 in cell culture and hamsters. Antivir Res 221:105778

Alata W, Yogi A, Brunette E, Delaney CE, van Faassen H, Hussack G, Iqbal U, Kemmerich K, Haqqani AS, Moreno MJ et al (2022) Targeting insulin-like growth factor-1 receptor (IGF1R) for brain delivery of biologics. FASEB J 36:e22208

Alexander E, Leong KW (2024) Discovery of nanobodies: a comprehensive review of their applications and potential over the past five years. J Nanobiotechnol 22:661

Arbabi Ghahroudi M, Desmyter A, Wyns L, Hamers R, Muyldermans S (1997) Selection and identification of single domain antibody fragments from camel heavy-chain antibodies. FEBS Lett 414:521–526

Arbabi-Ghahroudi M (2017) Camelid single-domain antibodies: historical perspective and future outlook. Front Immunol 8:1589

Arbabi-Ghahroudi M (2022) Camelid single-domain antibodies: promises and challenges as life-saving treatments. Int J Mol Sci 23(9):5009

Baker M (2005) Upping the ante on antibodies. Nat Biotechnol 23:1065–1072

Barderas R, Benito-Pena E (2019) The 2018 Nobel Prize in Chemistry: phage display of peptides and antibodies. Anal Bioanal Chem 411:2475–2479

Berland L, Kim L, Abousaway O, Mines A, Mishra S, Clark L, Hofman P, Rashidian M (2021) Nanobodies for medical imaging: about ready for prime time? Biomolecules 11(5):637

Bird RE, Hardman KD, Jacobson JW, Johnson S, Kaufman BM, Lee SM, Lee T, Pope SH, Riordan GS, Whitlow M (1988) Single-chain antigen-binding proteins. Science 242:423–426

Bradley ME, Dombrecht B, Manini J, Willis J, Vlerick D, De Taeye S, Van den Heede K, Roobrouck A, Grot E, Kent TC et al (2015) Potent and efficacious inhibition of CXCR2 signaling by biparatopic nanobodies combining two distinct modes of action. Mol Pharmacol 87:251–262

David MA, Jones DR, Tayebi M (2014) Potential candidate camelid antibodies for the treatment of protein-misfolding diseases. J Neuroimmunol 272:76–85

de Marco A (2020) Recombinant expression of nanobodies and nanobody-derived immunoreagents. Protein Expr Purif 172:105645

Debie P, Lafont C, Defrise M, Hansen I, van Willigen DM, van Leeuwen FWB, Gijsbers R, D'Huyvetter M, Devoogdt N, Lahoutte T et al (2020) Size and affinity kinetics of nanobodies influence targeting and penetration of solid tumours. J Control Release 317:34–42

Decanniere K, Desmyter A, Lauwereys M, Ghahroudi MA, Muyldermans S, Wyns L (1999) A single-domain antibody fragment in complex with RNase A: non-canonical loop structures and nanomolar affinity using two CDR loops. Structure 7:361–370

Desmyter A, Transue TR, Ghahroudi MA, Thi MH, Poortmans F, Hamers R, Muyldermans S, Wyns L (1996) Crystal structure of a camel single-domain VH antibody fragment in complex with lysozyme. Nat Struct Biol 3:803–811

Desmyter A, Decanniere K, Muyldermans S, Wyns L (2001) Antigen specificity and high affinity binding provided by one single loop of a camel single-domain antibody. J Biol Chem 276:26285–26290

Desmyter A, Spinelli S, Payan F, Lauwereys M, Wyns L, Muyldermans S, Cambillau C (2002) Three camelid VHH domains in complex with porcine pancreatic alpha-amylase. Inhibition and versatility of binding topology. J Biol Chem 277:23645–23650

Dumoulin M, Conrath K, Van Meirhaeghe A, Meersman F, Heremans K, Frenken LG, Muyldermans S, Wyns L, Matagne A (2002) Single-domain antibody fragments with high conformational stability. Protein Sci 11:500–515

Eyer L, Hruska K (2012) Single-domain antibody fragments derived from heavy-chain antibodies. Vet Med 57:439–513

Farrington GK, Caram-Salas N, Haqqani AS, Brunette E, Eldredge J, Pepinsky B, Antognetti G, Baumann E, Ding W, Garber E et al (2014) A novel platform for engineering blood-brain barrier-crossing bispecific biologics. FASEB J 28:4764–4778

Fleischman JB, Pain RH, Porter RR (1962) Reduction of gamma-globulins. Arch Biochem Biophys Suppl 1:174–180

Fridy PC, Rout MP, Ketaren NE (2024) Nanobodies: from high-throughput identification to therapeutic development. Mol Cell Proteomics 23(12):100865. https://doi.org/10.1016/j.mcpro.2024.100865

Goldman ER, Anderson GP, Liu JL, Delehanty JB, Sherwood LJ, Osborn LE, Cummins LB, Hayhurst A (2006) Facile generation of heat-stable antiviral and antitoxin single domain antibodies from a semisynthetic llama library. Anal Chem 78:8245–8255

Griffiths AD (1993) Production of human antibodies using bacteriophage. Curr Opin Immunol 5:263–267

Guttler T, Aksu M, Dickmanns A, Stegmann KM, Gregor K, Rees R, Taxer W, Rymarenko O, Schunemann J, Dienemann C et al (2021) Neutralization of SARS-CoV-2 by highly potent, hyperthermostable, and mutation-tolerant nanobodies. EMBO J 40:e107985

Hamers-Casterman C, Atarhouch T, Muyldermans S, Robinson G, Hamers C, Songa EB, Bendahman N, Hamers R (1993) Naturally occurring antibodies devoid of light chains. Nature 363:446–448

Harmsen MM, Ruuls RC, Nijman IJ, Niewold TA, Frenken LG, de Geus B (2000) Llama heavy-chain V regions consist of at least four distinct subfamilies revealing novel sequence features. Mol Immunol 37:579–590

Hartung F, Kruwel T, Shi X, Pfizenmaier K, Kontermann R, Chames P, Alves F, Pardo LA (2020) A novel anti-Kv10.1 nanobody fused to single-chain TRAIL enhances apoptosis induction in cancer cells. Front Pharmacol 11:686

Henry KA, Tanha J, Hussack G (2015) Identification of cross-reactive single-domain antibodies against serum albumin using next-generation DNA sequencing. Protein Eng Des Sel 28:379–383

Hmila I, Saerens D, Ben Abderrazek R, Vincke C, Abidi N, Benlasfar Z, Govaert J, El Ayeb M, Bouhaouala-Zahar B, Muyldermans S (2010) A bispecific nanobody to provide full protection against lethal scorpion envenoming. FASEB J 24:3479–3489

Holt LJ, Herring C, Jespers LS, Woolven BP, Tomlinson IM (2003) Domain antibodies: proteins for therapy. Trends Biotechnol 21:484–490

Hoogenboom HR (2005) Selecting and screening recombinant antibody libraries. Nat Biotechnol 23:1105–1116

Huen J, Yan Z, Iwashkiw J, Dubey S, Gimenez MC, Ortiz ME, Patel SV, Jones MD, Riazi A, Terebiznik M et al (2019) A novel single domain antibody targeting FliC flagellin of *Salmonella enterica* for effective inhibition of host cell invasion. Front Microbiol 10:2665

Hussack G, Riazi A, Ryan S, van Faassen H, MacKenzie R, Tanha J, Arbabi-Ghahroudi M (2014) Protease-resistant single-domain antibodies inhibit *Campylobacter jejuni* motility. Protein Eng Des Sel 27:191–198

Ishiwatari-Ogata C, Kyuuma M, Ogata H, Yamakawa M, Iwata K, Ochi M, Hori M, Miyata N, Fujii Y (2022) Ozoralizumab, a humanized anti-TNFalpha NANOBODY((R)) compound, exhibits efficacy not only at the onset of arthritis in a human TNF transgenic mouse but also during secondary failure of administration of an anti-TNFalpha IgG. Front Immunol 13:853008

Janeway CA Jr et al (2016) Janeway's immunobiology. Garland Science, New York

Jespers LS, Roberts A, Mahler SM, Winter G, Hoogenboom HR (1994) Guiding the selection of human antibodies from phage display repertoires to a single epitope of an antigen. Biotechnology (N Y) 12:899–903

Jin S, Sun Y, Liang X, Gu X, Ning J, Xu Y, Chen S, Pan L (2022) Emerging new therapeutic antibody derivatives for cancer treatment. Signal Transduct Target Ther 7:39

Jin BK, Odongo S, Radwanska M, Magez S (2023) NANOBODIES(R): a review of diagnostic and therapeutic applications. Int J Mol Sci 24(6):5994

Kirchhofer A, Helma J, Schmidthals K, Frauer C, Cui S, Karcher A, Pellis M, Muyldermans S, Casas-Delucchi CS, Cardoso MC et al (2010) Modulation of protein properties in living cells using nanobodies. Nat Struct Mol Biol 17:133–138

Kohler G, Milstein C (1975) Continuous cultures of fused cells secreting antibody of predefined specificity. Nature 256:495–497

Ladenson RC, Crimmins DL, Landt Y, Ladenson JH (2006) Isolation and characterization of a thermally stable recombinant anti-caffeine heavy-chain antibody fragment. Anal Chem 78:4501–4508

Lauwereys M, Arbabi Ghahroudi M, Desmyter A, Kinne J, Holzer W, De Genst E, Wyns L, Muyldermans S (1998) Potent enzyme inhibitors derived from dromedary heavy-chain antibodies. EMBO J 17:3512–3520

Liu Y, Huang H (2018) Expression of single-domain antibody in different systems. Appl Microbiol Biotechnol 102:539–551

Liu B, Yang D (2022) Easily established and multifunctional synthetic nanobody libraries as research tools. Int J Mol Sci 23(3):1482

Lowman HB, Wells JA (1993) Affinity maturation of human growth hormone by monovalent phage display. J Mol Biol 234:564–578

Martina M, Banderali U, Yogi A, Arbabi Ghahroudi M, Liu H, Sulea T, Durocher Y, Hussack G, van Faassen H, Chakravarty B et al (2024) A novel antigen design strategy to isolate single-domain antibodies that target human Nav1.7 and reduce pain in animal models. Adv Sci (Weinh) 11:e2405432

Maussang D, Mujic-Delic A, Descamps FJ, Stortelers C, Vanlandschoot P, Stigter-van Walsum M, Vischer HF, van Roy M, Vosjan M, Gonzalez-Pajuelo M et al (2013) Llama-derived single variable domains (nanobodies) directed against chemokine receptor CXCR7 reduce head and neck cancer cell growth in vivo. J Biol Chem 288:29562–29572

Mazor Y, Sachsenmeier KF, Yang C, Hansen A, Filderman J, Mulgrew K, Wu H, Dall'Acqua WF (2017) Enhanced tumor-targeting selectivity by modulating bispecific antibody binding affinity and format valence. Sci Rep 7:40098

McCafferty J, Griffiths AD, Winter G, Chiswell DJ (1990) Phage antibodies: filamentous phage displaying antibody variable domains. Nature 348:552–554

McGonigal K, Tanha J, Palazov E, Li S, Gueorguieva-Owens D, Pandey S (2009) Isolation and functional characterization of single domain antibody modulators of Caspase-3 and apoptosis. Appl Biochem Biotechnol 157:226–236

Medina Perez VM, Baselga M, Schuhmacher AJ (2024) Single-domain antibodies as antibody-drug conjugates: from promise to practice—a systematic review. Cancers (Basel) 16(15):2681

Monegal A, Ami D, Martinelli C, Huang H, Aliprandi M, Capasso P, Francavilla C, Ossolengo G, de Marco A (2009) Immunological applications of single-domain llama recombinant antibodies isolated from a naive library. Protein Eng Des Sel 22:273–280

Morrison C (2019) Nanobody approval gives domain antibodies a boost. Nat Rev Drug Discov 18:485–487

Moutel S, Bery N, Bernard V, Keller L, Lemesre E, de Marco A, Ligat L, Rain JC, Favre G, Olichon A et al (2016) NaLi-H1: a universal synthetic library of humanized nanobodies providing highly functional antibodies and intrabodies. elife 5:e16228

Mullard A (2020) FDA approves first BCMA-targeted therapeutic. Nat Rev Drug Discov 19:659

Murakami T, Kumachi S, Matsunaga Y, Sato M, Wakabayashi-Nakao K, Masaki H, Yonehara R, Motohashi M, Nemoto N, Tsuchiya M (2022) Construction of a humanized artificial VHH library reproducing structural features of camelid VHHs for therapeutics. Antibodies (Basel) 11(1):10

Muruganandam A, Tanha J, Narang S, Stanimirovic D (2002) Selection of phage-displayed llama single-domain antibodies that transmigrate across human blood-brain barrier endothelium. FASEB J 16:240–242

Muyldermans S (2013) Nanobodies: natural single-domain antibodies. Annu Rev Biochem 82:775–797

Muyldermans S, Atarhouch T, Saldanha J, Barbosa JA, Hamers R (1994) Sequence and structure of VH domain from naturally occurring camel heavy chain immunoglobulins lacking light chains. Protein Eng 7:1129–1135

Nachit M, Montemagno C, Clerc R, Ahmadi M, Briand F, Bacot S, Devoogdt N, Serdjebi C, Ghezzi C, Sulpice T et al (2023) Molecular imaging of liver inflammation using an anti-VCAM-1 nanobody. Nat Commun 14:1062

Nguyen VK, Muyldermans S, Hamers R (1998) The specific variable domain of camel heavy-chain antibodies is encoded in the germline. J Mol Biol 275:413–418

Nguyen VK, Hamers R, Wyns L, Muyldermans S (2000) Camel heavy-chain antibodies: diverse germline V(H)H and specific mechanisms enlarge the antigen-binding repertoire. EMBO J 19:921–930

Nguyen VK, Su C, Muyldermans S, van der Loo W (2002) Heavy-chain antibodies in Camelidae; a case of evolutionary innovation. Immunogenetics 54:39–47

Nguyen VK, Zou X, Lauwereys M, Brys L, Bruggemann M, Muyldermans S (2003) Heavy-chain only antibodies derived from dromedary are secreted and displayed by mouse B cells. Immunology 109:93–101

Oosterlaken M, Rogliardo A, Lipina T, Lafon PA, Tsitokana ME, Keck M, Cahuzac H, Prieu-Serandon P, Diem S, Derieux C et al (2025) Nanobody therapy rescues behavioural deficits of NMDA receptor hypofunction. Nature 645:262–270

Perez JM, Renisio JG, Prompers JJ, van Platerink CJ, Cambillau C, Darbon H, Frenken LG (2001) Thermal unfolding of a llama antibody fragment: a two-state reversible process. Biochemistry 40:74–83

Peyvandi F, Scully M, Kremer Hovinga JA, Cataland S, Knobl P, Wu H, Artoni A, Westwood JP, Mansouri Taleghani M, Jilma B et al (2016) Caplacizumab for acquired thrombotic thrombocytopenic purpura. N Engl J Med 374:511–522

Prantner AM, Yin C, Kamat K, Sharma K, Lowenthal AC, Madrid PB, Scholler N (2018) Molecular imaging of mesothelin-expressing ovarian cancer with a human and mouse cross-reactive nanobody. Mol Pharm 15:1403–1411

Riazi A, Strong PC, Coleman R, Chen W, Hirama T, van Faassen H, Henry M, Logan SM, Szymanski CM, Mackenzie R et al (2013) Pentavalent single-domain antibodies reduce *Campylobacter jejuni* motility and colonization in chickens. PLoS One 8:e83928

Rossey I, Gilman MS, Kabeche SC, Sedeyn K, Wrapp D, Kanekiyo M, Chen M, Mas V, Spitaels J, Melero JA et al (2017) Potent single-domain antibodies that arrest respiratory syncytial virus fusion protein in its prefusion state. Nat Commun 8:14158

Rossotti MA, Belanger K, Henry KA, Tanha J (2022) Immunogenicity and humanization of single-domain antibodies. FEBS J 289:4304–4327

Saerens D, Ghassabeh GH, Muyldermans S (2008) Single-domain antibodies as building blocks for novel therapeutics. Curr Opin Pharmacol 8:600–608

Safarzadeh Kozani P, Naseri A, Mirarefin SMJ, Salem F, Nikbakht M, Evazi Bakhshi S, Safarzadeh Kozani P (2022) Nanobody-based CAR-T cells for cancer immunotherapy. Biomark Res 10:24

Schepens B, Ibanez LI, De Baets S, Hultberg A, Bogaert P, De Bleser P, Vervalle F, Verrips T, Melero J, Vandevelde W et al (2011) Nanobodies(R) specific for respiratory syncytial virus fusion protein protect against infection by inhibition of fusion. J Infect Dis 204:1692–1701

Skerra A, Pluckthun A (1988) Assembly of a functional immunoglobulin Fv fragment in *Escherichia coli*. Science 240:1038–1041

Soetens E, Ballegeer M, Saelens X (2020) An inside job: applications of intracellular single domain antibodies. Biomolecules 10(12):1663

Soleimanizadeh A, Dinter H, Schindowski K (2021) Central nervous system delivery of antibodies and their single-domain antibodies and variable fragment derivatives with focus on intranasal nose to brain administration. Antibodies (Basel) 10(4):47

Spinelli S, Frenken LG, Hermans P, Verrips T, Brown K, Tegoni M, Cambillau C (2000) Camelid heavy-chain variable domains provide efficient combining sites to haptens. Biochemistry 39:1217–1222

Spinelli S, Desmyter A, Frenken L, Verrips T, Tegoni M, Cambillau C (2004) Domain swapping of a llama VHH domain builds a crystal-wide beta-sheet structure. FEBS Lett 564:35–40

Steeland S, Vandenbroucke RE, Libert C (2016) Nanobodies as therapeutics: big opportunities for small antibodies. Drug Discov Today 21:1076–1113

Stijlemans B, Conrath K, Cortez-Retamozo V, Van Xong H, Wyns L, Senter P, Revets H, De Baetselier P, Muyldermans S, Magez S (2004) Efficient targeting of conserved cryptic epitopes of infectious agents by single domain antibodies. African trypanosomes as paradigm. J Biol Chem 279:1256–1261

Sulea T (2022) Humanization of camelid single-domain antibodies. Methods Mol Biol 2446:299–312

Ta DT, Guedens W, Vranken T, Vanschoenbeek K, Steen Redeker E, Michiels L, Adriaensens P (2016) Enhanced biosensor platforms for detecting the atherosclerotic biomarker VCAM1 based on bioconjugation with uniformly oriented VCAM1-targeting nanobodies. Biosensors (Basel) 6(3):34

Tapia-Galisteo A, Alvarez-Vallina L, Sanz L (2023) Bi- and trispecific immune cell engagers for immunotherapy of hematological malignancies. J Hematol Oncol 16:83

van Faassen H, Ryan S, Henry KA, Raphael S, Yang Q, Rossotti MA, Brunette E, Jiang S, Haqqani AS, Sulea T et al (2020) Serum albumin-binding V(H) Hs with variable pH sensitivities enable tailored half-life extension of biologics. FASEB J 34:8155–8171

Van Roy M, Ververken C, Beirnaert E, Hoefman S, Kolkman J, Vierboom M, Breedveld E, t Hart B, Poelmans S, Bontinck L et al (2015) The preclinical pharmacology of the high affinity anti-IL-6R Nanobody(R) ALX-0061 supports its clinical development in rheumatoid arthritis. Arthritis Res Ther 17:135

Verheesen P, ten Haaft MR, Lindner N, Verrips CT, de Haard JJ (2003) Beneficial properties of single-domain antibody fragments for application in immunoaffinity purification and immuno-perfusion chromatography. Biochim Biophys Acta 1624:21–28

Vincke C, Loris R, Saerens D, Martinez-Rodriguez S, Muyldermans S, Conrath K (2009) General strategy to humanize a camelid single-domain antibody and identification of a universal humanized nanobody scaffold. J Biol Chem 284:3273–3284

Vu KB, Ghahroudi MA, Wyns L, Muyldermans S (1997) Comparison of llama VH sequences from conventional and heavy chain antibodies. Mol Immunol 34:1121–1131

Wang J, Kang G, Yuan H, Cao X, Huang H, de Marco A (2021) Research progress and applications of multivalent, multispecific and modified nanobodies for disease treatment. Front Immunol 12:838082

Ward ES, Gussow D, Griffiths AD, Jones PT, Winter G (1989) Binding activities of a repertoire of single immunoglobulin variable domains secreted from *Escherichia coli*. Nature 341:544–546

Wesolowski J, Alzogaray V, Reyelt J, Unger M, Juarez K, Urrutia M, Cauerhff A, Danquah W, Rissiek B, Scheuplein F et al (2009) Single domain antibodies: promising experimental and therapeutic tools in infection and immunity. Med Microbiol Immunol 198:157–174

Winter G, Griffiths AD, Hawkins RE, Hoogenboom HR (1994) Making antibodies by phage display technology. Annu Rev Immunol 12:433–455

Xu J, Xu K, Jung S, Conte A, Lieberman J, Muecksch F, Lorenzi JCC, Park S, Schmidt F, Wang Z et al (2021) Nanobodies from camelid mice and llamas neutralize SARS-CoV-2 variants. Nature 595:278–282

Yan J, Wang P, Zhu M, Li G, Romao E, Xiong S, Wan Y (2015) Characterization and applications of nanobodies against human procalcitonin selected from a novel naive nanobody phage display library. J Nanobiotechnol 13:33

Yang EY, Shah K (2020) Nanobodies: next generation of cancer diagnostics and therapeutics. Front Oncol 10:1182

Yu YJ, Zhang Y, Kenrick M, Hoyte K, Luk W, Lu Y, Atwal J, Elliott JM, Prabhu S, Watts RJ et al (2011) Boosting brain uptake of a therapeutic antibody by reducing its affinity for a transcytosis target. Sci Transl Med 3:84ra44

Zhao Y, Li Y, Wu X, Li L, Liu J, Wang Y, Liu Y, Li Q, Wang Z (2020) Identification of anti-CD16a single domain antibodies and their application in bispecific antibodies. Cancer Biol Ther 21:72–80

Zhu H, Ding Y (2025) Nanobodies: from discovery to AI-driven design. Biology (Basel) 14(5):547

Zou X, Smith JA, Nguyen VK, Ren L, Luyten K, Muyldermans S, Bruggemann M (2005) Expression of a dromedary heavy chain-only antibody and B cell development in the mouse. J Immunol 175:3769–3779

Camels for Milk: Innovations and Health Value in Production

7

Gaukhar Konuspayeva, Bilal Aslam, and Sulaiman Aljasir

Abstract

Camel milk is increasingly associated with potential health benefits, although many claims lack strong scientific validation. While it contains bioactive components that may support nutrition and health, it should primarily be considered a food rather than a therapeutic product. Its unique composition, including beneficial fatty acids, absence of β-lactoglobulin, and presence of immune-related proteins, distinguishes it from cow milk and contributes to its growing popularity. The camel dairy sector is expanding rapidly, driven by consumer interest and new market opportunities, despite production challenges in arid environments. Advances in processing have enabled the development of products such as pasteurized milk, fermented products, cheese, and milk powder, though adaptations from bovine technologies are often required. This chapter highlights camel milk's composition, technological innovations, and ongoing challenges in processing and market development. Particular attention is given to the role of One Health in addressing antimicrobial resistance, zoonotic risks, and environmental sustainability.

Keywords

One Health · Food safety · Milk composition · Camel milk processing · Climate change

---

G. Konuspayeva (✉)
Biology and Biotechnology Faculty, Department of Biotechnology, Farabi University, Almaty, Kazakhstan

B. Aslam · S. Aljasir
Department of Veterinary Preventive Medicine, College of Veterinary Medicine, Qassim University, Buraydah, Saudi Arabia
e-mail: b.aslam@qu.edu.sa; S.ALJASIR@qu.edu.sa

S. A. Gordine et al. (eds.), *Camels and One Health*,
https://doi.org/10.1007/978-981-92-2246-9_7

7.1 Introduction

Over the past two decades, interest in camel production has increased markedly, leading to a rapid expansion of camel breeding in several countries, including outside of the camel's countries of origin. Globally, the relative growth rate of camel populations surpasses that of most other domestic herbivores, except for goats (Faye 2020). Some authors have also described a genuine surge of enthusiasm for camel-derived products, driven by three major factors (Faye 2018):

- First, climate change, reflected in increasingly irregular rainfall and the progressive aridification of many regions, has contributed to the extension of camel husbandry into new areas, particularly across the semiarid zones of Africa (Jones and Thornton 2009; Faye et al. 2012).
- Second, the increasing integration of livestock production into market economies has facilitated the development of trade in camel products, especially milk. This commercialization has been supported by the establishment of organized milk collection networks and processing facilities that supply local, regional, and even international markets with various dairy products (Faye 2016).
- Third, camel products (including milk, meat, and fiber) are widely attributed with (true or perceived) nutritional and medicinal benefits, which are increasing their consumption for both food and nonfood purposes, such as cosmetics and textiles (Yadav et al. 2015).

One of the most significant transformations in camel milk production in recent decades has been the emergence of intensive dairy farming systems, which are characterized by the cessation of traditional herd mobility (Chamekh et al. 2020). Central Asian countries during the USSR period, and more recently the Gulf States, have been pioneers in developing such systems. Dairy intensification within the camel sector involves managing camels for milk production in a manner analogous to modern dairy cattle operations, incorporating the full suite of practices that have underpinned the rapid advancement of the dairy industry in Western countries. These include the use of reproductive biotechnologies (such as artificial insemination and embryo transfer), machine milking, systematic disease prevention, controlled feeding strategies, herd sedentarization, performance monitoring, and genetic selection.

This chapter outlines the status of camel milk production and describes the main recent innovations in this sector. It summarizes developments in how camel milk is produced, processed, and brought to market, highlighting the key changes that have taken place in production practices, processing approaches, and marketing strategies.

7.2 Advances in Camel Milk Production

The global camel dairy sector is gradually expanding, primarily due to the growing demand for camel milk in response to its (both real and perceived) health benefits. Detailed experimental studies exploring breeding strategies and good farming practices that are aligned with modernized standards may provide invaluable insights to strengthen this industry across the globe. Likewise, a sustainable camel dairy industry requires an integrated approach across the entire value chain that considers the human-animal-environment interface, incorporating the One Health concept.

Currently, camel dairy is gaining global attention due to the increasing demand for alternative dairy sources (Kumar et al. 2016). The recent surge in camel milk demand is associated with its unique health benefits, making it a promising functional food (Arain et al. 2023). Furthermore, bioactive components and the medicinal potential of camel milk are key drivers for the establishment of camel dairy markets worldwide (Alhaj et al. 2024).

Recent economic assessments have highlighted the global growth potential of the camel dairy industry. According to reports, it was valued at USD 14 billion in 2023 and is projected to reach USD 25 billion by 2032, with a compound annual growth rate (CAGR) of 6%). Similarly, another estimate suggests that in 2024, the global camel dairy market was valued at USD 14.3 billion, with a projected CAGR of 9%, reaching USD 24 billion by 2030. This economic trajectory positions camel dairy as a potentially strong competitor within the global dairy industry.

Although demand for camel milk is rising, the dominant global dairy sector presents considerable challenges to the growth of the camel dairy industry (Ali et al. 2019). Key limiting factors include inadequate infrastructure, limited digitalization, and low production levels (Fig. 7.1). Additionally, environmental and climatic conditions in the MENA (Middle East and North Africa) region further contribute to these constraints (Kena 2022). As a result, global distribution and market expansion remain complex and logistically demanding (Oselu et al. 2022).

7.3 Health Significance

Safeguarding human health and food safety remains the focal point of a sustainable environment. Hence, a sustainable supply chain requires wholesome camel dairy products for human consumption that are free from pathogens and all other possible contaminants. However, relevant unsafe practices, such as the consumption of unpasteurized milk, milk adulteration, and the presence of antibiotic residues, may cause negative health effects and threaten the sustainability of the industry.

Health impacts of adulterated or residue-containing camel milk have been well documented, including mutagenic, teratogenic, and carcinogenic effects (Fig. 7.2) (Mishra et al. 2025). It is vital to monitor for the presence of antimicrobial residues and milk adulteration to guarantee food safety and sustainability. In this regard, global organizations like *Codex Alimentarius* and the European Union have crafted regulations stating maximum residue limits (MRL) for antimicrobial residues, which are essential to ensure global food safety and public health (European Commission 2007; Bevilacqua 2025).

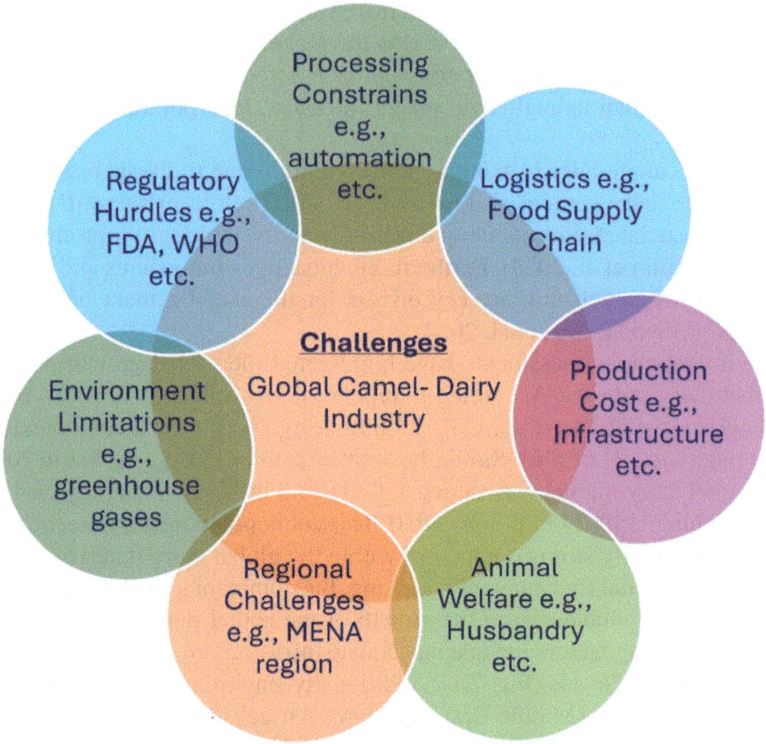

Fig. 7.1 Challenges for the global camel dairy industry

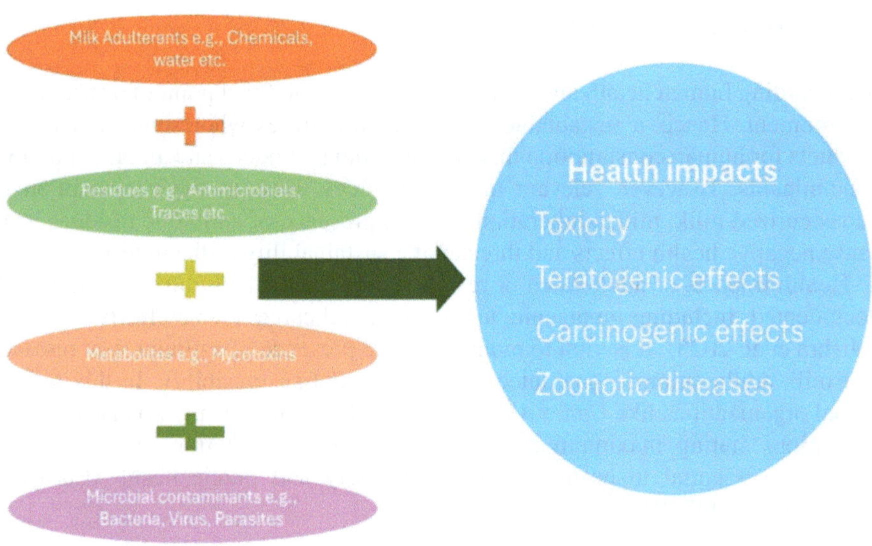

Fig. 7.2 Drivers associated with the health impacts of contaminated camel milk

The consumption of raw camel milk is associated with health risks, particularly in regions like the Gulf or MENA, where the consumption of camel milk is high compared with other parts of the world (Ismail et al. 2022). Studies from the UAE have documented the presence of antimicrobial residues in camel milk that are associated with toxic health effects. Additionally, such residues persist in the environment, posing a potential additional threat to sustainability (Morsi et al. 2024).

7.3.1 Husbandry and Animal Welfare

Good farming practices and animal welfare are crucial considerations within the One Health framework, as they have both direct and indirect consequences for human, animal, and environmental health. The International Finance Corporation has highlighted that animal welfare is essential to boost business viability and profitability, and importantly, animal welfare standards are also needed to meet the expectations of consumers (Ryan 2012). At present, there is a degree of public apprehension regarding the potential ethical consequences of intensive farm practices on animal welfare (Fernandes et al. 2019). In recent years, demand for ethical production systems along with animal welfare standards has increased across the globe. In a comparative survey conducted in Europe, the expressed concern of animal welfare at a farm level had increased to 57% by 2015, up from 7.8% in 2006. Subsequent studies have affirmed the public's perceived importance of animal welfare, demonstrating that such concerns have become more relevant with time across the globe (Eurobarometer 2007).

The sustainability aspects of animal welfare are well recognized, as have been outlined in the recommendations of the UNC-WFS (UN Committee on World Food Security) on sustainable agricultural development (Aredes 2023). Additionally, the health standards from the World Organization for Animal Health outline five freedoms for the welfare of terrestrial animals (freedom from hunger/thirst, fear/distress, physical discomfort, pain/disease, and freedom to express natural behaviors) (Mellor 2016). Improved animal welfare as per these standards, along with capacity building initiatives and voluntary programs are considered crucial for sustainable development in the agricultural sector (Pingault et al. 2016).

Contemporary husbandry practices in camel dairies are associated with increased milk production (Breulmann and Böer 2010). These efficiency gains, however, do not necessarily come at the expense of animal welfare. For instance, digitalization and automation are now applied and practiced in various regions such as Saudi Arabia, the UAE, and Central Asia, with the potential to support improved management, health monitoring, and resource efficiency. Under such intensive husbandry conditions, milk yields can be up to 20 liters/day with lactation lasting 18–20 months, highlighting the productive potential achievable under well-managed conditions. Farms such as these that utilize modern husbandry management are now serving consumers through national and international markets (Gossner et al. 2016).

7.3.2 Infectious Disease and AMR in Camel Milk Production: A One Health Perspective

Camel farming, along with other animal agriculture, is closely linked to the One Health framework. Current farm animal practices directly impact animal health and welfare, including infection management and prophylactic measures. Animal disease dynamics and antimicrobial resistance (AMR) that are associated with modern practices are a threat to global food security.

For instance, the use of antimicrobials as feed additives may create selection pressure and drive the emergence of AMR and resistant strains. Correspondingly, farm waste (e.g., biological waste, wastewater, effluent, etc.) is associated with various zoonotic pathogens (e.g., Shiga toxin-producing *E. coli* (STEC), methicillin-resistant *S. aureus* (MRSA), MERS-CoV, Brucellosis, etc.) (Fig. 7.3). Additionally, infections such as mastitis are also a challenge that is associated with milk production loss. Recently, a case report from the UAE described the transmission of skin diseases (i.e., sarcoptic mange) from camel to camel and camel to human (Naveenkumar et al. 2025). Furthermore, food products are considered one of the main routes for human infection, particularly bacterial pathogens. In the first report published by WHO on foodborne infection, it was reported that out of 600 million annual recorded cases, the death toll was 420,000 (Hoffmann et al. 2025).

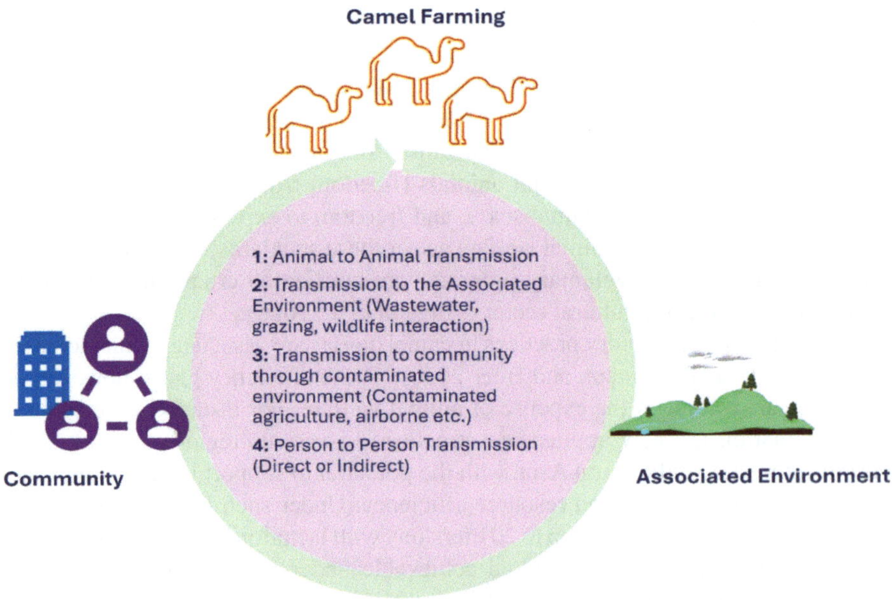

Fig. 7.3 Disease transmission dynamics at animal-environment-human interface

7.3.3 Breeding Advancements

Farm management and contemporary breeding strategies are essential to achieve the maximum potential in the camel dairy industry. In this regard, milk yields can vary significantly, and there is no specific "dairy camel" exclusively bred to maximize milk yield and consistency. Genomic selection is a practical tool for breed selection, which helps to identify camels with the best production potential. Assisted reproductive techniques (ARTs) such as artificial insemination (AI), ovulation synchronization, and in vitro fertilization (IVF) are also important for sustainable camel dairy production. In addition, modern husbandry practices, such as intensive or semi-intensive farming that uses nontraditional feeding systems, and health-tech digital herd-monitoring systems to record the health and performance of camels, can help to improve milk yields. Similarly, using state-of-the-art equipment (e.g., automated milking machines and data software) is essential for selection programs and production, which ultimately determines the economics and sustainability of this industry.

For the genetic advancement of camels for use in the dairy industry, several challenges remain to be addressed. These may be tackled with high-throughput methods used for genomic selection. Currently, single-gene tests are being used for selection in other farm animals, which may be exploited in camel genomics, as has previously been used for color coat genes (Almathen et al. 2016). Likewise, Best Linear Unbiased Prediction (BLUP) may be used to observe Estimated Breeding Values (EBVs). Genomic selection (GS) may accelerate the rate of genetic gain because of long generation intervals in camels. Moreover, GS may result in precise EBVs and decreased progeny-testing costs (Kess et al. 2016).

AI is also a considerable hurdle for the genetic advancement of camel dairy. Disseminating superior genetics is made difficult due to the challenges of performing artificial insemination, which is complicated by a few associated factors, such as the collection and handling of semen. Additionally, the induced ovulatory nature of female camels makes it both unique and difficult. A potential tool known as FWsynch, which synchronizes follicular development by using gonadotropin-releasing hormone (GnRH) and prostaglandin F2 alpha (PGF2α), was newly developed (Manjunatha et al. 2015). Such procedures may be adopted at the research level under the direction of governments, but these tools are difficult for small-scale camel owners to manage.

In the age of genetics, the significance of phenotypes should not be ignored, as well-defined phenotypes are used to study genetic traits (Gonzalez-Recio et al. 2014). Though the developed world has camel-breeding societies, many developing countries lack such practices. In addition, these camel dairies have inadequate data on fertility traits and milking records, which would be useful for improving milk production (Faye 2018). This builds a barrier for genomic advancement and requires considerable teamwork from researchers and other stakeholders.

7.3.4 Sustainability

While climate change poses a threat to camel dairy production (Fig. 7.4), camel milk production offers environmental advantages in the face of climate change, such as camels' resilience to hot climates and reduced greenhouse gas emissions compared to livestock farming. Climate change is also contributing to the rising demand for camel milk, both domestically and in regional export markets, catalyzing the expansion of value chains that extend from remote pastoral settlements to urban processors and consumers. In many production hubs, camel milk has become a reliable and sometimes superior source of cash income compared to cattle, particularly during drought periods when other livestock underperform. These dynamics demonstrate how climate pressures can stimulate innovation, diversify income sources, and shift the socio-economic architecture of camel production.

Additional beneficial environmental impacts include improved land use, lower nutrient loading (nitrogen/phosphorus), reduced eutrophication potential, and support for biodiversity. However, some challenges (e.g., overgrazing) are harmful to the desert ecosystem, which may be addressed through modern grazing management and holistic One Health strategies. Likewise, One Health-driven wastewater management strategies are also crucial in intensive contemporary farm practices to minimize the environmental impact of production.

At present, climate change is a pressing challenge impeding the sustainable development of livestock production and the camel dairy industry (Kena 2022). Though data is available on livestock livelihood and production, the understanding of climate change and its impact on animals, especially camels as dairy animals, is still an understudied research domain. A few studies have reported the deleterious effects on camels of climate change and extreme weather events (e.g., rainfall shortage, droughts, etc.). Observed effects include reduced milk and meat production. Animal deaths and decreased consumption of dairy products have also been recorded, which causes huge economic losses to the industry (Wako et al. 2017). These deleterious effects of climate change are more impactful due to the fact that value chains are fragmented with structural inefficiencies such as limited cold-chain infrastructure, inadequate transport networks, and inconsistent quality-control systems. Moreover, spoilage increases during droughts when milk yields fluctuate, and seasonal surpluses often exceed the capacity of local processors (Ait El Alia et al. 2025).

During extended drought conditions, camels produce less milk. Dehydration is one of the leading factors resulting in decreased milk yields in heat-stressed camels. In addition to decreased water quantity, other contributing factors include feed availability, heat stress, and water quality (George 2024). Dry weather periods greatly disturb the composition of camel milk. Some studies have reported that certain components (e.g., milk fat) vary by season. It has been observed that during the summer season, milk fat content is lower, while in winter, protein content is higher. Lactose levels are also observed to be higher during the winter season (Babiker and El-Zubeir 2024).

Fig. 7.4 Impact of climate change on the global camel dairy industry

Due to the limited data availability, the impact of climate change on camel dairy production remains vague and needs comprehensive investigation from various regions of the world, particularly countries with significant camel populations such as Saudi Arabia and the UAE. Further study of the association between climate change, camel dairy, and global dynamics will be crucial to review and design appropriate control interventions. A better understanding of farmers' perceptions of climate change and its relationship to animal production will also be important for the formulation and implementation of effective strategies to improve camel dairy production.

The effects of climate change on camel dairy farming remain largely unresearched due to a shortage of available data. Thorough scientific studies are needed across various regions, especially in nations where camels play a major economic role, such as the Middle East and Central Asia. Beyond direct impacts on milk production, it will also be important to examine how these environmental shifts affect the people involved with the industry. Climate change not only affects animal productivity but also influences the socioeconomic structures and livelihoods of communities dependent on these systems. Gaining a clearer picture of how climate change connects camel dairy operations to wider economic and social patterns will be needed. Another critical piece of the puzzle deserving closer examination is the perception of farmers themselves regarding climate change and how it affects their livestock. These personal viewpoints matter enormously as they directly influence which adaptation methods farmers will choose, and how successfully they will implement them.

7.3.5 Regulatory Policies

Alongside good farm practices, adequate economic and regulatory policies are important for the development of a sustainable camel dairy industry. In this regard, technological advancements should be employed that develop organized value chains to ensure sustainability. Regulatory challenges also need to be addressed to achieve the maximum potential of this industry. As per the recommendations of FAO, the development of global standards for the quality and specificities of camel milk and dairy products would strengthen consumer trust across the globe. To protect global trade, guidance should include topics such as farm settings, production, marketing, and methods to ensure the unique traits of camel milk and dairy products are preserved, while utilizing established Codex standards on dairy products (Codex 2007). While global standards still need to be defined, a few government organizations at a regional level (i.e., GCC Standardization Organization) have developed standards for camel milk (i.e., GSO 1970:2009 and GSO 174:2021). Likewise, the African Organization for Standardization (ARSO) is also working to streamline the standards of camel dairy, which will be beneficial for the trade of these products across borders.

Furthermore, government policies that encourage investment in the camel dairy sector will increase stakeholders' trust and investment, which would support research and development, capacity building, and the adoption of novel practices and tools. Camel dairy has the potential to contribute substantially to achieving relevant Sustainable Development Goals (Fig. 7.4). Globally, governmental and inter-governmental bodies, along with the International Dairy Federation (IDF) and the International Union of Food Science and Technology (IUFoST), will be crucial for the development of Codex standards. Such entities have observer status with Codex and may be included in the process of standard elaboration.

7.4 Camel Milk Composition and Nutrition

Camel milk contains several bioactive components that may contribute to dietary value and potential health effects, but it should primarily be regarded as a food product rather than a medical treatment. Although the gross composition of camel milk is broadly similar to bovine milk, its detailed molecular and biochemical profile exhibits substantial differences. These compositional distinctions may underline the specific health benefits attributed to regular consumption of camel milk (Konuspayeva 2019).

An analysis of 121 publications from 1905 to 2019 from all around the world shows that camel milk contains, on average, 3.68 ± 1.00% fat, 3.28 ± 0.59% total protein, 4.47 ± 0.66% lactose, and 0.81 ± 0.19% ash. The dry matter content ranges from 8.25% to 16.70%, with a mean value of 12.2 ± 1.62%. In the compilation of those references, 15 were issued from Central Asia, 20 from South Asia, 24 from the Middle East, 27 from East Africa, 27 from North Africa, one from West Africa, and one from Europe, with six being undetermined. The samples from Central Asia were significantly richer in fat (4.87 ± 0.77%) and in total protein (3.86 ± 0.57%) than in other regions. This difference is due to differences in milk composition between the two species of camels—*Camelus bactrianus* and *Camelus dromedarius*—and hybrids of the two (which are typical in that region) (Faye et al. 2008). The milk composition for these three types of camel is summarized (Table 7.1).

Although the gross composition of camel milk is generally similar to that of cow milk, certain components within its detailed composition—such as fatty acids, caseins and other proteins, vitamins, and both major and trace minerals—may differ

Table 7.1 Differences in the gross composition of Bactrian, dromedary, and hybrid milk in animals reared in the same farms from Kazakhstan (Faye et al. 2008)

Parameters	Bactrian	Dromedary	Hybrid	*P* value
Fat, %	6.67 ± 2.93	5.94 ± 2.26	6.09 ± 1.81	0.04
Lactose, %	2.77 ± 0.96	3.12 ± 0.92	3.04 ± 0.60	0.45
Protein, %	5.23 ± 1.17	4.76 ± 1.13	5.15 ± 1.59	0.11
Vitamin C, (mg/L)	177 ± 109	152 ± 91	133 ± 133	<0.001
Ca, (g/L)	1.303 ± 0.287	1.163 ± 0.273	1.257 ± 0.268	0.003
P, (g/L)	1.075 ± 0.177	0.915 ± 0.190	1.067 ± 0.273	<0.001
Fe, (mg/L)	2.11 ± 1.63	1.93 ± 1.06	2.01 ± 0.78	0.55

significantly in concentration or quality (Al haj and Al Kanhal 2010). A comprehensive review of these elements would necessitate an entire book.

Camel milk contains biologically active proteins and peptides (Hsieh et al. 2015; Mati et al. 2017), with whey proteins composing 20% and offering immunomodulatory (Legrand et al. 2004), anticarcinogenic (Habib et al. 2013), antibacterial, and antifungal benefits (Kanwar et al. 2015). Pattern-recognition proteins like peptidoglycan recognition protein (PGRP) help regulate immune responses and participate in bacterial killing; PGRP is found at low levels in ruminant milks but has been identified in camel and porcine mammary secretions (Kappeler et al. 2004). Lactoferrin and lysozyme C target Gram-negative and Gram-positive bacteria, respectively, with lysozyme C showing less affinity than PGRP (Dziarski et al. 2012). Lactophorin supports immune protection in lactating camels and their young by inhibiting pathogen growth (Girardet et al. 2000). Camel milk also includes whey acidic protein (WAP), whose function is unclear, though related proteins act as protease inhibitors (Hennighausen and Sippel 1982; Bouchard et al. 2006).

Regarding the fat content of camel milk, the fatty acid profile differs significantly from that of cow's milk, with camel milk containing predominantly monounsaturated fatty acids (MUFA) (Karray et al. 2004).

7.5 Advances in Processing Camel Milk

This section focuses on four main dairy products that have varying market success-pasteurize— milk, fermented milk, camel cheese, and camel milk powder. The same processing methods are frequently used for camel milk as for bovine milk, but their suitability varies. To date, many researchers and processors have relied on cow's milk indicators and conditions by default. Sometimes, existing bovine milk processing methods are suitable for camel milk, but other times they are not. Pasteurization is a process used not only for milk but also for many other foods. Appropriate pasteurization parameters must be defined, with each product requiring specific parameters to be adapted to its raw composition. Despite similar gross composition, individual components differ in their properties and thermal behavior (Konuspayeva and Faye 2021; Lajnaf et al. 2020). For example, alkaline phosphatase is not a reliable pasteurization indicator for camel milk (Lorenzen et al. 2011). Regarding ultrahigh treatment (UHT) processing, camel milk is considerably more complex (Karray et al. 2004; Laleye et al. 2008; Alhaj et al. 2011). Camel milk does not withstand ultra-high-temperature treatment due to the distinct structure and relative proportions of its protein fractions compared with those of cow's milk (Hattem et al. 2011). At present, the only feasible approach for producing UHT camel milk is through reconstituted milk, i.e., milk obtained from powder. However, this raises important concerns regarding the preservation of its functional and nutritional properties. During the drying process, several bioactive components are susceptible to degradation, and upon reconstitution followed by secondary heat treatment (including UHT), some of these components may be irreversibly lost (Alhaj et al. 2011; Ho et al. 2019; Zouari et al. 2020).

A comparable issue arises with camel cream. Owing to its proportionally higher MUFA concentration, camel milk has a higher susceptibility to oxidation. Oxidative processes may therefore proceed more rapidly in camel milk than in cow's milk. Cream from camel milk has been documented only by Kashaninejad and Razavi (2020) and was produced as an experimental batch by a private company in the UAE, but at this point, it is not available for sale commercially.

Fermented camel milk is, however, entirely different. Fermentation is one of the oldest methods of food preservation, and camel milk has traditionally been conserved in fermented form by most nomadic societies because it was the easiest way to prolong the preservation of this highly perishable product in the absence of a cold chain. This practice yields characteristic sensory properties in terms of taste, aroma, and texture that vary between regions. Since fermentation is driven by the indigenous microflora, geographical differences in microbial communities produce distinct end products (Ayyash et al. 2020; Bilal et al. 2024). This diversity underpins regional cultures, consumption practices, and consumer preferences. The microflora of fermented camel milk therefore constitutes a large research topic that requires further study (Belkheir et al. 2020; Mortazavi et al. 2021).

Many products may be classified as fermented camel milk, yet they can be fundamentally different in both technological and sensory terms. The case of *camel-milk yogurt* is illustrative (Kamal-Eldin et al. 2020). To date, only weak, unstable gel textures have been obtained using camel milk. When conventional yogurt technology is applied, a liquid product is typically produced. This can be classified as drinkable yogurt or, more generally, as fermented milk, and is often better accepted by urban consumers. Gelled camel-milk yogurts reported in the literature have generally been produced using gelling agents (e.g., alginate, agar-agar, starch, pectin) (Chen et al. 2019; Sobti et al. 2021; Buchilina and Aryana 2021). Unfortunately, such formulations sometimes contain less than 50% camel milk and should therefore be regarded primarily as products containing camel milk rather than as authentic camel-milk dairy products (Atwaa et al. 2020; Kamal-Eldin et al. 2020).

The modern development of *camel milk cheese* spans less than 40 years (Ramet 1989; Jones-Abeiderrahmane 2013). Due to the low content of κ-casein, traditional coagulation of camel milk was neither practical nor profitable (Kappeler et al. 2006). Since 2010, camel chymosin obtained by genetic engineering has become available, enabling the emergence of new products from camel milk (Konuspayeva et al. 2014a, b; Jensen et al. 2013; Mbye et al. 2020). Today, various types of cheeses are being developed in different regions of the world. However, not all cheese types have yet been tested with camel milk (Hailu et al. 2014). Camel milk first entered international cheese competitions in 2023, and many new products are expected to appear in the coming years.

An interesting aspect of camel milk cheese-making is the differing focus of private companies and scientists. The private sector tends to explore new types of products, while researchers often concentrate on classical trials such as alternative coagulants, sometimes without considering consumer preferences (El-Hatmi et al. 2020; Hailu et al. 2014; Mohamed et al. 2013). Another noteworthy point is the behavior of camel milk curd with starter cultures (Konuspayeva et al. 2013;

Moynihan et al. 2014; Konuspayeva et al. 2017). For certain cheese types, the use of specific starters in camel milk cheese making produces different aromas and textures to the initial cheese product (Konuspayeva et al. 2014a, b).

Camel milk powder is another product that is better known in the production and commercial sectors than in research. It should be emphasized that camel milk powder has significant potential to boost the camel milk industry (Zouari et al. 2020; Ho et al. 2019; Deshwal et al. 2020). Since its introduction, many countries without camel milk production have started to consume camel milk powder and related products. In camel milk-producing countries, powder production is often developed with the main goals of export and for the generation of higher-value products (Konuspayeva et al. 2022). The availability of camel milk powder has also enabled the creation of new product formats, such as chocolate, cappuccino ("camelccino"), and specialty coffee beverages and cosmetics (Faye 2018).

7.5.1 Marketing Strategies of Camel Milk

Until recently, camel milk was not marketed (Jones-Abeiderrahmane 2013) and was part of the "gift economy" (Faye and Corniaux 2024). Its very recent introduction into national and international markets faces several constraints that limit its ability to compete with milk from other dairy species.

The high market price of camel milk is partly due to its attributed health benefits but is also directly linked to the low productivity of dairy camels. Camels produce a modest milk yield, calving every 2 years, and are often difficult to milk. Additionally, cisternal milk represents less than 10% of total camel milk volume. As a result, one liter of camel milk generally costs between two and 20 times more than cow's milk, depending on the country of sale.

There is often a geographical separation between milk production areas in arid pastoral zones and consumption areas. This situation is typical in central Asia, the Sahel, and North Africa, leading to high transaction (transport) costs or, due to difficulties in maintaining appropriate hygiene and the cold chain, to the marketing of poor-quality milk.

The preference of many consumers in the Middle East and African countries is for raw camel milk, to which health benefits are traditionally attributed. Many of these consumers remain uncertain as to the health benefits of pasteurized and powdered milk, and these have not yet been scientifically proven. Except for some Central Asian and Gulf countries, most camel dairies are small-scale, processing very limited volumes of milk. This makes them highly dependent on strong seasonal variability in both production and consumption (e.g., during Ramadan). The small size of these facilities also limits investment in modern equipment that is better suited to large volumes. Milk powder production in particular needs investment in specific equipment. Only some small processors succeed in diversifying the production of camel milk products that can be easily marketed.

The distribution of camel milk products relies largely on small shops or informal vendors. Sales through specialized retail outlets are generally limited to facilities associated with large dairies, while supermarkets require both large and regular volumes and quality controls that are difficult for small dairies to provide. In several developing countries, competition is often observed between the formal and informal sectors. Informally, producers sell their milk directly to consumers in the absence of sanitary controls, while the formal sector processes milk and milk products through collection centers, often resulting in higher prices for consumers.

Although negotiations are ongoing, the absence of an international standard for camel milk is hampering the development of an international market, particularly in industrialized countries (Faye et al. 2025). Regarding processed products, the lack of standardized processing methods leads to the marketing of inconsistent products, often with tastes and aromas that may vary over time, which can confuse consumers and decrease consumption opportunities. Quality control measures for camel milk are still lacking standard procedures for the detection of specific parameters. Examples of such parameters include freezing point analysis to detect fraud by the addition of water, or the detection of β-lactoglobulin to identify the presence of other (bovine or goat) milk. There is also no officially recognized "organic" label for camel milk products (though producers sometimes claim one), and the conditions required to obtain such a label are not defined by any organizations. It will be important to develop appropriate reference values and sufficiently adapt quality control techniques to increase the camel milk market.

Marketing strategies for camel milk should be built on three major pillars: (i) the development of quality control systems specifically adapted to camel milk and its derivatives; (ii) the establishment of a broad distribution network including structures of very different scales, ranging from small shops to supermarkets; and (iii) the provision of reliable, scientifically sound information on the health effects of regular camel milk consumption.

7.6 Conclusion

The One Health framework provides a relevant and practical basis for the sustainable development of the camel dairy sector. It contributes to animal health and herd productivity while also supporting environmental sustainability by protecting biodiversity and limiting ecosystem damage. When applied through good farm practices, the One Health approach can support a climate-resilient and equitable camel dairy industry and ensure the production of safe camel milk for human consumption. Camel milk should therefore be considered first and foremost as a food product. It is a high-quality resource that is well adapted to arid environments. The use of technological and scientific innovations can further improve its production and processing and help expand its role as a food. Such an approach should be given priority over traditional health claims that are not supported by scientific evidence. Several practical processing approaches already exist and could be adopted by the private sector in different countries. In addition, governmental guidance and compliance with Codex standards are important to support sustainability and enable international trade of camel dairy products.

References

Ait El Alia O, Zine-Eddine Y, Chaji S, Boukrouh S, Boutoial K, Faye B (2025) Global camel milk industry: a comprehensive overview of production, consumption trends, market evolution, and value chain efficiency. Small Rumin Res 243:107441

Al Haj OA, Al Kanhal HA (2010) Compositional, technological and nutritional aspects of dromedary camel milk. Int Dairy J 20:811–821

Alhaj OA, Metwalli AM, Elsayed A (2011) Heat stability of camel milk proteins after sterilisation process. J Camel Pract Res 18:277–282

Alhaj OA, Abodoleh G, Abu Jadayil S, Irshaid N, Mehta BM, Faye B, Jahrami HA (2024) A systematic review and meta-analysis of the antihypertensive effects of camel milk hydrolysates. Int J Dairy Technol 77(4):1029–1043

Ali W, Akyol E, Ceyhan A, Dilawar S, Firdous A, ul Qasim MZ, Ahmad MM (2019) Milk production and composition in camel and its beneficial uses: a review. Turk J Agric Food Sci Technol 7(12):2142–2147

Almathen F, Charruau P, Mohandesan E, Mwacharo JM, Orozco-terWengel P, Pitt D, Abdussamad AM, Uerpmann M, Uerpmann HP, De Cupere B, Magee P (2016) Ancient and modern DNA reveal dynamics of domestication and cross-continental dispersal of the dromedary. Proc Natl Acad Sci 113(24):6707–6712

Arain MA, Khaskheli GB, Shah AH, Marghazani IB, Barham GS, Shah QA, Khand FM, Buzdar JA, Soomro F, Fazlani SA (2023) Nutritional significance and promising therapeutic/medicinal application of camel milk as a functional food in human and animals: a comprehensive review. Anim Biotechnol 34(6):1988–2005

Aredes MA (2023) The UN Committee on World Food Security: an opportunity to counter the ongoing corporate capture of FAO. World Nutr 14(3):79–82

Atwaa EH, Hassan MAA, Ramadan MF (2020) Production of probiotic stirred yoghurt from camel milk and oat milk. J Food Dairy Sci Mansoura Univ 11:259–264

Ayyash M, Abu-Jdayil B, Itsaranuwat P, Almazrouei N, Galiwango E, Esposito G, Hunashal Y, Hamed F, Najjar Z (2020) Exopolysaccharide produced by the potential probiotic *Lactococcus garvieae* C47: structural characteristics, rheological properties, bioactivities and impact on fermented camel milk. Food Chem 333:127418

Babiker WIA, El-Zubeir IEM (2024) Impact of husbandry, stages of lactation and parity number on milk yield and chemical composition of dromedary camel milk. Emir J Food Agric 26(4):333–341

Belkheir B, Zadi-Karam H, Karam NE, Carballo J, Centeno JA (2020) Effects of selected mesophilic Lactobacillus strains obtained from camel milk on the volatile and sensory profiles of a model short-ripened pressed cows' milk cheese. Int Dairy J 109:104738

Bevilacqua D (2025) The codex alimentarius commission. In Elgar concise encyclopedia of food law (pp. 38–42). Edward Elgar Publishing

Bilal Z, Akhmetsadykova S, Baubekova A, Tormo H, Faye B, Konuspayeva G (2024) The main features and microbiota diversity of fermented camel milk. Foods 2024(13):1985. https://doi.org/10.3390/foods13131985

Bouchard D, Morisset D, Bourbonnais Y, Tremblay GM (2006) Proteins with whey-acidic-protein motifs and cancer. Lancet Oncol 7(2):167–174. https://doi.org/10.1016/S1470-2045(06)70579-4. PMID: 16455481

Breulmann M, Böer B (2010) Camel farms: a new idea to help desert ecosystems recover. Int J Environ Rural Dev 44(2):39–40

Buchilina A, Aryana K (2021) Physicochemical and microbiological characteristics of camel milk yogurt as influenced by monk fruit sweetener. J Dairy Sci 104:1484–1493

Chamekh L, Calvo M, Khorchani T, Castro-Gómez P, Hammadi M, Fontecha J, Yahyaoui MH (2020) Impact of management system and lactation stage on fatty acid composition of camel milk. J Food Compos Anal 87:103418. https://doi.org/10.1016/j.jfca.2020.103418

Chen C, Wang P, Zhang N, Zhang W, Ren F (2019) Improving the textural properties of camel milk acid gel by treatment with trisodium citrate and transglutaminase. LWT 103:53–59

Deshwal DK, Kumar Singh A, Kumar D, Sharma H (2020) Effect of spray and freeze drying on physico-chemical, functional, moisture sorption and morphological characteristics of camel milk powder. LWT 134:110117

Dziarski R, Kashyap DR, Gupta D (2012) Mammalian peptidoglycan recognition proteins kill bacteria by activating two-component systems and modulate microbiome and inflammation. Microb Drug Resist 18(3):280–285. https://doi.org/10.1089/mdr.2012.0002. PMID: 22432705

El Hatmi, H.; Jrad, Z.; Mkadem,W.; Chahbani, A.; Oussaief, O.; Ben Zid, M.; Nouha, M.; Zaidi, S.; Khorchani, S.; Belguith, K.; et al. 2020. Fortification of soft cheese made from ultrafiltered dromedary milk with *Allium roseum* powder: effects on textural, radical scavenging, phenolic profile and sensory characteristics. LWT 2020, 132, 109885

European Commission (2007) Attitudes of EU citizens towards animal welfare, Report; Special Eurobarometer 270. European Commission, Brussels, p 82

Faye B (2016) The camel, new challenges for a sustainable development. Trop Anim Health Prod 48:689–692. https://doi.org/10.1007/s11250-016-0995-8

Faye B (2018) The enthusiasm for camel production. Edit Emir J Food Agric 30(4):249–250

Faye B (2020) How many large camelids in the world? A synthetic analysis of the world camel demographic changes. Pastor Policy Res Prac 10:25. https://doi.org/10.1186/s13570-020-00176-z

Faye B, Konuspayeva G, Messad S, Loiseau G (2008) Discriminant milk components of Bactrian camel (*Camelus bactrianus*), dromedary (*Camelus dromedarius*) and hybrids. Dairy Sci Technol 88(6):607–617. https://doi.org/10.1051/dst:2008008

Faye B, Chaibou M, Vias G (2012) Integrated impact of climate change and socioeconomic development on the evolution of camel farming systems. Br J Environ Clim Chang 2(3):227–244. https://doi.org/10.9734/BJECC/2012/1548

Faye B, Corniaux C (2024) Le lait de chamelle au risque de l'économie politique : de l'économie du don à l'économie marchande. Revue d'Elevage et de Medecine Veterinaire des Pays Tropicaux 77:37263

Faye B, Konuspayeva G, Kiselev S, Nagy P (2025) Review and documentation of adulteration in camel milk. In: Proceeding of symposium "supporting the development of the camel milk sector", 24–25 September 2024, in Journal of Global Food Regulatory Science Society, pp 22–25

Fernandes J, Blache D, Maloney SK, Martin GB, Venus B, Walker FR, Head B, Tilbrook A (2019) Addressing animal welfare through collaborative stakeholder networks. Agriculture 9(6):132

George AS (2024) Camel milk production as an adaptation to climate change induced drought in East Africa. Partners Univ Multidiscip Res J 1:109–126

Girardet JM, Saulnier F, Gaillard JL, Ramet JP, Humbert G (2000) Camel (*Camelus dromedarius*) milk PP3: evidence for an insertion in the amino-terminal sequence of the camel milk whey protein. Biochem Cell Biol 78(1):19–26. PMID: 10735560

Gonzalez-Recio O, Coffey MP, Pryce JE (2014) On the value of the phenotypes in the genomic era. J Dairy Sci 97(12):7905–7915

Gossner C, Danielson N, Gervelmeyer A, Berthe F, Faye B, Kaasik Aaslav K, Adlhoch C, Zeller H, Penttinen P, Coulombier D (2016) Human–dromedary camel interactions and the risk of acquiring zoonotic Middle East respiratory syndrome coronavirus infection. Zoonoses Public Health 63(1):1–9

Habib HM, Ibrahim WH, Schneider-Stock R, Hassan HM (2013) Camel milk lactoferrin reduces the proliferation of colorectal cancer cells and exerts antioxidant and DNA damage inhibitory activities. Food Chem 141(1):148–152. https://doi.org/10.1016/j.foodchem.2013.03.039. PMID: 23768340

Hailu Y, Seifu E, Yilma Z (2014) Physicochemical properties and consumer acceptability of soft unripened cheese made from camel milk using crude extract of ginger (*Zingiber officinale*) as coagulant. Afr J Food Sci 8:87–91

Hattem HE, Manal AN, Hanna SS, Elham AA (2011) A study on the effect of thermal treatment on composition and some properties of camel milk. J Brew Distill 2:51–55

Hennighausen LG, Sippel AE (1982) Mouse whey acidic protein is a novel member of the family of `four-disulfide core' proteins. Nucleic Acids Res 10(8):2677–2684

Ho TM, Chan S, Yago AJE, Shravya R, Bhandari BR, Bansal N (2019) Changes in physico-chemical properties of spray-dried camel milk powder over accelerated storage. Food Chem 295:224–233

Hoffmann S, White AE, McQueen RB, Ahn JW, Gunn-Sandell LB, Scallan Walter EJ (2025) Economic burden of foodborne illnesses acquired in the United States. Foodborne Pathog Dis 22(1):4–14

Hsieh CC, Herna Ândez-Ledesma B, FernaÂndez-TomeÂ S, Weinborn V, Barile D, De Moura Bell JMLN (2015) Milk proteins, peptides, and oligosaccharides: effects against the 21st century disorders. Biomed Res Int 2015:146840

Ismail LC, Osaili TM, Mohamad MN, Zakaria H, Ali A, Tarek A, Ashfaq A, Al Abdouli MA, Saleh ST, Al Daour R, AlRajaby R (2022) Camel milk consumption patterns and perceptions in the UAE: a cross-sectional study. J Nutr Sci 11:e59

Jensen JL, Molgaard A, Poulsen JN, Harboe MK, Simonsen JB, Lorentzen AM, Hjerno K, van den Brink JM, Qvist KB, Larsen S (2013) Camel and bovine chymosin: the relationship between their structures and cheese-making properties. Acta Crystallogr D Biol Crystallogr D69:901–913

Jones PG, Thornton PK (2009) Croppers to livestock keepers: livelihood transitions to 2050 in Africa due to climate change. Environ Sci Pol 12(4):427–437. https://doi.org/10.1016/j.envsci.2008.08.006

Jones-Abeiderrahmane N (2013) Camel cheese: seemed like a good idea. Publ Nancy Jones-Abeiderrahmane, Nouakchott Mauritania, 387 p

Kamal-Eldin A, Alhammadi A, Gharsallaoui A, Hamed F, Ghnimi S (2020) Physicochemical, rheological, and micro-structural properties of yogurts produced from mixtures of camel and bovine milks. NFS J 19:26–33

Kanwar JR et al (2015) Multifunctional iron bound lactoferrin and nanomedicinal approaches to enhance its bioactive functions. Molecules 20(6):9703–9731. https://doi.org/10.3390/molecules20069703. PMID: 26016555

Kappeler SR, Heuberger C, Farah Z, Puhan Z (2004) Expression of the peptidoglycan recognition protein, PGRP, in the lactating mammary gland. J Dairy Sci 87(8):2660–2668. https://doi.org/10.3168/jds.S0022-0302(04)73392-5. PMID: 15328291

Kappeler SR, van den Brink HJ, Rahbek-Nielsen H, Farah Z, Puhan Z, Hansen EB, Johansen E (2006) Characterizations of recombinant camel chymosin reveals superior properties for the coagulation of bovine and camel milk. Biochem Biophys Res Commun 342:647–654

Karray N, Lopez C, Lesieur P, Ollivon M (2004) Dromedary milk fat, thermal and structural properties 1. Crystalline forms obtained by slow cooling. Lait 80:399–416

Kashaninejad M, Razavi SMA (2020) Influence of thermosonication treatment on the average size of fat globules, emulsion stability, rheological properties and color of camel milk cream. LWT 132:109852

Kena D (2022) Review on camel production and marketing status in Ethiopia. Pastoralism 12(1):38

Kess T, Gross J, Harper F, Boulding EG (2016) Low-cost ddRAD method of SNP discovery and genotyping applied to the periwinkle Littorina saxatilis. J Molluscan Stud 82(1):104–109

Konuspayeva G (2019) Camel milk composition and Nutritional Value. In: Al-Haj OA, Faye B, Agrawal RP (eds) Handbook of research on health and environmental benefits of camel products. IGI Global, pp 15–40. https://doi.org/10.4018/978-1-7998-1604-1.ch002

Konuspayeva G, Faye B (2021) Recent advances in camel milk processing. Animals 11(4):1045. https://doi.org/10.3390/ani11041045

Konuspayeva G, Camier B, Gaucheron F, Faye B (2014a) Some parameters to process camel milk into cheese. Emir J Food Agric 26:354–358

Konuspayeva G, Shomamrey M, Aleliwi N, Camier B, Faye B (2014b) New trials to make camel mozzarella. Rev KazNU Biol Ser №1/2 60:138–140

Konuspayeva G, Camier B, Aleilawi N, Al-Shumeimyri M, Al-Hammad K, Algruin K, Alshammari F, Beaucher E, Faye B (2017) Manufacture of dry- and brine-salted soft camel cheeses for the camel dairy industry. Int J Dairy Technol 68:92–101. https://doi.org/10.1111/1471-0307.12319

Konuspayeva G, Faye B, Duteurtre G (2022) Online camel milk trade: new players, new markets (Update) [translation and update of the original French article published under. https://doi.org/10.19182/remvt.36746]. Rev Elev Med Vet Pays Trop 75(4). https://doi.org/10.19182/remvt.37041

Kumar D, Verma AK, Chatli MK, Singh R, Kumar P, Mehta N, Malav OP (2016) Camel milk: alternative milk for human consumption and its health benefits. Nutr Food Sci 46(2):217–227

Lajnaf R, Zouari A, Trigui I, Attia H, Ayadi MA (2020) Effect of different heating temperatures on foaming properties of camel milk proteins: a comparison with bovine milk proteins. Int Dairy J 104:104643

Laleye LC, Jobe B, Wasesa AAH (2008) Comparative study on heat stability and functionality of camel and bovine milk whey proteins. J Dairy Sci 91:4527–4534

Legrand D, Elass E, Pierce A, Mazurier J (2004) Lactoferrin and host defence: an overview of its immuno-modulating and anti-inflammatory properties. Biometals 17(3):225–229. PMID: 15222469

Lorenzen PC, Wernery R, Johnson B, Jose S, Wernery U (2011) Evaluation of indigenous enzyme activities in raw and pasteurized camel milk. Small Rumin Res 97:79–82

Manjunatha BM, Al-Bulushi S, Pratap N (2015) Synchronisation of the follicular wave with GnRH and PGF2α analogue for a timed breeding programme in dromedary camels (*Camelus dromedarius*). Anim Reprod Sci 160:23–29

Mati A, Senoussi-Ghezali C, Si Ahmed Zennia S, Almi-Sebbane D, El-Hatmi H, Girardet JM (2017) Dromedary camel milk proteins, a source of peptides having biological activities a review. Int Dairy J 73:25–37

Mbye M, Sobti B, Al Nuami MK, Al Shamsi Y, Al Khateri L, Al Saedi R, Saeed M, Ramachandran T, Hamed F, Kamal-Eldin A (2020) Physicochemical properties, sensory quality, and coagulation behavior of camel versus bovine milk soft unripened cheeses. NFS J 2020(20):28–36

Mellor DJ (2016) Moving beyond the "five freedoms" by updating the "five provisions" and introducing aligned "animal welfare aims". Animals 6(10):59

Mishra A, Dhyani P, Goyal C, Kumari S, Yadav K, Kumar A, Saharan BS (2025) Process optimization and characterization of ragi-based probiotic camel milk drink for enhanced functionality. Food Biomacromol 2(4):437–450

Mohamed AE, Babiker IA, Mohamed TE (2013) Preparation of fresh soft cheese from dromedary camel milk using acid and heat method. Res Opin Anim Vet Sci 3:289–292

Morsi R, Ghoudi K, Ayyash MM, Jiang X, Meetani MA (2024) Detection of 11 carbamate pesticide residues in raw and pasteurized camel milk samples using liquid chromatography tandem mass spectrometry: method development, method validation, and health risk assessment. J Dairy Sci 107(4):1916–1927

Mortazavi SM, Jalali H, Hamidreza Ziaolhagh S (2021) Production of a probiotic camel milk enriched with pomegranate peel powder. Iran Food Sci Technol Res J 16:123–132

Moynihan AC, Govindasamy-Lucey S, Jaeggi JJ, Johson ME, Lucey JA, McSweeny LH (2014) Effect of camel chymosin on the texture functionality and sensory properties of low-moisture part-skim Mozzarella cheese. J Dairy Sci 97:85–96

Naveenkumar V, Thangapandiyan M, Sakthivel V, Kalyaan US, Duraisamy C (2025) Sarcoptic mange in a camel (*Camelus dromedarius*): a case report emphasizing zoonotic concerns and preventive measures. Int J Vet Sci Anim Husb 10(6):32–34

Oselu S, Ebere R, Huka G, Musalia L, Marete E, Mathara JM, Mwobobia F, Arimi JM (2022) Production and characterisation of camel milk yoghurt containing different types of stabilising agents. Heliyon 8(11):e11816

Pingault N, Caron P, Kalafatic C, Allahoury A, Fresco LO, Kennedy E, Khan M, Kliksberg B, Mei F, Murphy S, Saeid Noori Naeini M (2016) Sustainable agricultural development for food security and nutrition: what roles for livestock? A report by the High Level Panel of Experts on Food Security and Nutrition of the Committee on World Food Security

Ramet JP (1989) Cheese making capacity of camel milk. Rev Elev Med Vet Pays Trop 42:105–111
Ryan O (2012) Animal welfare and economic development: a financial institution perspective. In: Animals, ethics and trade. Routledge, pp 238–247
Session T (2007) Joint FAO/WHO food standards programme Codex Alimentarius Commission. Codex
Sobti B, Aljneibi AHA, Seraidy HAA, Alnaqbi AAH, Al Zain B, Ramachandran T, Hamed F, Kamal-Eldin A (2021) Short communication: the effect of pectin and sodium alginate on labans made from camel milk and bovine milk. J Dairy Sci 104(5):5279–5284
Wako G, Tadesse M, Angassa A (2017) Camel management as an adaptive strategy to climate change by pastoralists in Southern Ethiopia. Ecol Process 6(1):1–2
Yadav AK, Kumar R, Priyadarshini L, Singh J (2015) Composition and medicinal properties of camel milk: a review. Asian J Dairy Food Res 34(2):83–91. https://doi.org/10.5958/0976-0563.2015.00018.4
Zouari A, Schuck P, Gaucheron F, Triki M, Delaplace G, Gauzelin-Gaiani C, Lopez C, Attia H, Ayadi MA (2020) Microstructure and chemical composition of camel and cow milk powders' surface. LWT 2020(117):108693

Camels in Their Environment: Symbiosis and the Extremes

8

Abdelmadjid Chehma and Bernard Faye

Abstract

This chapter examines the remarkable adaptations that enable camels to survive and remain productive in arid and semiarid environments. It describes the extreme climatic conditions of desert ecosystems characterized by high temperatures, low and irregular rainfall, and sparse vegetation, and their effects on plant life and forage quality. The chapter highlights the camel's unique anatomical, physiological, and behavioral adaptations, including thermoregulation, water conservation, and the efficient utilization of low-quality, fibrous feed resources. These adaptations allow camels to maintain homeostasis, withstand dehydration, and exploit widely dispersed resources. Beyond survival, the camel plays a significant ecological role by promoting vegetation regeneration, dispersing seeds, reducing grazing pressure through its ambulatory feeding behavior, and contributing less to greenhouse gas emissions compared to other ruminants. The chapter also discusses camel health within environmental and management contexts, emphasizing the importance of ecopathological approaches. Finally, it considers the implications of intensifying camel production systems and highlights the need to balance productivity with ecological sustainability in fragile desert ecosystems.

A. Chehma (✉)
Laboratoire des bioressources sahariennes, Préservation et valorisation, Ouargla University, Ouargla, Algeria

University of Ghardaïa, Bounoura, Algeria
e-mail: chehma.abdelmadjid@univ-ghardaia.edu.dz

B. Faye
CIRAD-SELMET (French Agricultural Research Centre for International Development), Montpellier, France

S. A. Gordine et al. (eds.), *Camels and One Health*,
https://doi.org/10.1007/978-981-92-2246-9_8

Keywords

Camel adaptation · Desert ecology · Thermoregulation · Ecopathology · Biodiversity · Water conservation

8.1 Introduction

The camel is endowed with evolutionary adaptations that enable it to survive in arid and semiarid lands. These physiological and anatomical traits ensure that camels can make the most of the very meagre availability of mediocre quality forage. The camel's rangelands are constrained not only by scarcity of water and pasture but also by extreme temperature variation and a high disease burden. Thanks to its physiological adaptations, the camel is the only livestock species capable of producing and reproducing in this very hostile environment. What's more, it is a species that lives in harmony with its desert ecosystem, with its ambulatory feeding behavior contributing to the preservation and proliferation of desert plants.

8.2 The Characteristics of the Camel's Desert Environment

Camels live in a very harsh and restrictive desert landscape. This environment is characterized by very high temperatures, very low and irregular rainfall, intense luminosity, high evaporation, and wide temperature variations that are accentuated by the hot, dry winds (Dubief 1963; Ozenda 1991). The inadequacy of desert rainfall is accompanied by highly irregular rainfall patterns with considerable interannual variability, leading to periods of drought (Alahacoon et al. 2021). Average annual temperatures are high, with absolute maximums in July–August that can exceed 50 °C. Soil surface temperatures can exceed 70 °C, but below the ground, the temperature rapidly falls and then plateaus. The temperature can drop to freezing in the northern Sahara, in the northern Arabian Peninsula, and in Central Asia, as well as, of course, in the mountains (Zhou and Wang 2016). However, minimum temperatures in January range from 2 to 9 °C in the semiarid African Sahel (Le Houerou 1990). Because of the low cloud cover in the atmosphere, the amount of sunlight is relatively high, which has a drying effect by raising the temperature (Ozenda 1991). In the desert, sunshine lasts a very long time (9–12 h per day) with the ultraviolet index ranging from 6 (high) to 12 (extreme) (Dubief 1963). These climactic characteristics impose strong physiological constraints on both the vegetation and animals that live in such a hostile environment.

8.2.1 The Impact of the Desert Climate on Vegetation

The plant cover in the desert is mainly composed of xerophytes, which possess specific adaptations that enable them to resist drought and high evapotranspiration. The low and irregular desert rainfall results in very sparse plant cover with relatively poor nutritional value. These extreme conditions are not favorable for the proliferation of plants. However, in some areas, such as wadis, more abundant biomass is present. This can be observed through the presence of trees, primarily from the *Acacia* species.

Generally, two categories of plants can be distinguished in the desert landscape (Ozenda 1991; Chehma 2019):

- Ephemeral plants, also known as *achebs*, only appear after the rainy season and complete their vegetative cycle before the soil has dried out. These plants therefore occupy the land sporadically, and even then, only fleetingly (Chehma et al. 2008; Malika et al. 2019).
- Permanent or perennial plants possess a set of morphological and anatomical adaptations that allow for increased water absorption and a reduced evaporative surface. This type of vegetation remains constant and is less subject to seasonal variation. It therefore constitutes the only fodder that is reliably available for camels, even in the summer (Chehma and Youcef 2009; Berg and Steinberger 2012).

The chemical composition and nutritional value of the plants grazed by camels reflect the conditions of the desert environment in which they live (Faye 1997; Chehma et al. 2005; Madouh 2022). These plants limit evaporation rate as much as possible through a reduced leaf/stem ratio (both in number and in surface area), through the formation of a thick cuticle over the stomata, as well as the lignification of support tissues (Chehma et al. 2010). Similarly, the generally low levels of total nitrogen (which rumen microorganisms can convert to protein) can be attributed to the drought adaptation strategy of these plants. Though these desert plants contain a large concentration of soluble proteins, enzymes, and chlorophyll within their leaves, these leaves are small in size and low in number (Peguero-Pina et al. 2020). Due to these specific adaptations to drought, desert plants are therefore characterized by their richness in parietal compounds, a high degree of lignification, and relatively poor protein content. They can therefore only be utilized by camels, thanks to their unique digestive abilities (Ouajd and Kamel 2009; Chehma 2019; Louhaichi et al. 2021).

The survival and growth of desert plants depend largely on the germination of seeds. In the extreme conditions of arid areas, seeds do not germinate if rainfall is not sufficient for seedling survival. To survive for long periods of time without water (sometimes several years), the seeds from desert plants possess specific dormancy mechanisms. This prevents germination until favorable conditions arise (Suleiman et al. 2023), such as sufficient soil humidity and optimal temperature (De Queiroz and Meyer 2023). The longevity of these seeds results in a long-life soil seed bank, representing an adaptative strategy of desert vegetation in the face of this unpredictable environment (Ali et al. 2022).

8.2.2 Adaptations of Camels to Their Desert Environment

Camels face many challenges, including wide day-night temperature variability, scarcity of water, and dispersed, low-nutritional-value food resources. The anatomical and physiological features of the camel demonstrate the adaptation of this unique animal to these environmental challenges.

8.2.2.1 Adaptations to Heat and Temperature Fluctuation

Most desert animals bury themselves in the ground during the hot hours of the day, a behavior that is impossible for camels. Thus, the camel has developed other strategies to limit heat stress. Several millennia of presence in desert areas have shaped this animal anatomically to withstand high temperatures and the wide variations of temperature over a 24-h period. Anatomically and structurally, camels have long legs to move the body away from the hot ground. When standing in a stooped position, the sternal pad keeps the abdomen slightly above the ground, allowing air to circulate under the body to dissipate heat. In cold weather, the squatting position gives a compact shape to the whole body, limiting the impact of the cold. Moreover, the camel's skin and hair are adapted to insulate against heat by reflecting the sun's burning rays, thereby helping to regulate temperature. Its skin is thick, especially in areas most exposed to the sun (on the back and neck), and contains pigments that absorb and disperse some of the ultraviolet rays (Gebreyohanes and Assen 2017). During winter, camels develop thick coats (especially Bactrians who live in the cold desert) to protect themselves from the cold. Their hair (or wool in the case of Bactrians) falls out by itself when temperatures increase in the spring. This is replaced in summer with new, finer, lighter hair that is better suited to the heat. Bactrian wool is particularly known for its superior insulating and anti-UV properties compared with other wool species (Hasi et al. 2020). The light coat colors that dominate the different camel breeds also contribute to better reflection of the sun's rays. Additionally, its feet have large footpads that distribute weight evenly. This makes it easier to move over sandy terrain, and thanks to the thick footpad, it is more resistant to the high soil temperatures (Bornstein 1990; Kebir et al. 2024).

From a behavioral point of view, camels act to conserve as much water as possible. When grazing in areas where trees are present, they prefer to browse the tree leaves rather than grasses, to take advantage of the shade of the trees while feeding. They reserve the steppe grassland areas for the beginning and end of the day (Faye and Tisserand 1989). Camels reduce their activity during the hottest periods of the day and ingest very large quantities of water when it is available (El-Zeiny 2010; Bekele et al. 2013). They rest grouped together in a sitting position, with their legs tucked under their bodies, thus reducing the surface area exposed to the hot sand. They also orient their bodies with their heads facing the sun. This minimizes the surface area exposed to the sun, further reducing evaporative water loss (Gebreyohanes and Assen 2017; Fatnassi and Padalino 2024).

The fat reserve of the hump also plays an essential role in thermoregulation due to its concentration on the animal's back (Chilliard et al. 2005; Djenane and Aider 2024). The concentration of fat at this one point (the hump) allows for decreased fat cover in the other parts of the body, especially under the skin, which facilitates the dissipation of heat. The camel is also able to vary its internal temperature. When the ambient temperature drops, especially during the night, the internal temperature of the dromedary can decrease to 34 °C. During the hottest hours of the day, the camel's rectal temperature can reach 42 °C without compromising the animal's health. Thanks to this mechanism of adaptive heterothermy, camels can limit sweating to maintain a stable temperature, which avoids losing water (Bekele et al. 2013; Hoter et al. 2019; Mohammed et al. 2025). Moreover, this mechanism, which would be fatal for most mammals, is a powerful instrument of adaptation to the wide variations in day-night temperature (Robertshaw and Zine-Filali 1995). Thus, an increase of 6 °C in body temperature can save five liters of water.

"Heat stroke," which can be fatal, occurs when increasing external temperatures exceed an animal's thermoregulatory mechanisms. The thyroid gland plays a central role in this process by increasing its activity in response to external heat. The subsequent metabolic heat production continues to increase until it bypasses the animal's capacity for heat regulation. In camels, however, an increase in internal temperature is conversely associated with a general decrease in metabolism, notably in oxygen consumption (Omidi et al. 2024). This means that camels can suppress internal heat production during the hottest part of the day. It has been shown that the thyroid gland increases bromide secretion, which suppresses thyroid activity and thus slows overall metabolism (Etzion et al. 1987). At the molecular level, dromedaries have developed specific heat shock proteins that protect their cellular proteins from denaturation under high heat stress, playing a crucial role in maintaining cellular homeostasis (Hoter et al. 2019; Dowidar et al. 2022).

8.2.2.2 Adaptations to Water Scarcity

The adaptive mechanisms to a hot climate described above involve physiological mechanisms that also contribute to saving water. The camel's ability to resist thirst is well-known. Indeed, it can tolerate losing up to 30% of its body weight in water (Perk 1963; Ali et al. 2023). This resistance is exceptional both in the case of severe dehydration for several weeks and also during rapid rehydration. In such cases, camels can ingest large quantities of water very quickly after a long session of water deprivation without disturbing their metabolism.

In the face of dehydration, the camel has developed two main mechanisms for the maintenance of homeostasis: (i) the reduction of water loss through water-conserving processes and (ii) the regulation of vital parameters (maintenance of homeostasis) through maximization of metabolic waste excretion (Bengoumi and Faye 2002).

Both external and internal anatomical properties of the camel contribute to its ability to limit water losses. These include:

- The nostrils can close to protect against sand and conserve moisture (Gebreyohanes and Assen 2017; Ridouh et al. 2023).
- The nasal ducts are long and convoluted, contributing to water conservation by reducing respiratory losses (Mohammed et al. 2025; Ridouh et al. 2023).
- Its characteristic mouth with thick and flexible lips, robust and well-arranged teeth, and thick and resistant gums with increased salivation. This enables the camel to select, grasp, moisten, and crush its desert fodder plants, which are often very dry and thorny (Kebir et al. 2024). A normally hydrated camel produces a large amount of saliva (about 20 L/day per gland!), but in the dehydrated animal, saliva production can drop to 1 L/day without affecting intake (Almansour et al. 2024).

- A hump composed mainly of adipose tissue, which stores fat that can be metabolized into energy and water during periods of scarcity.
- Specialized kidneys with a larger medullary part and a larger diameter of the renal structures. This favors water reabsorption and produces highly concentrated urine, thus minimizing water loss (El-Hady et al. 2023). Urine volumes of dehydrated camels can be 0.1% of the animal's weight, compared with 2% in sheep (Mokhtar et al. 1989). Dehydrated camel urine is twice as concentrated as seawater—at 42 °C, the dromedary loses about 20 g of water/kg of body weight (BW)/day, compared to more than 40 g in sheep. These renal properties in camels are regulated by hormones such as arginine vasopressin and oxytocin to promote water reabsorption in the kidneys (Bengoumi and Faye 2002; Lin et al. 2022; El-Hady et al. 2023).
- Unique pulmonary vascular structures, such as constricted arteries and muscle pads in the veins, which facilitate the hemodynamic, humeral, and thermoregulatory functions crucial to maintaining lung function in hostile desert environments (Mohamedien et al. 2022).

The main storage organ for ingested water is the stomach. A dehydrated camel will maintain a high proportion of water, especially in the C1 compartment (rumen), which ensures a smooth passage of ingesta into the other gastric compartments and into the intestine, thus maintaining its appetite. Massive water reabsorption takes place at the terminal level of the intestines, resulting in the excretion of particularly dry droppings. Indeed, the percentage of dry matter in dromedary excreta is about 50%, compared to 30% for small ruminants and 15% for cattle. In dehydrated dromedaries, the fecal dry matter rate increases slightly (55%), whereas it does not change in cattle. Ultimately, camels lose seven times less water through the fecal route than cattle (Ali et al. 2023).

Blood also plays a pivotal role in the conservation of water. Camel erythrocytes (red blood cells) are oval-shaped, have a longer lifespan (90–150 days) than other mammals, and can resist osmotic stress to maintain efficient oxygen transport, even in the event of dehydration (Mohammed et al. 2025). Thanks to the mobilization of water from tissues and the regulation of blood pressure, it can tolerate a significant reduction in blood volume without compromising vital functions (Bornstein 1990; Ali et al. 2023; Mohammed et al. 2025). In most mammal species, dehydration provokes an increase in blood viscosity, leading to a rise in internal temperature, which is often fatal for the animal (described as heat stroke). Conversely, the dromedary's blood remains fluid, enabling it to maintain its heat transfer function from the periphery (which is cooler due to evaporation) to the heart (Windberger et al. 2018).

The camel is remarkable also in its ability to be rapidly rehydrated without deleterious effect. In most mammals, a rapid and large consumption of water causes an uncontrolled influx of water into the bloodstream, which leads to a dilution of the extracellular milieu (hypotonicity). To compensate for the difference between this hypotonic milieu and the hypertonic internal cytoplasm of the red blood cells, the latter swell with water until they burst (hemolysis). As described above, the dromedary has red blood cells that are extremely resistant to hypotonicity, due to the composition of its cell wall (Al-Qarawi and Mousa 2004). After rehydration, the return to a normal, hydrated cellular and metabolic state takes about 4 h for blood cells and kidney functions. Globally, compared to other species, the water turnover in dromedaries is particularly low (50–100 ml/kg/d vs 70–230 ml/kg/d for zebu), although the percentage of body water is higher: 72 vs. 60–65% in other domestic herbivore species (Wilson 2012).

8.2.2.3 Adaptations to Low-Nutritive Feed Resources

Arid environments are characterized by limited food resources, wide spatial dispersion, and high seasonal (and even interannual) rainfall variability. In such a context, large camelids only survive thanks to: (i) its feeding behavior and (ii) its exceptional digestive physiology.

Regarding their feeding behavior, camels are browsers rather than grazers. They consume a wide variety of plants in their ecosystems, from the shortest grasses to the stems and leaves of trees accessible by their long necks. In Sudan, for example, it has been observed that the camel can ingest between 27 and 28 different plant species according to season, while cattle are limited to 2–8 different plants in the same season (Rutagwenda et al. 1990). In this respect, they are similar to goats and clearly distinguishable from cattle and sheep. Mastication of spinous, highly lignified desert plants is possible thanks to the hard, strong teeth of the camel, which can crush sclerophyllous branches and stems with their incisors and molars. Moreover, the internal side of the cheeks is covered with a horny layer so as to avoid any injuries from the spinous branches. When browsing, the camel has a nongregarious and ambulatory behavior, which means the camel herd never concentrates all together in the same place and they move continuously to browse. These movements can be more than 20 km per day (and even up to 50 km in the case of highly dispersed resources or distant waterholes), avoiding localized pressure on the available resources. On average, these animals can spend 8 h on rangelands foraging for food.

Good digestion of cellulose is possible due to several factors. These include the presence of cellulolytic bacteria in the stomach, strong secretion of hydrochloric acid in the C4 compartment, strong amylolytic activity allowing better digestion of starches, greater length of the intestine, and greater papillary surface. Through this highly efficient digestive system, the camel can extract water and nutrients from dry, fibrous vegetation. It can survive without food for several days using the reserves in its hump and its great capacity to metabolize lipids, which is essential for energy production during periods of fasting. Camels also have the ability to regulate protein metabolism by minimizing nitrogen waste and reducing water excretion. Additionally, they possess a remarkable capacity to regulate mineral balance, even in conditions of deficiency or excess. This is essential for the maintenance of cellular function and general health in the desert environment (Bornstein 1990; Faye and Bengoumi 2000; Chehma and Faye 2011; Mohammed et al. 2025). The camel's immune system is also well adapted to pathogens common in desert environments, enhancing its ability to survive in conditions where water and food resources are limited (Ali et al. 2019; Futas et al. 2019; Conca et al. 2023). The dromedary camel has demonstrated its ability to control mineral deficits through increasing absorption and minimizing losses. It can increase organ storage and also increase intestinal mineral absorption in times of deficit. It is also able to prevent mineral losses, when necessary, by reducing their metabolism through metalloenzymes, such as glutathione-peroxidase, sodium oxide dismutase, and ceruloplasmin (Bengoumi et al. 1998, 1999; Essamadi et al. 1998; Faye et al. 1999b).

Through its physiology and behavior, the camel has adaptive features for water conservation, thermoregulation, and efficient digestion, enabling it to survive in desert environments with limited resources. Importantly, the camel could also play an important ecological role in its environment, contributing to the preservation of its desert ecosystem.

8.3 The Ecological Role of the Camel in Preserving Its Environment

The camel, thanks to its physiological adaptations and feeding behavior, lives in harmony with its environment. It plays a role in the balance of its desert ecosystem and in the preservation and proliferation of spontaneous desert plant cover. Therefore, camels can be considered to play a significant ecological role.

The camel has little impact on global warming compared with other farmed ruminants, as it is less methanogenic. In fact, per unit of body surface area, dromedaries emit 30–50% less methane (a greenhouse gas) than ruminants such as cows. This is due to its digestion, which enables it to use food energy efficiently, with less water and feed consumption (Dittmann et al. 2014). Moreover, due to its low number (e.g., only 42 million head worldwide compared to the 1.5 billion cattle), its footprint is very light compared to animals reared in intensive agricultural systems. Even when utilized in intensive systems (developed particularly in the Middle East), the impact on water resources is lower compared to dairy cows (Faye 2015).

Unlike other farmed ruminants with hooves, the dromedary has a light footprint that minimizes soil erosion and degradation, thanks to the morphology of its feet. The large elastic pads distribute weight evenly, avoiding sinking and crushing of the soil, and thus limiting volatilization and dispersion of the thin layer of topsoil in the desert environment (Dagg 1974; Hadi 2016; Kebir et al. 2024). Notably, the pressure on soil has been evaluated at 2.6 Newton/cm^2 compared with 8.86 Newton/cm^2 for cows.

The ambulatory grazing mode of camels, even when forage is abundant, respects the balance and diversity of the vegetation in its environment. In fact, camels contribute to maintaining pastoral biodiversity by avoiding over-exploitation of specific plants and by diversifying their diet, grazing on almost all plants, even those considered toxic to other livestock species. They are highly selective regarding species and plant parts, never ceasing to wander between the plants and taking only small bites from each plant. As they walk, passing from one plant to another, they prune the plants, helping to regenerate them. They do not overgraze or uproot any type of vegetation and can reach the upper layers of plant formations, selectively grazing parts of the plant (leaves, stems, flowers, seeds) depending on availability. The duration of grazing each plant is linked to its architecture and its size, so that camels graze the most abundant species in large quantities and leave the less abundant ones.

Thanks to its selective, ingestive, and digestive aptitudes, the camel is able to satisfy its own nutritional requirements (energy and protein) regardless of the vegetation present, the season, and the type of rangeland (Towhidi et al. 2011; Chehma and Faye 2015). Unlike sheep and goats, dromedaries graze in a dispersed, nongregarious way, which reduces pressure on vegetation and contributes to preserving the natural environment (Mahma et al. 2019; Mallem et al. 2024). They exploit woody and thorny plants rejected by other herbivores, allowing the maintenance and regeneration of certain plant species capable of stabilizing and fixing sand dunes, thus preventing silting. Due to their ability to withstand water deprivation for very long periods, they can graze in places far from waterholes, unlike other herbivores, which are forced to stay in closer proximity to water. In a context where water reserves are beginning to be a global problem, this represents a major ecological advantage. Their dispersed grazing pattern results in a better distribution of the grazing area, which is highly beneficial to the maintenance and regeneration of the very fragile desert plant cover. Additionally, they consume relatively little feed (Slimani et al. 2013; Chehma et al. 2023).

More recently, it has been shown (in the Canary Islands, Spain) that within a 1 km radius of camel farms, wildlife biodiversity is greater than in areas further away (Gil Sanchez et al. 2025). Wild animal species around camel livestock farms in this study included wild mammals, birds, arachnids, and mollusks. This suggests that the sustainable management of camel farms could support the preservation of local agroecosystems and the survival of coexisting wild animal species, reinforcing their potential ecological role.

Camels also play an important role in dispersing the seeds of wild desert plants through endozoochory (where ingested seeds are dispersed through fecal droppings). This contributes to the maintenance and improvement of plant cover in these harsh desert ecosystems. Seeds that fall to the ground by autochory (ejected by the plant over a short distance) can remain dormant for several years. As a result, their offspring may be weaker due to the increased risk of mortality and the loss of potential generations due to predation, especially by ants. In desert environments, most seeds are small and easily carried by the wind to distant locations unsuitable for germination, which can hinder their survival and proliferation. This contrasts with seeds found in camel droppings, which have a much better chance of surviving and becoming viable seedlings. Of note, camels are the only livestock species that can exploit this desert resource, and due to their feeding behavior, they are able to distribute almost all available seed species. Although these seeds undergo a long process of ingestion, rumination, and microbial and chemical digestion by the camel, most of the seeds of the grazed species remain in good condition. This demonstrates that, despite the camel's complex digestion as well as the potential nutrient value of seeds, it allows seeds to pass through the gastrointestinal tract intact for fecal excretion without being damaged. Interestingly, this process actually enables the seeds to acquire their germinative power. The seeds of spontaneous desert species have mechanisms that adapt them to their environment, resulting in hard seed coats and impermeability to water. This induces dormancy and makes germination difficult under natural conditions. The acidic scarification of the seeds in the camel's digestive tract and chemical treatment by their digestive enzymes, such as diastases, have a positive effect on their germination capacity. Studies have shown that the seeds of certain plant species can only germinate in camel droppings, particularly those with thick seed coats (Trabelsi et al. 2023). By covering very long distances, camels contribute significantly to the dispersal of seeds over the widest possible area, thereby promoting the preservation, proliferation, and genetic diversity of rangeland flora (Wilson 1992). Their droppings provide a favorable environment for the survival and protection of seeds against the climatic constraints of the desert as well as protection from predation (Chehma et al. 2023). Camel droppings also constitute an important source of organic matter to promote the growth of seedlings, which are packaged and preserved until the appearance of favorable germination conditions, such as rainfall (Janzen 1983; Neffati 2008; Trabelsi et al. 2012; Kouadri et al. 2024).

8.4 Environment and Camel Diseases

The camel is globally less sensitive to the main infectious diseases affecting cattle and other ruminants (see the chapter on camel diseases), thanks to its immunological make-up based on unique "nanobody" immunoglobulins (see the chapter on nanobodies) (Al-Numair et al. 2022). However, camel health can be impacted by external and internal parasitosis, as well as some nutritional and metabolic diseases linked to farming practices and deleterious environmental conditions. For the investigation of these disorders, applying an ecopathological approach (see also Chap. 1) can be useful (Faye et al. 1999a). This investigates the interactions between animal factors and herd husbandry factors (such as feeding, housing, veterinary practices, and the microbiological environment), as well as other "nonanimal" factors (practices). This holistic approach is aimed at explaining an animal's health status within the overall dynamic of a livestock production system (Fortane 2023). These ecopathological studies are integrated with epidemiology to form a systemic approach. This involves "the development of a preliminary conceptual model, sampling based on the structure of the livestock production system, the establishment of a field study by a multidisciplinary team, the organization and management of the animal health and production information, data analysis, the distribution of results to all participants and the development of a preventive medicine program" (Faye et al. 1999a).

Such an approach is well adapted to multifactorial diseases (Benaissa and Iglesias Pastrana 2024). Some of these diseases can have infectious origins (such as mastitis, genital infections, respiratory syndromes, and diarrhea), but in most cases, complex disorders are linked to dietary imbalances (deficiency or toxicity) or metabolic failure (e.g., thyroid, kidney, or liver dysfunction). They are often detectable only by blood biochemistry analysis (Faye and Bengoumi 2018) and longitudinal surveys (e.g., infertility). The ecopathological approach has been applied primarily to investigate multifactorial diseases such as camel calf diarrhea (Bengoumi et al. 2004) and respiratory syndrome (Ahmed and Musa 2015).

A trend to more intensive farming involves dietary changes (a higher concentration of protein and energy in feeds), reduced animal mobility, and adoption of biotechnology (through machine milking and innovative reproduction technologies). These practices could change disease patterns, resulting in a higher prevalence of mastitis, infertility, and metabolic diseases (Nagy et al. 2022). Moreover, the geographical expansion of camel farming (Faye et al. 2012), especially in Africa and Western countries, exposes camels to more humid environments. This could also increase the risk of multi-parasitism and emerging diseases. This concern is supported by recent outbreaks of unknown etiology (Shadomy et al. 2025).

8.5 Conclusion: From Desert to Stable?

The unique adaptations of camels enable them to serve as a vital component in arid and semiarid lands (Faye 2011). This occurs at three main levels: (i) the camel's anatomical and physiological characteristics allow for the use of arid zones without excessive pressure, especially with extensive farming systems; (ii) desert areas can be regarded as productive zones for high quality animal proteins (camel milk and meat), thus increasing the value of agriculture in these regions; and (iii) through its other functions, the camel brings further value to local populations (wool, carrying and riding, leisure and tourism, transfer of desert seeds). The further development of the camel sector could contribute to combating desertification and to the sustainable development of arid lands. However, the introduction of high-added-value camel products in national and international markets could stimulate the development of the camel industry and important changes in camel farming. Consequently, the emergence of new camel farming systems, based on more intensive inputs, has the potential to modify the fragile balance between camel and environment. It will be important to examine the sustainability of these new farming systems to promote more ecologically responsible production practices. This is particularly important regarding the use of less water-consuming fodder (compared to the currently used irrigated grasses and cereals), and for the respect of the physiological welfare of an animal that is more adapted to wide spaces than to confinement in barns.

References

Ahmed ME, Musa MT (2015) Characterization of bacteria isolated from dromedary camels affected with pneumonia for the first time in Sudan. Ann Res Rev Biol 7(1):61–67. https://doi.org/10.9734/ARRB/2015/16744

Alahacoon N, Edirisinghe M, Simwanda M, Perera ENC, Nyirenda VR, Ranagalage M (2021) Rainfall variability and trends over the African continent using TAMSAT data (1983–2020): towards climate change resilience and adaptation. Remote Sens 14(1):96. https://doi.org/10.3390/rs14010096

Ali A, Baby B, Vijayan R (2019) From desert to medicine: a review of camel genomics and therapeutic products. Front Genet 10:17. https://doi.org/10.3389/FGENE.2019.00017

Ali AH, Marghany MR, Atito E, Barutçular C, Kamel NM, Ahmed MM, El-Sayed MA (2022) Desert plants seeds morphology and germination strategy. Int J Conserv Sci 13(4):1249–1260

Ali MA, Damir H, Adem MA, Ali OM, Amir N, Shah AAM, Muhairi SSA, Abdouli KOS, Khawaja JR, Fagieri TA, Adam A, Elkhouly AA, Marri ZJA, Jamali M, Murphy D, Adem A (2023) Effects of long-term dehydration on stress markers, blood parameters, and tissue morphology in the dromedary camel (*Camelus dromedarius*). Front Vet Sci. https://doi.org/10.3389/fvets.2023.1236425

Almansour M, Jarrar B, Faye B, Al-Doaiss A, Shati A, Meriane D (2024) The salivary glands of the camel: an element of adaptation to desert conditions and mitigation of climate change impacts. Jordan J Biol Sci 17(1):99–108. https://doi.org/10.54319/jjbs/170109

Al-Qarawi AA, Mousa HM (2004) Lipid concentrations in erythrocyte membranes in normal, starved, dehydrated and rehydrated camels (*Camelus dromedarius*), and in normal sheep (*Ovis aries*) and goats (*Capra hircus*). J Arid Environ 59(4):675–683. https://doi.org/10.1016/j.jaridenv.2004.02.003

Al-Numair NS, Theyab A, Alzahrani F, Shams AM, Al-Anazi IO, Oyouni AAA, Al-Amer OM, Mavromatis C, Saadeldin IM, Abdali WA, Hawsawi YM (2022) Camels' biological fluids contained nanobodies: promising avenue in cancer therapy. Cancer Cell Int 22(1):279. https://doi.org/10.1186/s12935-022-02696-7

Bekele T, Olsson KA, Olsson U, Dahlborn K (2013) Physiological and behavioral responses to different watering intervals in lactating camels (*Camelus dromedarius*). Am J Phys Regul Integr Comp Phys 305(6). https://doi.org/10.1152/AJPREGU.00015.2013

Benaissa MH, Iglesias Pastrana C (2024) Good health: recognition and prevention of disease and pain in dromedary camels. In: Dromedary camel behavior and welfare: camel friendly management practices, vol 24. Springer, pp 111–143. https://doi.org/10.1007/978-3-031-48600-5_6

Bengoumi M, Faye B (2002) Adaptation du dromadaire à la déshydratation. Sécheresse 13(2):121–129. https://www.jle.com/en/revues/sec/e-docs/adaptation_du_dromadaire_a_la_deshydratation_230117/article.phtml

Bengoumi M, Essamadi AK, Tressol JC, Faye B (1998) Comparative study of copper and zinc metabolism in cattle and camel. Biol Trace Elem Res 63:81–94. https://doi.org/10.1007/BF02778867

M, Moutaoukil F, De La Farge F, Faye B (1999) Thyroidal status of the dromedary camel: effect of some physiological factors. J Camel Pract Res 6(1):41–43

Bengoumi M, Berrada J, Hidane K, Faye B (2004) Diarrhées du chamelon: écopathologie, physiopathologie et prévention. In: Lait de chamelle pour l'Afrique. FAO Publ, Rome, pp 137–146

Berg N, Steinberger Y (2012) The role of perennial plants in preserving annual plant complexity in a desert ecosystem. Geoderma 185:6–11. https://doi.org/10.1016/j.geoderma.2012.03.023

Bornstein S (1990) The ship of the desert. The dromedary camel (*Camelus dromedarius*), a domesticated animal species well adapted to extreme conditions of aridness and heat. Rangifer 10(3):231–236. https://doi.org/10.7557/2.10.3.860

Chehma A (2019) Catalogue des plantes spontanées du Sahara septentrional algérien, 2éme édition. Edition « Editions universitaires Européennes», 156 p

Chehma A, Faye B (2011) Facultés digestives du dromadaire face aux contraintes alimentaire du milieu saharien. Rev Bioresour 1(1):26–30. ISSN: 2170-1806. http://www.ouargla-univ.dz/Pagesweb/PressUniversitaire/doc/08%20Bio%20recources/B0101/B010104.pdf

Chehma A, Faye B (2015) Nutritional characteristics of the camel's rangeland in Algeria. Veterinaria 42(2):42–43. http://publications.cirad.fr/une_notice.php?dk=577436

Chehma A, Youcef F (2009) Variations saisonnières des caractéristiques floristiques et de la composition chimique des parcours sahariens du sud est algérien. Sécheresse 20(4):373–381. http://www.jle.com/e-docs/00/04/51/6A/article.Phtml

Chehma A, Djebar MR, Hadjaiji F, Rouabeh (2005) Etude floristique spatiotemporelle des parcours sahariens du Sud-Est algérien. Rev Sécheresse 16(4):275–285. http://www.jle.com/fr/revues/agro_biotech/sec/e-docs/00/04/14/CB/article.md

Chehma A, Bouzegag I, Chehma Y (2008) Productivité de la phytomasse éphémère des parcours camelins du Sahara septentrional algérien. Fourrages 194:253–256

Chehma A, Faye B, Bastianelli D (2010) Valeurs nutritionnelles des plantes vivaces des parcours sahariens algériens pour dromadaires. Fourrages 204(2010):263–268. http://www.afpf-asso.org/index/action/page/id/33/title/Les-articles/article/1818

Chehma A, Mahma H, Trabelsi H, Huguenin J (2023) Place of the dromedary in the ecological balance of its Saharan ecosystem. In: The 10th international seminar of veterinary medicine: camelids in Algeria & Maghreb. https://doi.org/10.3390/blsf2023022007

Chilliard Y, Bengoumi M, Delavaud C, Faulconnier Y, Faye B (2005) Body lipids and adaptation of camel to food and water shortage: new data on adipocyte size and plasma leptin. FAO AGRIS—International System for Agricultural Science and Technology, pp 135–145. https://agritrop.cirad.fr/525435/

Conca W, Salih S, Al-Rabiah R, Parhar R, Abd-Elnaeim M, Al-Hindas H, Tinson A, Collison K, Kishore U, Al-Mohanna F (2023) The immunoglobulin A isotype of the Arabian camel (*Camelus dromedarius*) preserves the dualistic structure of unconventional single-domain and canonical heavy chains. bioRxiv. https://doi.org/10.1101/2023.09.04.556198

Dagg AI (1974) The locomotion of the camel (*Camelus dromedarius*). J Zool 174(1):67–78. https://doi.org/10.1111/J.1469-7998.1974.TB03144.X

De Queiroz T, Meyer SE (2023) Long-term seed bank persistence in a stochastic desert environment. Ecol Evol 13(3):e9924. https://doi.org/10.1002/ece3.9924

Dittmann MT, Runge U, Lang RA, Moser D, Galeffi C, Kreuzer M, Clauss M (2014) Methane emission by camelids. PLoS One 9(4). https://doi.org/10.1371/JOURNAL.PONE.0094363

Djenane D, Aider M (2024) The one-humped camel: the animal of future, potential alternative red meat, technological suitability and future perspectives. F1000Research. https://doi.org/10.12688/f1000research.125246.2

Dowidar YA, Zeidan AEB, Badr MR, Mohammed WS, Shehab A (2022) Evolutionary relationship between Arabian camel breeds based on sequencing of heat-stress genes under Egyptian conditions. Int J Vet Sci 11(4):533–538. https://doi.org/10.47278/journal.ijvs/2021.133

Dubief J (1963) Le climat du Sahara, vol 2. Mémoires de l'Institut de Recherches Sahariennes, Université d'Alger

El-Hady E, Behairy A, Goda NIA, Abdelbaset-Ismail A, Ahmed AS, Al-Doaiss AA, Abd El-Rahim IHA, Alshehri M, Aref M (2023) Comparative physiological, morphological, histological, and AQP2 immunohistochemical analysis of the Arabian camels (*Camelus dromedarius*) and oxen kidney: effects of adaptation to arid environments. Front Anim Sci 4:1078159. https://doi.org/10.3389/fanim.2023.1078159

El-Zeiny WT (2010) Some aspects of thermo-regulation in water deprived camels under the semi-arid conditions of the north-western coastal desert in Egypt. J Camel Pract Res 17(2):245–256. https://www.cabdirect.org/cabdirect/abstract/20113238304

Essamadi K, Bengoumi M, Tressol JC, Chacornac JP, Faye B (1998) Comparative relationship of plasma copper concentration and ceruloplasmin activity of camel and cow. Trends Comp Biochem Physiol 5:211–220

Etzion Z, Alfassi Z, Lavi N, Yagil R (1987) Halide concentrations in camel plasma in various states of hydration. Biol Trace Elem Res 12:411–418. https://doi.org/10.1007/BF02796697

Fatnassi M, Padalino B (2024) Behaviour: behavioural repertoire and behavioural needs of camels. Springer, pp 145–168. https://doi.org/10.1007/978-3-031-48600-5_7

Faye B (1997) Le guide de l'élevage du dromadaire. Ed. Sanofi, Libourne (France), 118 pp

Faye B (2011) Combating desertification: the added value of the camel farming. Ann Arid Zone 50(3&4):1–11

Faye B (2015) Camelids and sustainable development. In: Kazakhstan A, Konuspayeva G (eds) Proceedings of 4th conference of ISOCARD, "Silk Road Camel: the Camelids, main stakes for sustainable development", June 8–12, 2015. Special Issue of Scientific and Practical Journal Veterinariya #2(42), Key-note presentation, pp 38–43

Faye B, Bengoumi M (2000) Le dromadaire face à la sous-nutrition minérale: un aspect méconnu de son adaptabilité aux conditions désertiques. Sécheresse 11(3):155–160

Faye B, Bengoumi M (2018) Camel clinical biochemistry and hematology, vol 48. Springer, Cham, pp 689–692. https://doi.org/10.1007/978-3-319-95562-9_1
Faye B, Tisserand JL (1989) Problèmes de la détermination de la valeur alimentaire des fourrages prélevés par le dromadaire. Séminaire sur la nutrition et l'alimentation du dromadaire, Ouargla, Algérie. Opt Méditerr Sér A Semin 2:61–65
Faye B, Waltner-Toews D, McDermott J (1999a) From ecopathology to agroecosystem health. Prev Vet Med 39(2):111–128. https://doi.org/10.1016/S0167-5877(98)00149-4
Faye B, Bengoumi M, Tressol JC (1999b) Comparative trace-element excretion in camel and cow. J Camel Pract Res 6:19–25
Faye B, Chaibou M, Vias G (2012) Integrated impact of climate change and socioeconomic development on the evolution of camel farming systems. Br J Environ Clim Change 2(3):227–244. https://doi.org/10.9734/BJECC/2012/1548
Fortané N (2023) Rise and fall of ecopathology: the thwarted development of a preventive approach to animal health (France, 1970s-1980s). In: From breeding & feeding to medicalization: animal farming, veterinarization and consumers in twentieth-century Western Europe, pp 337–360. https://doi.org/10.1484/M.RURHE-EB.5.135032
Futas J, Oppelt J, Jelinek A, Elbers JP, Wijacki J, Knoll A, Burger PA, Horin P (2019) Natural killer cell receptor genes in camels: another mammalian model. Front Genet 10:620. https://doi.org/10.3389/FGENE.2019.00620
Gebreyohanes M, Assen A (2017) Adaptation mechanisms of camels (*Camelus dromedarius*) for desert environment: a review. J Vet Sci Technol 8(6):1–5. https://doi.org/10.4172/2157-7579.1000486
Gil Sánchez JM, Iglesias Pastrana C, Velado Alonso E, Fresno Baquero MDR, Delgado Bermejo JV, Camacho Vallejo ME (2025) The ecological landscape of camel farming: impacts on wild animal diversity in the Canary Islands. CABI Agric Biosci 6(1):0069
Hadi YA (2016) A study on the strain and function of one-humped dromedary footpad. J Biotechnol Biomater. https://doi.org/10.4172/2155-952X.C1.059
Hasi S, Amu G, Zhang W (2020) Camel hair structure, properties, and commercial products. In: AlHaj O, Faye B, Agrawal RD (eds) Handbook of research on health and environmental benefits of camel products. IGI Global, Hershey, pp 328–347
Hoter A, Rizk S, Naim HY (2019) Cellular and molecular adaptation of Arabian camel to heat stress. Front Genet 10:588. https://doi.org/10.3389/FGENE.2019.00588
Janzen DH (1983) Dispersal of seeds by vertebrate guts. In: Futuyma DJ, Slatkin M (eds) Coevolution. Sinauer Associates, pp 232–262
Kebir N, Berber N, Zahzeh M (2024) Anatomical and physiological properties of the dromedary: a potential sustainability alternative and a vital asset in the era of climate change. J Anim Behav Biometeorol 12(4):2024031. https://doi.org/10.31893/jabb.2024031
Kouadri A, Trabelsi H, Chehma A (2024) Effet du passage des graines de *Retama raetam* (Fabaceae) par le tractus digestif du dromadaire sur leur propriétés morpho-physiologiques. Rev Bioressour 15(spécial):19–21
Le Houérou HN (1990) The grazing land ecosystems of the African Sahel. Springer, Berlin
Lin P, Gillard B, Pauza AG, Iraizoz FA, Ali MA, Mecawi AS, Djazouli Alim FZ, Romanova EV, Burger PA, Greenwood MP, Adem A, Murphy D (2022) Transcriptomic plasticity of the hypothalamic osmoregulatory control centre of the Arabian dromedary camel. Commun Biol 5(1):1008. https://doi.org/10.1038/s42003-022-03857-0
Louhaichi M, Gamoun M, Hassan S, Abdallah MA (2021) Characterizing biomass yield and nutritional value of selected indigenous range species from arid Tunisia. Plants 10(10):2031. https://doi.org/10.3390/plants10102031
Madouh TA (2022) Eco-physiological responses of native desert plant species to drought and nutritional levels: case of Kuwait. Front Environ Sci 10:785517. https://doi.org/10.3389/fenvs.2022.785517
Mahma H, Chehma A, Huguenin J (2019) Etude du comportement alimentaire journalier du dromadaire (*Camelus dromedarius*) dans son milieu naturel. Fourrages 240:341–347

Malika B, Didi OEHM, Hadjer A, Aïssa A (2019) Diversity study of the spontaneous flora in Touggourt area (Sahara Septentrional Algerian). Int J Biosci 14(2):350–360. https://doi.org/10.12692/ijb/14.2.350-360

Mallem K, Chehma A, Huguenin j. (2024) Activités alimentaires journalières de la chamelle gestante dans son milieu naturel. Rev Elev Med Vet Pays Trop 2024(77):37447. https://revues.cirad.fr/index.php/REMVT/article/view/37447/37623

Mohamedien D, Mohamed SA, Salem A, El-Bab MF (2022) Mapping the distribution of special structured pulmonary vasculatures in the dromedary camel (*Camelus dromedarius*): histological and histochemical study. Microsc Microanal 28:858–868. https://doi.org/10.1017/S1431927622000381

Mohammed AE-NA, Almuyidi A, Almarri HA, Alkhalifah H, Alhmad A, ElBeshbishy HA, AlAli KH, AlHuwaish O, AlKhawaher M (2025) Unique characteristics of camel body systems: adaptation to harsh conditions, productive and reproductive performances: a review. Indian J Anim Res. https://doi.org/10.18805/ijar.bf-1922

Mokhtar MM, El-Younis AA, Azamel AA (1989) Physiological responses in camels, sheep and goats under hydrated and dehydrated conditions. In: Galal ESE, Aboul-Ela MB, Shafie MM (eds) Proceedings of the international symposium on the constraints and possibilities of ruminant production in the dry subtropics. EAAP Publication (Netherlands), issue 38, pp 193–195, ISBN 90-220-0949-1

Nagy PP, Skidmore JA, Juhasz J (2022) Intensification of camel farming and milk production with special emphasis on animal health, welfare, and the biotechnology of reproduction. Anim Front 12(4):35–45

Neffati M (2008) Domestication des plantes spontanées autochtones à usages multiples en zones arides et désertiques. Guide pratique de collecte, de conditionnement et de germination de leurs semences. Institut des Régions Arides, Médenine, Tunisie

Omidi A, Nazifi S, Rasekh M, Zare N (2024) Heat-shock proteins, oxidative stress, and antioxidants in one-humped camels. Trop Anim Health Prod 56(1):29. https://doi.org/10.1007/s11250-023-03876-x

Ouajd S, Kamel B (2009) Physiological particularities of dromedary (*Camelus dromedarius*) and experimental implications. Scand J Lab Anim Sci 36(1):19–29. https://doi.org/10.23675/sjlas.v36i1.165

Ozenda P (1991) Flore et végétation du Sahara (3e édition mise à jour et augmentée). Éditions du Centre National de la Recherche Scientifique, Paris

Peguero-Pina JJ, Vilagrosa A, Alonso-Forn D, Ferrio JP, Sancho-Knapik D, Gil-Pelegrín E (2020) Living in drylands: functional adaptations of trees and shrubs to cope with high temperatures and water scarcity. Forests 11(10):1028. https://doi.org/10.3390/f11101028

Perk K (1963) The camel's erythrocyte. Nature 200(4903):272–273. https://doi.org/10.1038/200272A0

Ridouh R, Tekkouk-Zemmouchi F, Guintard C (2023) Principal anatomy particularities in dromedary compared to ox: digestive and respiratory systems. Biol Life Sci Forum 22(1):13. https://doi.org/10.3390/blsf2023022013

Robertshaw D, Zine-Filali R (1995) Thermoregulation and water balance in the camel: a comparison with other ruminant species. In: von Engelhardt W, Leonhard-Marek S, Breves G, Giesecke D (eds) Proceedings 8th international symposium on ruminant physiology. Delmar Publishers, Albany, pp 563–578

Rutagwenda T, Lechner-Doll M, Schwartz HJ, Schultka W, Von Engelhardt W (1990) Dietary preference and degradability of forage on a semiarid thornbush savannah by indigenous ruminants, camels and donkeys. Anim Feed Sci Technol 31(3–4):179–192. https://doi.org/10.1016/0377-8401(90)90123-P

Shadomy SV, Ebrahim SH, Guagliardo SAJ, Sánchez-González L, Zureick K, Sinclair JR, Schneider DA, Walker AT, Payne DC, Vieira AR, Heitzinger K, Lenhart A, Oakley LP, Clemente J, Basler C, Beard CB, Armstrong PA, Burke H (2025) Emerging and re-emerging disease threats in the Middle East and North Africa region—One Health approaches and potential strategies. Eur J Pub Health 35(Suppl_1):i3–i5. https://doi.org/10.1093/eurpub/ckae122

Slimani N, Chehma A, Faye B, Huguenin J (2013) Diet and feeding behavior of dromedaries in their natural desert habitat in Algeria. Livest Res Rural Dev 25:Article #213. Retrieved December 1, 2013, from http://www.lrrd.org/lrrd25/12/slim25213.htm

Suleiman MK, Bhatt A, Jacob S, Thomas RR, Sivadasan MT (2023) Seed longevity in desert species and the possibility of forming a persistent soil seed bank. Sustainability 15(22):15904. https://doi.org/10.3390/su152215904

Towhidi A, Saberifar T, Dirandeh E (2011) Nutritive value of some herbage for dromedary camels in the central arid zone of Iran. Trop Anim Health Prod 43(3):617–622. https://doi.org/10.1007/S11250-010-9741-9

Trabelsi H, Senoussi A, Chehma A, Faye B (2012) The contribution of the dromedary in the spontaneous plant seeds transfer in the Northern Algerian Sahara. J Life Sci 6(3):300–303. http://www.davidpublishing.com

Trabelsi H, Senoussi A, Chehma A, Faye B, Mohamed Elhafed Kherraze MA (2023) Camel potentiality in survival and germination of wild pastoral species: the case of Fabaceae in Sahara rangelands of Algeria. J Arid Environ 216:105015. https://www.sciencedirect.com/journal/journal-of-arid-environments/vol/216/suppl/C

Wilson MF (1992) The ecology of seed dispersal. In: Fenner M (ed) Seed: the ecology of regeneration in plant communities. CAB International, Wallingford, pp 61–85

Wilson RT (2012) Ecophysiology of the Camelidae and desert ruminants. Springer, Berlin

Windberger U, Auer R, Plasenzotti R, Eloff S, Skidmore JA (2018) Temperature dependency of whole blood viscosity and red cell properties in desert ungulates: studies on scimitar-horned oryx and dromedary camel. Clin Hemorheol Microcirc 69(4):533–543. https://doi.org/10.3233/CH-189204

Zhou C, Wang K (2016) Land surface temperature over global deserts: means, variability, and trends. J Geophys Res Atmos 121(24):14–344. https://doi.org/10.1002/2016JD025410

Camels in a Changing Climate

9

Pauline Njoki Gitonga, Oscar Kipchirchir Koech,
and Dana Hyláková

Abstract

Climate change is transforming the ecological and socioeconomic systems that support camel-based livelihoods. This chapter analyses the direct and indirect impacts of climate change on camels (*Camelus dromedarius* and *Camelus bactrianus*) and their habitats across Africa, Asia, Australia, and Europe. Direct impacts include rising temperatures, erratic rainfall patterns, recurrent droughts, range degradation, and the emergence of vector-borne and zoonotic diseases. These biophysical stressors undermine the camels' inherent resilience and adaptability, negatively affecting their health, productivity, and mobility. Indirect impacts arise from socioeconomic transformations, including species substitution from cattle to camels, changes in pastoral mobility, peri-urbanization, and the commercialization of camel products. Gender roles are also evolving, with women taking on more roles in marketing camel products such as milk and meat, while youth serve as intermediaries who link markets. Global, regional, and national policy frameworks are gradually recognizing the strategic importance of

P. N. Gitonga (✉)
Centre for Epidemiological Modelling and Analysis (CEMA), The University of Nairobi, Nairobi, Kenya

Department of Biomedical Sciences, Colorado State University (CSU), Fort Collins, CO, USA

ASAL eXtension, Nairobi, Kenya

O. K. Koech
Department of Land Resource Management and Agricultural Technology (LARMAT), The University of Nairobi, Nairobi, Kenya
e-mail: oscarkip@uonbi.ac.ke

D. Hyláková
Veterinary Clinic Vyvet Vyškov, Vyškov, Czech Republic

Czech Association of Zoo and Wildlife Veterinarians (CAZWV), Brno, Czech Republic

S. A. Gordine et al. (eds.), *Camels and One Health*,
https://doi.org/10.1007/978-981-92-2246-9_9

camels for economic growth; however, significant gaps remain in research, value chain integration, and transboundary resource governance.

The chapter highlights a growing array of adaptation and resilience strategies that blend indigenous knowledge with modern science, including selective breeding, rangeland restoration, digital advisory platforms, and gender-inclusive cooperatives. These initiatives position camels as both vulnerable sentinels and vital actors in climate-impacted ecosystems. Securing the future of dryland livelihoods demands a paradigm shift that integrates camel systems into climate, food security, and biodiversity policies, underpinned by strengthened evidence-based research and participatory rangeland governance.

Keywor ds

Climate change · Camels (*Camelus dromedarius*, *Camelus bactrianus*) · Pastoralism · Dryland ecosystems · Rangeland degradation · Adaptation strategies · Resilience · Food security · One Health

9.1 Direct Impacts: Ecological and Health Consequences

Global mean surface temperatures have risen by approximately 1.1 degrees centigrade (°C) since the preindustrial period (1850–1900), with arid zones warming at a significantly faster rate than the global average (Mukherji 2023). In major camel-rearing regions such as North Africa, the Arabian Peninsula, and South Asia, summer temperatures now routinely exceed 45 °C (Malik et al. 2023). These escalating temperatures, combined with erratic rainfall and prolonged droughts, are accelerating aridification and transforming the ecological suitability of traditional rangelands (Vicente-Serrano et al. 2024). Such biophysical stressors not only push the camel's physiological limits but also fundamentally reshape the disease ecology (Kebir et al. 2024a) and the habitat stability of dryland ecosystems (Polley et al. 2017). This section analyses the direct impacts of climate change on camels (*Camelus dromedarius and Camelus bactrianus*) and their habitats worldwide.

9.1.1 Expanding Camel Range

Climate change is reshaping the ecological and socioeconomic systems that support camel-based livelihoods (Jassim and Sejian 2015). Camels are among the most climate-resilient domestic species; however, the accelerating pace of environmental change is beginning to exceed the thresholds of their physiological and ecological adaptability (Kebir et al. 2024b). These shifts are driving substantial changes in camel distribution, mobility, and population dynamics (Faye 2022). Rising temperatures, prolonged droughts, the loss of perennial water sources, and the expansion of aridification are transforming the geographical suitability of rangelands (Odongo et al. 2025; Adeel, Millennium Ecosystem Assessment, World Resources

Institute 2005). These changes are pushing camel herders to modify long-standing migration routes (Tefera and Abebe 2012). Historically confined to hot, arid zones, camel populations are now exhibiting both range contractions in degraded habitats and expansions into areas previously considered unsuitable (Faye 2022). Across North Africa, recurrent droughts have displaced camel herders southward into Sahelian zones (Sahoo 2020; Ashour et al. 2024). In Central Asia and northern India, warming winters have enabled Bactrian camels to move northward into higher latitudes (Faye 2024). Such shifts reflect not only environmental pressures but also emerging opportunities for camel husbandry in newly warming ecological niches. In East Africa and the Horn of Africa, camels have become a vital fallback species as cattle mortality increases under drought stress (Watson et al. 2016a; Bekele et al. 2025). This coping livestock husbandry strategy has led to the expansion of camel grazing territories into areas historically dominated by other livestock (Armara et al. 2025; Megersa et al. 2024). Globally, similar expansions are evident. Australia's wild camel population is approximately two million (Faye 2015; Hampton et al. 2016), and camel husbandry is reappearing in parts of Europe, where camels are valued for their drought resilience and ability to utilize marginal lands (Maitra et al. 2025).

Data from the Food and Agriculture Organization of the United Nations (FAO) indicate that the global camel population has nearly tripled since 1961, when it stood at approximately 13 million, reaching over 35 million by 2018 and an estimated 42 million by 2023 (Fig. 9.1), with projections suggesting it could reach 60 million by 2045 (FAO 2025a; Faye and Ratto 2022) if current trends persist. Africa accounts for more than 84% of the global dromedary camel population, with approximately 37 million animals. Countries such as Chad, Somalia, Sudan,

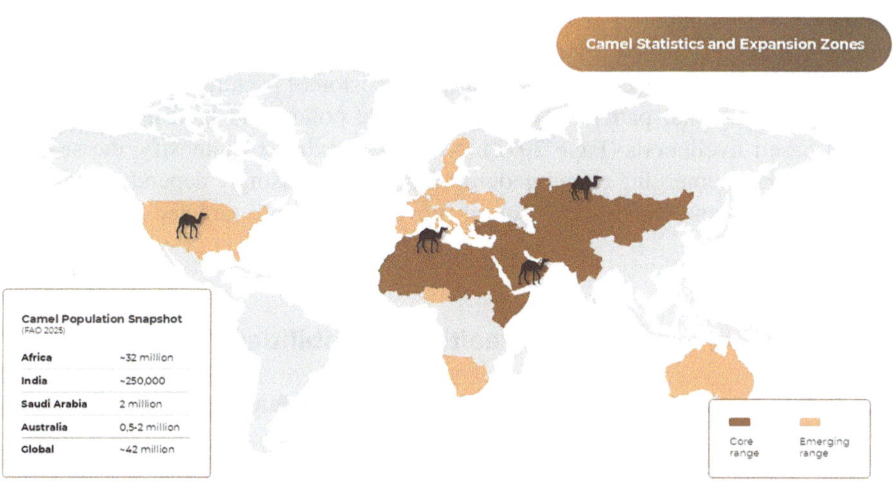

Fig. 9.1 A world map image showing the expansion pathways of camels that was adapted and designed by the authors using data from a FAO 2025 publication titled "Camelid biodiversity, population trends and geographic distribution". https://openknowledge.fao.org/handle/20.500.14283/cd5884en

and Kenya serve as major global strongholds for dromedary camels (Faye 2020). Despite global growth, some regions are experiencing a sharp decline in their dromedary camel populations. India's Rajasthan region, home to over 85% of India's camel population, has seen its camel population decline by more than 70% since the early 1990s. This is primarily due to non-climatic pressures such as shrinking grazing lands, restrictive forest and wildlife regulations, and legislative bans that have disrupted traditional pastoral trade systems (Choudhary 2019; Köhler-Rollefson and Rathore 2021). These examples reveal that climate change interacts with policy, land-use change, and socioeconomic dynamics in complex and uneven ways.

The expansion of camels into new marginal rangelands represents an important ecological shift. Although camels are primarily browsers, their introduction can still alter vegetation structure and modify plant community dynamics (Faye 2024; Gallacher and Hill 2006; Kuria et al. 2012). Selective browsing may reduce the regeneration of particular woody species (Box et al. 2016). These ecosystem shifts, even without heavy grazing pressure, may create conditions that favor opportunistic or invasive plants (Makoni 2020; Rojas-Sandoval and Acevedo-Rodríguez 2013). As a result, camel range expansion into fragile rangelands presents new management and conservation challenges that require careful monitoring. At the socio-ecological level, shifting migration patterns are straining customary land-tenure arrangements. Traditional nomadic routes based on predictable rainfall gradients are increasingly being disrupted by erratic precipitation, forcing herders to adopt opportunistic movement strategies. In regions such as the Horn of Africa and parts of Pakistan, camel caravans now travel longer distances in search of viable pasture, contributing to higher transit mortality and, at times, heightened conflict with sedentary crop farming communities (Faye 2015, 2024). The global picture of overall camel population growth reflects its remarkable resilience in a warming world. Yet this expansion often masks underlying vulnerabilities, such as ecological fragility in newly occupied territories, the erosion of pastoral mobility systems, and uneven regional patterns of decline, where policy and land use undermine camel-based livelihoods (Faye 2012). As climate extremes intensify, the sustainability of these emerging camel geographies will increasingly depend on adaptive land management, inclusive pastoral governance, and targeted conservation strategies.

9.1.2 Disease Ecology and Health Vulnerabilities

9.1.2.1 Climate-Driven Health Vulnerabilities in Natural Camel Ranges

Climate change is reshaping the health landscape of camel populations by intensifying heat exposure, altering rainfall patterns, and transforming the ecology of pathogens and vectors (Phillips 2025; Bona et al. 2022). Although dromedaries have the most advanced physiological adaptations to hot, arid environments, the magnitude and pace of current climate extremes are increasingly pushing these adaptive limits

(Tibary and El Allali 2020). The resulting vulnerabilities impact their productivity, immunity, and welfare. Camels are renowned for exceptional thermoregulation, tolerating body temperature fluctuations of between 34 and 41 °C and withstanding dehydration losses of up to 25% of their body weight. However, persistent heatwaves and chronic water scarcity undermine these mechanisms. Elevated temperatures increase evaporative water loss, depress feed intake, disrupt rumen function, and impair reproductive processes. Recent field studies in East Africa and the Arabian Peninsula show reductions in milk yield, conception rates, and calf survival during prolonged heat-stress events. Heat stress is accompanied by elevated cortisol and markers of oxidative stress (Tibary and El Allali 2020; Kandil et al. 2023). The data suggest that the camel's adaptive envelope is narrowing as thermal thresholds become more frequent and longer.

As temperatures rise and rainfall patterns shift, the ecology of camel diseases is modifying pathogen survival, altering the distribution and seasonality of vectors, and weakening the host immunity (Megersa et al. 2024). These three forces together amplify disease risk (Mekonnen 2018; Tidman et al. 2021). The warming world supports longer and more favorable activity periods for mosquitoes (Obolski et al. 2019), ticks (Nuttall 2022; Moreira et al. 2020; Cunze et al. 2022), and biting flies (Chalchisa et al. 2024), enabling pathogens to proliferate in places and seasons where they were once rare. At the same time, camels weakened by heat stress, malnutrition, or long migratory journeys become more susceptible to infections they once resisted (Kandeel and Al-Mubarak 2022; Othieno et al. 2022). Rift Valley fever (RVF) is perhaps the clearest illustration of a climate-sensitive camel disease. RVF outbreaks coincide with periods of unusually high rainfall that drive explosive *Aedes* mosquito populations. The 1997–1998 El Niño event in East Africa caused devastating RVF outbreaks marked by widespread camel mortality, catastrophic abortion storms, and severe human illness (Martin et al. 2008; Anyamba et al. 2001, 2002). Similar climatic anomalies have triggered outbreaks in Sudan, Saudi Arabia, and Mauritania, underscoring the inextricable link between environmental variability and zoonotic risk (Hassan and Jamil 2025; Khalafalla 2017, 2023). Middle East Respiratory Syndrome Coronavirus (MERS-CoV) further illustrates the complex interface between climate-driven camel expansion and zoonotic risk. While the virus is widespread in dromedary populations across Africa and the Middle East, an epidemiological paradox exists. Human cases remain primarily in the Arabian Peninsula despite high seroprevalence in African camels (Ogoti et al. 2024). This has been attributed to differences in camel-to-human transmission dynamics and the genetic diversity of viral clades across regions (Bowen et al. 2025; Ogoti et al. 2026; Zhou et al. 2023). As climate stress drives camels into closer proximity with human settlements and other livestock, the potential for viral evolution and cross-species spillover increases, necessitating a One Health approach to surveillance. Tick-borne diseases are expanding as well. Research from Kenya and Somalia shows that warming temperatures and sporadic surges in humidity have pushed *Hyalomma* ticks into new ecological zones and extended their active seasons (Makwatta et al. 2024; Alanazi et al. 2020; Getange et al. 2021). This shift has increased cases of Ehrlichiosis, a Rickettsial bacterial disease caused by *Ehrlichia species* (Collins et al. 2022). Some *Ehrlichia* pathogens, such as *Ehrlichia ruminantium,* the causative agent of heartwater disease, transmitted by *Amblyomma* ticks, can produce severe clinical disease, resulting in fever, respiratory distress, neurological symptoms, and, in severe cases, sudden death (Younan et al. 2021). Others, however, establish silent subclinical infections in camels, causing no obvious disease in the animal yet posing a significant risk of transmission to humans, particularly among herders and abattoir workers (Nuttall 2022).

Trypanosomiasis (Surra), spread by biting flies, is also intensifying as its vectors move into newly favorable habitats, turning marginal pastoral zones into disease hotspots (Molefe-Nyembe 2024; Meisner et al. 2022; Desquesnes et al. 2013). Heat stress and nutritional deficiency erode the camel's immune defenses, heightening vulnerability to bacterial and fungal diseases such as camel tuberculosis, pasteurellosis, dermatophytosis, and brucellosis (Khalafalla 2023; Rodriguez-Morales et al. 2021). The detection of a prion disease in Algerian dromedaries (Babelhadj et al. 2018) is a worrying discovery that raises urgent questions about how environmental stress, altered grazing behavior, and long-distance movements might expose camel populations to new pathogens or reveal previously unnoticed ones. Building a resilient future for camel-based livelihoods will require innovative and adaptive health systems that match the complexity of the challenges ahead (Watson et al. 2016b). The deployment of mobile veterinary units, equipped with multiplex rapid diagnostic platforms, will enable the real-time detection of co-circulating pathogens in herds traversing increasingly fragmented and distant grazing routes (Narayan et al. 2025). Climate-smart rangeland management and pasture restoration can help stabilize forage systems (Tulu et al. 2023; Nyongesa et al. 2023). Breeding programs that focus on heat- and disease-resilient camel ecotypes can drive long-term adaptability (Faye 2015; Gaouar et al. 2026; FAO 2025b). Community-led training in disease recognition, early reporting, and food hygiene practices can prevent large-scale outbreaks (Gitonga et al. 2020). Finally, the development of welfare standards tailored to mobile pastoralism can ensure that resilience is measured not only by survival but by the well-being of both animals and people (Padalino et al. 2024). As climate extremes accelerate, the fate of camel populations will depend on our capacity to weave together traditional pastoral knowledge with scientific innovation. In doing so, we can forge health systems that honor the camel's remarkable evolutionary legacy while strengthening the resilience of the communities that have depended on the camel for millennia.

9.1.2.2 Ex Situ Climate Adaptation and Its Implications for Camel Health and Welfare

The preceding discussion has focused on camels within their traditional arid ranges. However, climate-related health and welfare challenges also emerge when camels are kept ex situ, that is, in environments far removed from those that shaped their evolutionary adaptations. A case study contributed by co-author Dana Hyláková, based on her extensive experience breeding and managing Bactrian and dromedary camels in zoos and specialized breeding facilities in the Czech Republic and Slovakia, offers a compelling illustration of how contrasting climatic conditions can

Fig. 9.2 A photograph of a group of Bactrian camels (*Camelus bactrianus*) showing their dense winter coats, which are an adaptive morphology for temperate climates. (Photo credit: Dana Hyláková)

shape camel health and husbandry outcomes. In these settings, seasonal temperature fluctuations differ markedly from those of the deserts and steppes to which camels are naturally adapted (Maitra et al. 2025). These differences affect thermoregulation, nutritional balance, and disease susceptibility, revealing how limitations in housing design or insufficient thermal support can exacerbate musculoskeletal disorders, particularly in dromedaries, whose adaptations to subtropical heat do not translate well to long, cold Central European winters. Thus, ex situ camel management highlights the role of climate in shaping welfare outcomes, even for species renowned for their hardiness. The Bactrian camel, forged by the climatic extremes of Central Asia, possesses a specialized physiological toolkit for managing the thermal variability characteristic of Central Europe. Its robust frame and dense, insulating winter coat (Fig. 9.2) serve as a highly efficient evolutionary adaptation to environments, where temperatures regularly oscillate from intense summer heat to winter lows of −30 °C or below. In zoo park facilities across the Czech Republic and Slovakia, these animals demonstrate remarkable acclimatization, often thriving outdoors year-round, provided that basic shelters offer protection from strong winds and heavy precipitation. Because their natural physiology aligns so closely with the temperate zone, ex situ husbandry for Bactrians remains comparatively straightforward.

The same cannot be said for the dromedary. The one-humped camel has a smaller frame, sparse hair, and the ability to retain heat during the day, all of which are remarkable advantages in hot, arid environments. These adaptations become a

Fig. 9.3 A photograph of a climate-responsive facility design for dromedary camels (*Camelus dromedarius*). (Photo credit: Dana Hyláková)

physiological liability when ambient temperatures drop for extended periods. In Central Europe, winters stretch for months, and rain, low temperatures, and high humidity often mark spring and autumn. Over time, exposure to cold and damp conditions places sustained stress on the dromedary's musculoskeletal system. The inability to generate and retain adequate body heat forces behavioral adjustments that indirectly strain joints and connective tissue. In Czech and Slovak zoo facilities, keepers frequently report chronic arthritis and arthrosis in dromedaries kept without sufficient thermal support. These musculoskeletal conditions are rarely observed in Bactrian camels living under the same management. Once these disorders develop, they compromise mobility, reduce feeding efficiency, and may affect reproductive success. Dromedary calves raised in cold housing without adequate protection show slower growth and are more prone to illness. These health disparities highlight the importance of climate-responsive facility design. Dromedaries require deliberate thermal engineering, such as insulated and heated stables (Fig. 9.3), and dry bedding, such as sand, straw, or wood shavings (Fig. 9.4). The beddings help prevent conductive heat loss through the limbs as well as reduce the risk of pressure sores or slips.

Observational experience from several facilities shows that dromedaries readily choose heated indoor spaces during autumn and winter, often spending long periods

Fig. 9.4 A photograph showing the use of dry sandy bedding to enhance dromedary welfare at ZOOPARK Vyškov. (Photo credit: Dana Hyláková)

resting inside. In contrast, Bactrians continue to move comfortably between indoor and outdoor areas. The lessons emerging from Czech and Slovak Zoo and breeding facilities extend far beyond Central Europe. They illustrate a broader truth relevant to zoos, conservation centers, and ex situ breeding programs worldwide. Evolutionary adaptation sets fundamental thermal boundaries for each species, and exceeding those boundaries has cascading effects on health. Camels may be icons of resilience, but their resilience is not limitless. Welfare, disease susceptibility, and even reproductive success hinge on whether the constructed environment respects these evolutionary constraints. Beyond the direct challenges of thermoregulation, ex situ camel husbandry in temperate regions faces significant nutritional hurdles. Unlike the sparse, high-fiber vegetation of arid rangelands, the natural forage and supplementary feeds available in Europe are often characterized by a high concentration of protein and energy. This nutritional surplus, combined with reduced activity levels in confined or cold environments, predisposes camels to metabolic diseases such as obesity, insulin-resistant diabetes, and laminitis, which are rarely encountered in traditional desert systems (Hamid et al. 2026; Nagy et al. 2022). Consequently, the welfare of ex situ camels hinges not only on thermal engineering but also on the strict management of caloric intake to mirror their evolutionary adaptation to nutrient-poor environments.

9.1.3 Vegetation Dynamics and Rangeland Degradation

Rangeland vegetation is the ecological foundation of camel production systems, yet these landscapes are undergoing profound change due to climate stress. Rising temperatures and erratic rainfall have been linked to shifts in plant community composition, favoring drought-tolerant shrubs and unpalatable species at the expense of nutritious forbs and grasses (Polley et al. 2017). Invasive species such as *Prosopis juliflora* in East Africa (Eshetu 2024; Boy and Witt 2013) and *Calotropis procera* in South Asia (Gonçalves-Oliveira et al. 2022) have expanded rapidly, displacing native browse plants. While camels can consume certain invasive plants, overreliance on them reduces dietary diversity and may cause physical injuries, such as mouth sores from long thorns, and the accumulation of phytotoxins (Gebeyew et al. 2020), which are natural plant poisons that can damage a camel's liver or kidneys over time. While the rangeland may look green, the lack of plant diversity means camels are forced to eat plants that are low in essential nutrients. Environmentally, these invaders act like "water thieves," using deep taproots to suck up limited groundwater, which kills off the diverse local grasses that other animals and the soil need to stay healthy. Vegetation monitoring in Kenya reveals progressive bush encroachment and a decline in herbaceous cover over the past two decades (Kipkulei et al. 2025). These processes reduce carrying capacity and hinder herd mobility. Furthermore, the degradation of rangelands diminishes soil carbon sequestration, exacerbating local warming and erosion. The resulting feedback loop, where climate change accelerates degradation and degradation amplifies climate impacts, threatens the long-term sustainability of camel-based livelihoods. Remote sensing analyses corroborate these field observations, showing consistent declines in the Normalized Difference Vegetation Index (NDVI) across major camel belts (Xie et al. 2024). In the Sahel, NDVI reductions of 10–20% between 2000 and 2020 correspond to declines in the productivity of key forage plants such as *Acacia tortilis* (umbrella thorn acacia) and *Atriplex halimus* (Mediterranean saltbush) (Xie et al. 2024). This has forced herders to diversify grazing routes, often encroaching on wildlife reserves or agricultural zones, increasing the potential for conflict (Asongo et al. 2026) and disease transmission (Faye et al. 2012a).

9.2 Indirect Impacts: Socioeconomic Transformations

The previous sections examined the biological and environmental pressures on camels; the following discussion focuses on the human dimension of these ecosystems. The socioeconomic stability of these regions relies on the adaptive capacity of the pastoralists themselves. Rather than merely surviving, camel herders have developed proactive risk-management strategies to navigate intensifying climate stressors. As demonstrated in the Saharan and Sahelian regions, these adaptations include the spatial reorganization of herds into smaller, family-managed units (Correra et al. 2008), the massive expansion of traditional grazing ranges, the adoption of supplemental feeding such as wheat or millet to buffer against forage scarcity, and the diversification of household income into non-pastoral sectors like fishing or tourism (Jones and Thornton 2009).

9.2.1 Adaptive Livelihood Transitions

Climate change is reshaping not only the biological and ecological dimensions of camel production but also the socioeconomic systems in which camel-based livelihoods operate. As crop production in many semiarid regions becomes increasingly unreliable due to more frequent heat stress, drought, and rainfall variability. Rural households are shifting their livelihood focus toward livestock as a more dependable buffer against climate shocks. Analyses by Jones and Thornton (2009) indicate that large portions of sub-Saharan Africa will cross critical thresholds of crop viability by 2050, creating extensive "transition zones," where farmers may be compelled to abandon marginal cropping and rely more heavily on livestock for food security and income (Jones and Thornton 2009). In this context, climate change heightens the strategic importance of camel production, particularly in regions where resource scarcity, socioeconomic vulnerability, and constrained adaptation options converge.

9.2.2 Species Substitution and Pastoral Adaptation

Traditional rangelands have become less reliable due to recurrent droughts, which cause pasture failures and water scarcity. This is undermining the viability of cattle and small ruminant production systems and driving pastoral communities into adaptive transformations that extend far beyond husbandry practices. These indirect impacts influence species composition, labor roles, settlement patterns, market structures, and governance frameworks, collectively redefining the future of camel-dependent economies (Faye et al. 2012a). These transitions are visible across East Africa, the Horn of Africa, and arid parts of South Asia, where camels increasingly serve as a core asset for milk production, income generation, and household nutrition (Rahimi et al. 2022). This adaptive trend is equally pronounced across West Africa and the Sahel. In Mali, the dromedary is now recognized as a primary contributor to climate resilience, often replacing more vulnerable livestock species as pastoralists seek to stabilize food security under erratic rainfall patterns (Ibrahima et al. 2020). Similarly, in Niger, the combined impact of climatic and socioeconomic shifts has triggered a fundamental evolution in farming systems to adopt camel keeping as a modern pastoral survival strategy (Faye et al. 2013). These shifts represent more than a simple substitution of species; they signal deeper changes in economic strategies, risk management, and intergenerational knowledge. The decision to invest in camels requires new veterinary competencies, innovative grazing strategies, and greater engagement with emerging markets. In some regions, this transition has even stimulated demand for improved breeding stock and extension services tailored specifically to camel health and productivity. Thus, livelihood adaptation under climate stress is not merely reactive; it is a proactive restructuring of pastoral production systems to align with evolving environmental realities (Faye et al. 2012a).

9.2.3 Mobility Reconfiguration and Emerging Peri-Urban Interfaces

Pastoral mobility has long been the principal mechanism through which camel herders manage climatic variability (Faye 2018). However, shifting climatic patterns, resource degradation, conflict, and expanding land fragmentation are prompting a substantial reconfiguration of movement patterns (Kipkulei et al. 2025). Some herders now travel longer distances in search of grazing, crossing ecological boundaries that historically remained separate. Others, constrained by encroachment, fenced land, or shrinking rangelands, increasingly settle near peri-urban centers. These emergent peri-urban interfaces introduce new risks and opportunities. On one hand, proximity to markets, veterinary services, and water points can enhance economic engagement. On the other hand, higher livestock densities, exposure to waste-contaminated substrates, and interaction with diverse species assemblages increase the risk of disease spillover, environmental contamination, and welfare challenges. Camels that once roamed arid plains now traverse corridors marked by roads, settlements, and urban sprawl, altering their traditional movement ecology. These changes not only reshape pastoral production but also force communities to negotiate hybrid identities neither fully nomadic nor fully settled, as they adapt to new ecological and socioeconomic landscapes (Faye 2024).

9.2.4 Market Integration, Value Chains, and Economic Diversification

Climate-driven shifts in production systems are occurring alongside the rapid commercialization of camel products. Rising demand for camel milk and meat, both domestically and in regional export markets, is catalyzing the expansion of value chains that extend from remote pastoral settlements to urban processors and consumers (Mahamat Ahmat et al. 2023). In many production hubs, camel milk has become a reliable and sometimes superior source of cash income compared to cattle, particularly during drought periods when other livestock underperform. Yet these value chains remain marked by fragmentation and structural inefficiencies. Limited cold-chain infrastructure, inadequate transport networks, and inconsistent quality-control systems constrain market growth. Spoilage increases during droughts when milk yields fluctuate, and seasonal surpluses often exceed the capacity of local processors. Where they exist, regulatory systems struggle to keep pace with the expansion of informal trade. At the same time, new economic actors are emerging. Youth often serve as transporters, brokers, and information intermediaries, while women, especially in East African camel hubs (Nori 2023), play pivotal roles in milk bulking, hygiene, and retail markets. These dynamics demonstrate how climate pressures can stimulate innovation, diversify income sources, and shift the socioeconomic architecture of camel production.

9.2.5 Social Differentiation in Camel Economies: Gender and Generational Dynamics

Climate-induced changes in camel production are reshaping social relations within pastoral communities (Faye et al. 2012b). As camels gain prominence in household economies, women increasingly engage in camel milk handling, processing, and marketing, expanding their role in decision-making and income control. In some regions, women's cooperatives have become central to ensuring milk hygiene, organizing supply chains, and negotiating with traders and processors. This shift enhances women's economic visibility but also increases their labor burden, particularly during prolonged droughts when herds must be watered and milked under challenging conditions. Youth, meanwhile, are emerging as key agents of economic diversification. Many now operate as motorcycle milk couriers, digital service providers, or feed entrepreneurs (Gitonga 2017). These roles not only position them at the nexus of traditional pastoralism and modern market systems but also reflect broader generational aspirations. However, unequal access to capital, training, and decision-making structures can marginalize certain groups, highlighting the importance of socially inclusive adaptation strategies.

9.2.6 Governance Gaps in a Rapidly Transforming Camel Sector

Despite the growing economic and ecological importance of camels, policy and governance frameworks have been slow to adapt. National livestock policies in many countries continue to prioritise cattle, overlooking the unique veterinary, genetic, and market needs of camel systems. Research investments remain limited, leaving gaps in disease surveillance, breeding programs, and value-chain development. Transboundary issues such as shared grazing corridors, drought mobility, and cross-border disease transmission require coordinated regional governance mechanisms that are often absent or fragmented. The commercialization of camel milk and meat adds further complexity. Regulatory oversight of hygiene, processing standards, animal health certification, and product safety lags behind market expansion. Moreover, conflicts over land tenure, water access, and rangeland fragmentation constrain the mobility essential for climate-resilient camel husbandry. Without robust governance structures that integrate climate adaptation, animal health, and market development, pastoral households bear disproportionate risks. Yet the sector also presents significant opportunities. Effective policies rooted in participatory pastoral planning, improved veterinary infrastructure, harmonized rangeland governance, and targeted investment in camel value chains can help communities navigate climate variability while unlocking economic potential. Strengthening governance is therefore not only a matter of supporting camel-based livelihoods but also a critical component of building climate-resilient dryland economies (Faye 2024).

9.3 Conclusion

Climate change is reshaping both the ecological and socioeconomic foundations of camel-based systems. Rising temperatures, altered rainfall patterns, emerging diseases, and widespread rangeland degradation are eroding the environmental stability that once sustained camel mobility, nutrition, and health. Although camels remain among the most resilient domestic species, the pace of climate change increasingly pushes their physiological and ecological limits. Ex situ experiences from Central Europe further demonstrate that climate remains a fundamental determinant of camel health and welfare, reinforcing the importance of aligning husbandry systems with species-specific thermal and behavioral needs. At the same time, climate pressures are catalyzing far-reaching socioeconomic transitions. Camels are gaining prominence as cattle and small ruminants decline under drought stress, stimulating new livelihood strategies, shifting gender and generational roles, and accelerating the commercialization of camel products. Yet these opportunities unfold within governance systems that often lag, characterized by weak veterinary infrastructure, fragmented value chains, insufficient rangeland management, and policies that remain largely cattle-centric. Without supportive institutional frameworks, pastoral communities bear disproportionate climate risks despite their critical role in sustaining food security in dryland regions. Looking ahead, strengthening camel-based resilience will require integrated approaches that combine climate-smart rangeland management, one health disease surveillance, improved mobility, targeted investments in value-chain infrastructure, and land-governance policies. Research and innovation, especially in genetics, nutrition, welfare, and digital advisory tools, must reinforce, rather than replace, indigenous knowledge systems. If these efforts are aligned, camels can continue to serve as both vital contributors to dryland livelihoods and sentinels of environmental change, helping societies navigate an increasingly uncertain climate future.

References

Adeel Z, Millennium Ecosystem Assessment, World Resources Institute (eds) (2005) Ecosystems and human well-being: desertification synthesis; a report of the Millennium Ecosystem Assessment. World Resources Institute, Washington, DC, 26 p

Alanazi AD, Nguyen VL, Alyousif MS, Manoj RRS, Alouffi AS, Donato R et al (2020) Ticks and associated pathogens in camels (*Camelus dromedarius*) from Riyadh Province, Saudi Arabia. Parasit Vectors 13(1):110. https://doi.org/10.1186/s13071-020-3973-y. PubMed PMID: 32111251; PubMed Central PMCID: PMC7048137

Anyamba A, Linthicum K, Tucker C (2001) Climate-disease connections: Rift Valley fever in Kenya. Cadernos de saúde pública/Ministério da Saúde, Fundação Oswaldo Cruz, Escola Nacional de Saúde Pública 17(Suppl):133–140. https://doi.org/10.1590/S0102-311X2001000700022

Anyamba A, Linthicum K, Mahoney R, Tucker C, Kelley PW (2002) Mapping potential risk of Rift Valley fever outbreaks in African savannas using vegetation index time series data. Photogramm Eng Remote Sens 68:137–145

Armara G, Galwab AM, Koech OK, Galwab AM, Wasonga OV (2025) Determinants of livelihood choices in response to climate variability among the pastoral and agro-pastoral communities in Northern Kenya | Grassroots Journals. www.grassrootsjournals.org/ [Internet]; 2025 [cited 2025 July 9]. Available from https://grassrootsjournals.org/pp/03m00543.html#status

Ashour G, El-Bahrawy K, Abdel-Rahman S (2024) Climate change and camels: exploring adaptations to harsh conditions. Egypt J Camel Sci 2:63–72. https://doi.org/10.21608/ejcs.2025.336975.1023

Asongo T, Igbe JE, Ushie EA, Williams WO, Bisong PO (2026) Climate change: nomadic movement and the rise of rural banditry in Benue and Nasarawa States, Nigeria. Afr Secur 0(0):1–17. https://doi.org/10.1080/19392206.2026.2613176

Babelhadj B, Di Bari MA, Pirisinu L, Chiappini B, Gaouar SBS, Riccardi G et al (2018) Prion disease in dromedary camels, Algeria. Emerg Infect Dis 24(6):1029–1036. https://doi.org/10.3201/eid2406.172007. PubMed PMID: 29652245; PubMed Central PMCID: PMC6004840

Bekele B, Eshetu M, Wolkero T, Berhe T, Galmessa U, Gadissa S (2025) Perception of camel herders on climate change and variability in relation to camel production in Borana zone, Southern Ethiopia. Pastoralism 15:13986. https://doi.org/10.3389/past.2025.13986

Bona M, Blaňárová L, Stanko M, Mošanský L, Čepčeková E, Víchová B (2022) Impact of climate factors on the seasonal activity of ticks and temporal dynamics of tick-borne pathogens in an area with a large tick species diversity in Slovakia, Central Europe. Biologia 77(6):1619–1631. https://doi.org/10.1007/s11756-021-00902-x

Bowen RA, Hartwig A, Bruening A, Walker A, Peiris M (2025) Pathogenesis and transmissibility of Middle East respiratory syndrome coronaviruses of African origin in Alpacas. Viruses 17(11):1524. https://doi.org/10.3390/v17111524

Box JB, Nano CEM, McBurnie G, Waller DM, McConnell K, Brock C et al (2016) The impact of feral camels (*Camelus dromedarius*) on woody vegetation in arid Australia. Rangel J 38(2):181–190. https://doi.org/10.1071/RJ15073

Boy G, Witt A (2013) Invasive alien plants and their management in Africa. CABI Africa, Nairobi

Chalchisa A, Kumsa B, Gutema Wegi F (2024) Biting flies and associated pathogens in camels in Amibara District of Afar Region, Ethiopia. Vet Med Int 2024(1):5407898. https://doi.org/10.1155/2024/5407898

Choudhary M (2019) Changing area, shrinking spaces and struggling species: history of camels. Hist Sociol South Asia 13(2):93–111. https://doi.org/10.1177/22308075211043285

Collins M, Ngetich C, Owido M, Getange D, Harris R, Bargul J et al (2022) Detection of antibodies to *Ehrlichia* spp. in dromedary camels and co-grazing sheep in Northern Kenya using an *Ehrlichia ruminantium* polyclonal competitive ELISA. Microorganisms 10(5):916. https://doi.org/10.3390/microorganisms10050916

Correra A, Lefeuvre JC, Faye B (2008) Organisation spatiale et stratégie d'adaptation des nomades du parc national du Banc d'Arguin à la sécheresse. Sci Chang Planet Sécheresse 19(4):245–251. https://doi.org/10.1684/sec.2008.0150

Cunze S, Glock G, Kochmann J, Klimpel S (2022) Ticks on the move—climate change-induced range shifts of three tick species in Europe: current and future habitat suitability for Ixodes ricinus in comparison with Dermacentor reticulatus and Dermacentor marginatus. Parasitol Res 121(8):2241–2252. https://doi.org/10.1007/s00436-022-07556-x

Desquesnes M, Holzmuller P, Lai DH, Dargantes A, Lun ZR, Jittaplapong S (2013) Trypanosoma evansi and Surra: a review and perspectives on origin, history, distribution, taxonomy, morphology, hosts, and pathogenic effects. Biomed Res Int 2013:194176. https://doi.org/10.1155/2013/194176. PubMed PMID: 24024184; PubMed Central PMCID: PMC3760267

Eshetu AA (2024) A valuable or a curse resource? A systematic review on expansion, perception of local community, benefits and side effects of *Prosopis juliflora*. Front Conserv Sci 5:1491618. https://doi.org/10.3389/fcosc.2024.1491618

FAO (2025a) Camelid biodiversity, population trends and geographic distribution [Internet]. FAO; 2025 [cited 2025 Jul 6]. Available from https://openknowledge.fao.org/handle/20.500.14283/cd5884en

FAO (2025b) History of camelids [Internet]. 2025 [cited 2025 Jul 6]. FAO. Available from https://openknowledge.fao.org/handle/20.500.14283/cd5819en

Faye B (2012) Integrated impact of climate change and socioeconomic development on the evolution of camel farming systems. BJECC:227–244. https://doi.org/10.9734/BJECC/2012/1548
Faye B (2015) Role, distribution and perspective of camel breeding in the third millennium economies. Emir J Food Agric 27. https://doi.org/10.9755/ejfa.v27i4.19906
Faye B (2018) Recent advances in camelids biology, health and production. In: Proceedings of the 5th conference ISOCARD 2018 [conference_item] [Internet]. ISOCARD; 2018 [cited 2026 Mar 24]. What future for camel pastoralism in the world? Available from: https://agritrop.cirad.fr/591632/
Faye B (2020) How many large camelids in the world? A synthetic analysis of the world camel demographic changes. Pastoralism 10(1):25. https://doi.org/10.1186/s13570-020-00176-z
Faye B (2022) Is the camel conquering the world? Anim Front 12(4):8–16. https://doi.org/10.1093/af/vfac034
Faye B (2024) With camelids into a sustainable future: learning from pastoralist communities [book_section] [Internet]. Welt-Sichten; 2024 [cited 2025 Jul 10]. Camelid husbandry worldwide: its importance and evolving dynamics. Available from https://agritrop.cirad.fr/609442/
Faye B, Ratto MH (2022) Camelids: an old family spread over four continents. Anim Front 12(4):3–5. https://doi.org/10.1093/af/vfac045. PubMed PMID: null; PubMed Central PMCID: PMC9374518
Faye B, Chaibou M, Gilles V (2012a) Integrated impact of climate change and socioeconomic development on the evolution of camel farming systems. Br J Environ Clim Change 2:227–244. https://doi.org/10.9734/BJECC/2012/1548
Faye B, Chaibou M, Vias G (2012b) Integrated impact of climate change and socioeconomic development on the evolution of camel farming systems. Int J Environ Clim Chang 2(3):227–244. https://doi.org/10.9734/BJECC/2012/1548
Faye B, Vias-Franck G, Chaibou M (2013) autres repères autres paysages. Courrier de l'environnement de l'INRA n°63
Gallacher DJ, Hill JP (2006) Effects of camel grazing on the ecology of small perennial plants in the Dubai (UAE) inland desert. J Arid Environ 66(4):738–750. https://doi.org/10.1016/j.jaridenv.2005.12.007
Gaouar SBS, Meghelli I, Kaouadji Z, Meutchieye F, Gherissi DE (2026) African dromedary genetic resources, diversity and breeding systems. In: Ibeagha-Awemu EM, Peters SO, Djikeng A, Rege JEO (eds) African livestock genetic resources and sustainable breeding strategies: unlocking a treasure trove and guide for improved productivity [Internet]. Springer, Cham, pp 395–450. [cited 2025 Dec 9]. Available from https://doi.org/10.1007/978-3-031-92076-9_9
Gebeyew K, Abera B, Bajigo A, Gebresilassie G, Martínez Y, Adebowale T (2020) Indigenous medicinal uses, toxicity, and chemical composition of browsing plant used by camel in Ethiopia Somali Regional State: a survey. Trop Anim Health Prod 52(3):1459–1466. https://doi.org/10.1007/s11250-019-02152-1
Getange D, Bargul JL, Kanduma E, Collins M, Bodha B, Denge D et al (2021) Ticks and tick-borne pathogens associated with dromedary camels (*Camelus dromedarius*) in Northern Kenya. Microorganisms 9(7):7. https://doi.org/10.3390/microorganisms9071414
Gitonga P (2017) Camel milk value chain analysis and feasibility study_Wajir County, Kenya. supporting the responsible transfer and use of resources for the achievement of local economic development (LED) at county level
Gitonga PN, Bishop S, Watson C, Yigzaw A (2020) Livestock emergency guidelines and standards LEGS core standards and community-based animal health services [Internet]. 2020 [cited 2025 Oct 9]. Available from https://www.academia.edu/42974474/Livestock_Emergency_Guidelines_and_Standards_LEGS_Core_Standards_and_Community_Based_Animal_Health_Services
Gonçalves-Oliveira RC, Rodrigues HB, Benko-Iseppon AM (2022) Range distribution of the invasive alien species *Calotropis procera* in South America dry environments under climatic change scenarios. J Arid Environ 205:104819. https://doi.org/10.1016/j.jaridenv.2022.104819
Hamid MMA, Tharwat M, Ebeid TA, Alshanbari FA (2026) Nutritional disorders and metabolic adaptations in dromedary camels: insights into foregut fermentation and mineral balance.

Animals (Basel) 16(4):689. https://doi.org/10.3390/ani16040689. PubMed PMID: 41751150; PubMed Central PMCID: PMC12937329

Hampton J, Jones B, Perry A, Miller C, Hart Q (2016) Integrating animal welfare into wild herbivore management: lessons from the Australian Feral Camel Management Project. Rangel J 38(2):163–171. https://doi.org/10.1071/RJ15079

Hassan M, Jamil M (2025) The role of climate in burning the lines between human and animal diseases. In: Holistic health: one health in a changing world: climate, disease, policy, and innovation, pp 170–178. https://doi.org/10.47278/book.HH/2025.128

Ibrahima N, Aune J, Gry S, Harouna Y (2020) RÉPUBLIQUE DU MALI Un Peuple-Un But-Une Foi INSTITUT D'ÉCONOMIE RURALE Éditeurs scientifiques Adaptation de l'Agriculture et de l'Élevage au Changement Climatique au Mali Résultats et leçons apprises au Sahel

Jassim RA, Sejian V (2015) Review paper: climate change and camel production: impact and contribution. J Camelid Sci 8:1–17

Jones PG, Thornton PK (2009) Croppers to livestock keepers: livelihood transitions to 2050 in Africa due to climate change. Environ Sci Pol 12(4):427–437. https://doi.org/10.1016/j.envsci.2008.08.006

Kandeel M, Al-Mubarak AIA (2022) Camel viral diseases: current diagnostic, therapeutic, and preventive strategies. Front Vet Sci 9:915475. https://doi.org/10.3389/fvets.2022.915475

Kandil H, Wassif IM, Rabee A, Shokry MM, Khidr RES, Askar AR (2023) Camel, the animal of food security and climate change. Egypt J Camel Sci 1(1):1–8. https://doi.org/10.21608/ejcs.2023.220447.1008

Kebir N, Berber N, Meriem ZM (2024a) Anatomical and physiological properties of the dromedary: a potential sustainability alternative and a vital asset in the era of climate change. J Anim Behav Biometeorol 12:2024031. https://doi.org/10.31893/jabb.2024031

Kebir NE, Berber N, Zahzeh MR (2024b) Anatomical and physiological properties of the dromedary: a potential sustainability alternative and a vital asset in the era of climate change. J Anim Behav Biometeorol 12(4):2024031–2024031. https://doi.org/10.31893/jabb.2024031

Khalafalla AI (2017) Emerging infectious diseases in camelids. In: Bayry J (ed) Emerging and re-emerging infectious diseases of livestock [Internet]. Springer, Cham; 2017 [cited 2025 Jun 22]. pp 425–441. Available from https://doi.org/10.1007/978-3-319-47426-7_20

Khalafalla AI (2023) Zoonotic diseases transmitted from the camels. Front Vet Sci 10:1244833. https://doi.org/10.3389/fvets.2023.1244833

Kipkulei H, Rotich B, Ahmed A, Lameck A, Burudi J, Hounkpati K et al (2025) Land use/land cover dynamics in an arid and semi-arid landscape: a 24-year analysis of Baringo County, Kenya (2000–2024). Glob Earth Surface Process Change 4:100006. https://doi.org/10.1016/j.gespch.2025.100006

Köhler-Rollefson I, Rathore HS (2021) The case of the Kumbhalgarh Wildlife Sanctuary and Camel Pastoralism in Rajasthan (India). Sustainability 13(24):24. https://doi.org/10.3390/su132413914

Kuria SG, Tura IA, Amboga S, Walaga HK (2012) Forage species preferred by camels (*Camelus dromedarius*) and their nutritional composition in North Eastern Kenya. Livest Res Rural Dev 24:453–463

Mahamat Ahmat MA, Moulin CH, Koussou MO, Duteurtre G (2023) Le lait comme facteur de sécurisation des chameliers en zone périurbaine de N'Djamena au Tchad. Rev Elev Med Vet Pays Trop 76:1–10. https://doi.org/10.19182/remvt.37134

Maitra A, Iglesias Pastrana C, Faye B, Burger P, Ciani E (2025) Breeding of camels in Europe: between continuity and innovation. Appl Sci 15(3):3. https://doi.org/10.3390/app15031644

Makoni M (2020) Africa's invasive species problem. Lancet Planet Health 4(8):e317–e319. https://doi.org/10.1016/S2542-5196(20)30174-1

Makwatta JO, Ndegwa PN, Oyieke FA, Ahuya P, Masiga DK, Getahun MN (2024) Exploring the dynamic ticks-camel-pathogens interaction [Internet]. bioRxiv; 2024 [cited 2025 Dec 9]:2024.05.15.594365. Available from https://www.biorxiv.org/content/10.1101/2024.05.15.594365v1

Malik A, Stenchikov G, Mostamandi S, Parajuli S, Lelieveld J, Zittis G et al (2023) Accelerated historical and future warming over the Middle East and North Africa in response to the global temperature change. https://doi.org/10.22541/essoar.168500249.91012919/v1

Martin V, Chevalier V, Ceccato P, Anyamba A, Simone LD, Lubroth J et al (2008) The impact of climate change on the epidemiology and control of Rift Valley fever. Rev Sci Tech 27(2):413–426

Megersa B, Temesgen W, Amenu K, Asfaw W, Gizaw S, Yussuf B et al (2024) Camels in Ethiopia: an overview of demography, productivity, socio-economic value and diseases [Internet]. International Livestock Research Institute; 2024 [cited 2026 Mar 23]. Available from https://hdl.handle.net/10568/149289

Meisner J, Kato A, Lemerani M, Miaka EM, Taban AI, Wakefield J et al (2022) Livestock, pathogens, vectors, and their environment: a causal inference-based approach to estimating the pathway-specific effect of livestock on human African trypanosomiasis risk [Internet]. medRxiv; 2022 [cited 2025 Dec 10]:2022.01.13.22268994. https://doi.org/10.1101/2022.01.13.22268994. Available from https://www.medrxiv.org/content/10.1101/2022.01.13.22268994v1

Mekonnen N (2018) The impact of climatic change on animal disease ecology, Distribution and emergence: a review. Arch Appl Sci Res 10(1):14–20

Molefe-Nyembe NI (2024) An update on the African Trypanosomosis status in South Africa. In: Trypanosoma—recent advances and new perspectives [Internet]. IntechOpen; 2024 [cited 2025 Dec 10]. https://doi.org/10.5772/intechopen.115057. Available from https://www.intechopen.com/chapters/89543

Moreira RP, Costa AC, Gomes TF, de Oliveira Ferreira G (2020) Climate and climate-sensitive diseases in semi-arid regions: a systematic review. Int J Public Health 65(9):1749–1761. https://doi.org/10.1007/s00038-020-01464-6

Mukherji A (2023) Climate change 2023 synthesis report [Internet]. https://doi.org/10.59327/ipcc/ar6-9789291691647.001

Nagy PP, Skidmore JA, Juhasz J (2022) Intensification of camel farming and milk production with special emphasis on animal health, welfare, and the biotechnology of reproduction. Anim Front 12(4):35–45. https://doi.org/10.1093/af/vfac043. PubMed PMID: 35974793; PubMed Central PMCID: PMC9374475

Narayan M, Srivastava AK, Mohanty B, Sharma A (2025) From herds to markets: entrepreneurial innovations for economic resilience in camel pastoralism. J Camel Pract Res 32(2):155–166. https://doi.org/10.5958/2277-8934.2025.00023.7

Nori M (2023) High quality, high reliability: the dynamics of camel milk marketing in northern Kenya. Pastoralism 13(1):9. https://doi.org/10.1186/s13570-022-00265-1

Nuttall PA (2022) Climate change impacts on ticks and tick-borne infections. Biologia 77(6):1503–1512. https://doi.org/10.1007/s11756-021-00927-2

Nyongesa JM, Donatti CI, Kiptenai TK (2023) Community-based rangeland restoration for climate resilience and pastoral livelihoods in Chyulu, Kenya. In: Nishi M, Subramanian SM (eds) Ecosystem restoration through managing socio-ecological production landscapes and seascapes (SEPLS) [Internet]. Springer, Singapore, pp 35–59. [cited 2025 Dec 10]. Available from: https://doi.org/10.1007/978-981-99-1292-6_3

Obolski U, Perez PN, Villabona-Arenas CJ, Thézé J, Faria NR, Lourenço J (2019) MVSE: an R-package that estimates a climate-driven mosquito-borne viral suitability index. Methods Ecol Evol 10(8):1357–1370. https://doi.org/10.1111/2041-210X.13205

Odongo RA, Schrieks T, Streefkerk I, de Moel H, Busker T, Haer T et al (2025) Drought impacts and community adaptation: perspectives on the 2020–2023 drought in East Africa. Int J Disaster Risk Reduct 119:105309. https://doi.org/10.1016/j.ijdrr.2025.105309

Ogoti BM, Riitho V, Wildemann J, Mutono N, Tesch J, Rodon J et al (2024) Biphasic MERS-CoV incidence in nomadic dromedaries with putative transmission to humans, Kenya, 2022–2023. Emerg Infect Dis [Internet] 30(3):581–585. https://doi.org/10.3201/eid3003.231488

Ogoti B, Riitho V, Wildemann J, Mutono N, Mureithi M, Oyugi J et al (2026) Epidemiology and genomic features of MERS coronavirus in Africa: a systematic and meta-analysis review. Int J Infect Dis 165:108456. https://doi.org/10.1016/j.ijid.2026.108456

Othieno J, Njagi O, Masika S, Apamaku M, Tenge E, Mwasa B et al (2022) Knowledge, attitudes, and practices on camel respiratory diseases and conditions in Garissa and Isiolo, Kenya. Front

Vet Sci 9:1022146. https://doi.org/10.3389/fvets.2022.1022146. PubMed PMID: 36524228; PubMed Central PMCID: PMC9745045

Padalino B, Faraz A, Masebo NT, Waheed A, Ishaq HM, Tauqir NA et al (2024) Welfare assessment of dromedary camels kept under pastoralism in Pakistan. Front Vet Sci 11:1442628. https://doi.org/10.3389/fvets.2024.1442628

Phillips M (2025) The changing climate of vector-borne diseases [Internet], vol 9; [cited 2025 Jul 10]. Contagion, Winter 2024, Digital Edition. Available from https://www.contagionlive.com/view/the-changing-climate-of-vector-borne-diseases

Polley H, Bailey D, Nowak R, Stafford Smith M (2017) Ecological consequences of climate change on Rangelands. In: Rangeland systems, pp 229–260. https://doi.org/10.1007/978-3-319-46709-2_7

Rahimi J, Fillol E, Mutua JY, Cinardi G, Robinson TP, Notenbaert AMO et al (2022) A shift from cattle to camel and goat farming can sustain milk production with lower inputs and emissions in north sub-Saharan Africa's drylands. Nat Food 3(7):523–531. https://doi.org/10.1038/s43016-022-00543-6

Rodriguez-Morales AJ, Paniz-Mondolfi AE, Faccini-Martínez ÁA, Henao-Martínez AF, Ruiz-Saenz J, Martinez-Gutierrez M et al (2021) The constant threat of zoonotic and vector-borne emerging tropical diseases: living on the edge. Front Trop Dis 2:676905. https://doi.org/10.3389/fitd.2021.676905. PubMed PMID: 34010366; PubMed Central PMCID: PMC8132189

Rojas-Sandoval J, Acevedo-Rodríguez P (2013) Calotropis procera (apple of sodom). CABI Compendium, p CABI Compendium:16848. https://doi.org/10.1079/cabicompendium.16848

Sahoo A (2020) Camel: a fast declining animal species but can strive with its unique climate resilience and 'desert to medicine' application. EC Vet Sci 5:118–132

Tefera M, Abebe G (2012) Camel in Ethiopia. Ethiopian Veterinary Association

Tibary A, El Allali K (2020) Dromedary camel: a model of heat resistant livestock animal. Theriogenology 154:203–211. https://doi.org/10.1016/j.theriogenology.2020.05.046

Tidman R, Abela-Ridder B, de Castañeda RR (2021) The impact of climate change on neglected tropical diseases: a systematic review. Trans R Soc Trop Med Hyg 115(2):147–168. https://doi.org/10.1093/trstmh/traa192. PubMed PMID: 33508094; PubMed Central PMCID: PMC7842100

Tulu D, Gadissa S, Hundessa F, Kebede E (2023) Contribution of climate-smart forage and fodder production for sustainable livestock production and environment: lessons and challenges from Ethiopia. Adv Agric 2023(1):8067776. https://doi.org/10.1155/2023/8067776

Vicente-Serrano SM, Pricope NG, Toreti A, Morán-Tejeda E, Spinoni J, Ocampo-Melgar A et al (2024) The United Nations convention to combat desertification report on rising aridity trends globally and associated biological and agricultural implications. Glob Chang Biol 30(12):e70009. https://doi.org/10.1111/gcb.70009

Watson EE, Kochore HH, Dabasso BH (2016a) Camels and climate resilience: adaptation in northern Kenya. Hum Ecol 44(6):701–713. https://doi.org/10.1007/s10745-016-9858-1

Watson EE, Kochore HH, Dabasso BH (2016b) Camels and climate resilience: adaptation in Northern Kenya. Hum Ecol Interdiscip J 44(6):701–713. https://doi.org/10.1007/s10745-016-9858-1. PubMed PMID: 28018023; PubMed Central PMCID: PMC5143358

Xie Z, Game ET, Phinn SR, Adams MP, Bayarjargal Y, Pannell DJ et al (2024) A scalable big data approach for remotely tracking rangeland conditions. Commun Earth Environ 5(1):349. https://doi.org/10.1038/s43247-024-01516-2

Younan M, Ouso DO, Bodha B, Keitany EK, Wesonga HO, Sitawa R et al (2021) *Ehrlichia* spp. close to *Ehrlichia ruminantium*, *Ehrlichia canis*, and '*Candidatus Ehrlichia regneryi*' linked to heartwater-like disease in Kenyan camels (*Camelus dromedarius*). Trop Anim Health Prod 53(1):147. https://doi.org/10.1007/s11250-020-02524-y

Zhou Z, Ali A, Walelign E, Demissie GF, El Masry I, Abayneh T et al (2023) Genetic diversity and molecular epidemiology of Middle East respiratory syndrome coronavirus in dromedaries in Ethiopia, 2017-2020. Emerg Microbes Infect 12(1):e2164218. https://doi.org/10.1080/22221751.2022.2164218. PubMed PMID: 36620913; PubMed Central PMCID: PMC9888459

Camels in the Wild: An Outlook

10

Tungalag Ulambayar, Anna Jemmett,
Samantha Alex Gordine, and Jordan Hampton

Abstract

This chapter explores the deep and evolving relationship between camels, humans, and desert environments, highlighting both historical and contemporary perspectives. Focusing on two contrasting cases of camels living in the wild: *Camelus ferus* in Mongolia and the feral *Camelus dromedarius* population in Australia. It depicts their cultural relevance, ecological roles, and management challenges. Through these cases, the chapter illustrates how camels continue to influence human-environment interactions and discusses conservation, sustainability, and their relationship to One Health. These case studies highlight the different angles and complex considerations required for decision-making and strategy-setting in the field of camel conservation, as well as opportunities for managing camels that live in the wild.

Keywords

Conservation · Wild camel · Wildlife · Nomads · Environment · Ecosystem · Umbrella species

T. Ulambayar (✉)
Zoological Society Luujin, Ulaanbaatar, Mongolia
e-mail: tungalag.ulambayar@zs-compass.org

A. Jemmett
Wild Camel Protection Foundation, University of Kent, Kent, UK
e-mail: A.Jemmett-36@kent.ac.uk

S. A. Gordine
Lead Scientist, One Health Camel Program, Arcondis, Allschwil, Switzerland

J. Hampton
School of Veterinary Medicine, Harry Butler Institute, Murdoch University, Murdoch, WA, Australia

Faculty of Science, University of Melbourne, Parkville, VIC, Australia
e-mail: jordan.hampton@murdoch.edu.au

S. A. Gordine et al. (eds.), *Camels and One Health*,
https://doi.org/10.1007/978-981-92-2246-9_10

For centuries, camels have served humans in various ways and enabled humankind access to valuable resources in geographies defined by scarcity and harsh environmental conditions. This special relationship between camels, humans, and the environment might have been born out of human necessity. Yet, it has left a significant cultural, spiritual, and traditional imprint that is mutually shared across several geographies. The relationship between camels, humans, and their unique desert habitats is closely entwined and extends far beyond One Health outcomes. It has forged a deep personal relationship that bridges the past with the present and future. While few natively wild camels (*Camelus ferus*) still exist today, they continue to shape cultural identity as well as their environment. Their influence inspires a whole new generation of nomads, scientists, and everyone in between, driving conservation initiatives to save this sentinel species from extinction. About 8500 km south of the native home of the wild camel, the Gobi Desert, we can now find another camel living in the wild. Initially brought to Australia as a domesticated pack animal to help exploit desert environments, the dromedary has found a new home in the Australian outback.

To round out our exploration of the One Health relationship between camels, humans, and the unique environments they inhabit, we draw on two contrasting contexts from Mongolia and Australia. We do this from the perspective of two authors with a deep personal relationship to camels and explore the camels' heritage, different population management approaches, as well as current challenges and opportunities. We shine the light on two distinct camel populations: the wild camel (*Camelus ferus*) native to Mongolia and China, which has never been domesticated, and the formerly domesticated dromedary (*Camelus dromedarius*), which was set to roam free in Australia, hereafter referred to as the Australian dromedary or Australian feral camel.

10.1 A Personal Connection to Camels

10.1.1 Mongolia: Tungalag Ulambayar

In the Gobi's vast steppe, a lone wild camel treads through summer mirages in search of water and grazing, its survival entwined with nomadic herders and the desert ecosystem. This integrated vision recognizes that the Gobi's ecological balance, pastoral livelihoods, and human health are interdependent, shaped by cultural practices like mobile herding that foster resilience for both people and camels. A holistic perspective frames this interconnectedness, linking the vitality of wild camels, humans, and their environment (Gibbs 2014; Zinsstag et al. 2020). Sustainable conservation blends ecological, social, and cultural insights, shaping my journey from a Gobi herder girl to a conservation professional protecting the wild camel. This chapter explores different angles of One Health relative to my journey of blending cultural heritage with science to preserve a species vital to the Gobi's ecological and social fabric.

Fig. 10.1 A wild camel in the Gobi Desert. (Photo: T. Zoological Society Luujin)

The wild camel (Fig. 10.1), as a large migratory desert ungulate, requires vast ranges to access scarce food and water resources (Joly et al. 2019; Kaczensky et al. 2014; Yadamsuren et al. 2019). It is classified as endangered by the IUCN (Hare 2008; IUCN 2025) and has dwindled to just over 600 individuals in Mongolia (Jemmett 2023), being listed as *very rare* in Mongolia's Red Data Book (Shiirevdamba et al. 2013). This decline reflects global biodiversity loss, with a 73% drop in monitored wildlife populations over 50 years, including a 69% fall in terrestrial species. This is driven by habitat degradation, overexploitation, diseases, and climate change, which are exactly the challenges that intensify efforts to conserve the wild camel (WWF 2024). As an endemic umbrella species of the Gobi, the wild camel safeguards co-occurring species and protects the biodiversity of the Great Gobi through its vast habitat ranges (Hare 2008; Kaczensky et al. 2014). It maintains ecological balance through seed dispersal, soil distribution with wide hooves, water cycle regulation via trails, and browsing trees like poplar and saxaul (Burger et al. 2019; Xue et al. 2021a, b; Yadamsuren 2017). Its distinct adaptations, such as the ability to consume thorny shrubs, drink salty water, and migrate to remote pastures, reduce herbivore competition and optimize scarce desert resources (Jemmett et al. 2023; Xue et al. 2021a, b).

The Gobi Desert has one of the lowest levels of human influence globally and hosts a vibrant ecosystem rich in rare and endemic species, which makes it a UNESCO World Heritage Site (McCarthy et al. 2022). The Gobi was my childhood classroom, with its rolling hills, stony paths, and ravines woven with summer-green shrubs or winter snow. Lone elm trees stood guard, and summer mirages conjured phantom lakes. Raised by my grandparents, who are nomadic herders, I mastered pastoral skills early and helped them with their livestock as their first grandchild. Our old Bactrian camel (*Camelus bactrianus*), *which is* a domesticated and different species from the wild camel (*Camelus ferus*), carried water from distant wells or dried manure as essential fuel in the wood-scarce Gobi. His large eyes, framed by long lashes, calmly watched as I struggled to secure loads, taking time and testing his patience. Riding between his humps, I would nap to his steady gait, the Gobi unfolding around us: stony steppes, snowy hills, and scattered grazing livestock. In storms, I sheltered against his warm belly, confident in my nomadic know-how, which later fueled my passion to protect the wild camel's fragile habitat.

Visiting my uncle's camel herd, I saw Bactrian camels of every shade and wobbly calves. This exposure sharpened my skills, e.g., to shear wool for domestic use, race young camels to strengthen them, and train future herders. These cultural practices fostered ecological balance, sustaining human and camel well-being. Now, as a social scientist for rangelands and pastoralism and a wildlife conservation practitioner with the United Nations Development Programme (UNDP), the Zoological Society of London (ZSL), and the Zoological Society of Luujin, my work blends nomadic wisdom with science to conserve the wild camel. Mongolians revere the wild camel as an untamed symbol of the Gobi and call the wild camel *havtgai*, which means *flat* in Mongolian. This is due to the wild camel's unique flatter head shape compared to the domestic Bactrian camel (Jemmett et al. 2023; Dovchindorj 2017). Listed by ZSL as an Evolutionarily Distinct and Globally Endangered Species, it embodies cultural pride and ecological value (ZSL 2024). To a Mongolian herder, the population size of all extant wild camels is now far smaller than the livestock herd needed to sustain a herder family. This comparison underscores the urgency of its conservation.

10.1.2 Australia: Jordan Hampton

My paternal grandmother, Violet Zuleikha Hampton (née Mudgedeen), was the daughter of an Afghan cameleer. Cameleers came to Australia with their dromedary camels from what was then British India (now Pakistan and Afghanistan) during the late 1800s to the early 1900s (Jones and Kenny 2010). The Afghans were one of the few exceptions to the dichotomy of Australian people in the late 1800s, which were either Indigenous or British (Kabir 2005). Violet and her siblings (Fig. 10.2) grew up around Leonora in the arid inland *Goldfields* region of Western Australia, which is now a major mining region, as the colloquial name suggests. Like many families descending from Australia's Afghan cameleers, my family continues to regularly travel back to Marree (in remote South Australia), the site of Australia's first mosque, and the unofficial capital of the Australian Afghan community (Scriver 2004).

The reason for the Afghan cameleers coming to Australia in the first place relates to just how harsh and un-European inland Australia can be. As British colonists pushed inland from Australia's relatively temperate coastal strip, they found the terrain too harsh for their familiar pack animals: horses, donkeys, and bullocks (McKnight 1969). The first camels bound for Australia left Karachi in 1840, the same port that my great-grandfather farewelled, never to return, several decades

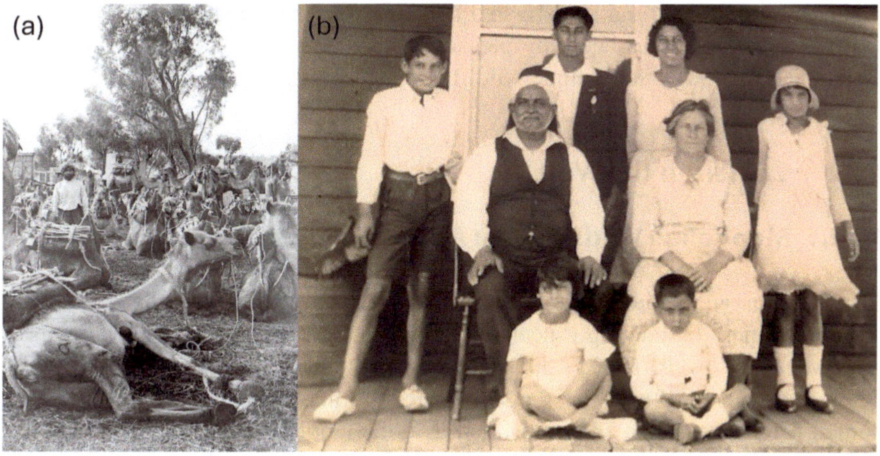

Fig. 10.2 The history of Afghan cameleers and their families in Australia: (**a**) a cameleer with resting camels at an unknown location in outback Australia, 1891 (State Library South Australia, n.d.), Photograph taken from https://collections.slsa.sa.gov.au/resource/B+45854 and (**b**) a photo of the Mudgedeen/Barradeen family in Leonora, Western Australia, in 1925. The author's grandmother is second from the right in the back row. (Photo: C. Barradeen)

later. The freight the camels carried could not be hauled effectively by any other means because the heat, lack of water, the camels' ability to forage on local native Australian vegetation, and in places the soft sand under hooves eliminated draft animals as alternatives (Stevens 2002). In a relatively short time, *cameleering* became a major industry in the goldfields of Western Australia.

If early-1900s Australia was a difficult time and place for the cameleers and their children in which to find a niche in a new continent, it was not much easier for the animals they brought with them. The era of camel transport in Australia abruptly snapped shut after 1920 when motor vehicles became widely available, rendering the beasts of burden obsolete (McKnight 1969). Rather than slaughtering their trusted camels when trucks and cars took over inland Australia, most cameleers simply set their companions free (Jones and Kenny 2010). Once camels were released into the rangelands of arid Australia in the 1930s, self-sustaining populations established totally outside of reliance on human husbandry, hence the free-ranging Australian feral camel herd was born.

Having grown up with this family history, I jumped at the chance as a young veterinarian to work on the species. My first paid job working with wildlife was figuring out how to capture free-ranging camels in the deserts of Western Australia to deploy satellite tracking collars (Boardman et al. 2014). From that starting point, I have spent much of the last 15 years of my career working on Australia's camels. In addition to establishing and refining capture methods, mostly involving the use of dart rifles and helicopters, I have assessed animal welfare outcomes for camel control methods (Hampton et al. 2014, 2016), examined their population genetics

(Spencer et al. 2015), and investigated the occurrence of plant toxins in their meat (Tan et al. 2016). It has been my pleasure to contribute to an improved scientific understanding of a highly unique wildlife population.

10.2 Camels in the Wild

10.2.1 Mongolian Wild Camels: An Endangered Heritage

The Russian geographer Nikolai Przhevalski first documented the wild camel (Fig. 10.3) in 1877, hunting it near Lake Lop and publishing a detailed description (Przewalski 1879). In 1900, Veniamin Ladygin confirmed its presence in the Altai Gobi (Ladygin 1900). Local herders, long familiar with this untamed species, hunted it for sustenance and embedded it in the Gobi cultural heritage (Bannikov 1975; Dementyev and Tsevegmid 1963; Sumikov 1937; Eregdendagva 1954). The wild camel was initially misclassified in Western literature as a wild progenitor or a feral version of the domesticated species. This was due to their morphological similarities, given that both species have two humps, overlap in range, and can hybridize to produce viable offspring (Przhevalsky 1878; IUCN 2025). This confusion echoed my childhood belief, raised among herders, that the wild camel was the ancestor of our domestic camels. Yet, Gobi herders always recognized the wild camel as a distinct and untamed species (Hare 1997).

Fig. 10.3 Wild camels in winter in the Gobi Desert (Photo: Ch. Khatantumur) at the WCPF captive breeding center at Zakhuin Us

Despite their apparent similarities, the wild camel (*Camelus ferus*) and the domestic Bactrian camel (*Camelus bactrianus*) are distinct species, separated by behavioral, morphological, and genetic differences. Behaviorally, the Bactrian camel is a tame beast of burden compared to the wild camel, which has never been domesticated and exhibits a natural fear of humans. Morphologically, the wild camel is smaller and more agile, with slimmer legs and smaller, pyramidal humps, and a flatter skull compared to the Bactrian camel (Ji et al. 2009). Genetically, high sequence divergence in nuclear loci (Chuluunbat et al. 2014; Jirimutu et al. 2012; Wu et al. 2014; Zhang et al. 2019) and mitochondrial DNA (Ji et al. 2009; Mohandesan et al. 2017; Silbermayr et al. 2010a, b) confirms the wild camel as a separate species, aligning Western science with local herders' long-held knowledge (Ji et al. 2009; Mohandesan et al. 2017; Silbermayr et al. 2010a, b). These differences illustrate the wild camel's unique ecological role in the Gobi, which needs to be protected.

The wild camels and domestic Bactrian camels diverged from a common ancestor approximately 0.7 million years ago, long before the Bactrian's domestication around 6000 years ago (Burger et al. 2019). This timeline, coupled with the genetic evidence, confirms that the wild camel is neither the progenitor nor a feral version of the Bactrian camel, but a distinct species, as herders like my grandparents always understood (Chuluunbat et al. 2014). Since the wild ancestors of both the Bactrian and dromedary camel are extinct, the wild camel stands as the last surviving wild member of the *Camelini* tribe, making its conservation critical to avoid its threatened extinction.

The wild camel's survival hinges on its ability to endure the harsh conditions of its limited habitat, confined to four locations: Gashuun Gobi, Lake Lop, and Taklamakan in China, and the Great Gobi A Strictly Protected Area (GGASPA) in Mongolia. Their habitat, shaped by four seasons with temperatures ranging from 40 °C in summer to −35 °C in winter and less than 50 mm annual precipitation, makes water a critical limiting factor (Jemmett 2023). Within GGASPA's 4.6 million hectares, approximately 50 water sources exist, of which only 30 are permanent, with saline springs (655–12,772 mg/l mineralization) containing high levels of iron, ammonium, arsenic, manganese, strontium, and uranium (Enkhjargal 2023). Sparse vegetation, driven by low precipitation, includes tall grasses (*Phragmites communis, Achnatherum splendens*), *Tamarix* shrubs, *Populus diversifolia* in oases, and drought-resistant shrubs like saxaul (*Haloxylon ammodendron*), *Caragana leucophloen, Reaumuria songarica, Salsola arbuscula, Calligonum mongolicum*, and *Ilyinia regelii*, which provide essential food for wild camels (Tulgat and Schaller 1992).

With its adaptations to the Gobi's harsh desert conditions forged over millennia, the wild camel's ecological roles are vital within a holistic perspective, supporting environmental stability and herder livelihoods. Its migrations, averaging 2821 km annually and 3–6.5 km daily, disperse seeds across desert steppes, fostering plant diversity (Jemmett 2023; Kaczensky et al. 2014). Its migration trails regulate the Gobi's water cycle, sustaining oases for other species, while grazing on shrubs and trees like saxaul and poplars prevents overgrowth, balancing pastures used by herders (Burger et al. 2019; Yadamsuren 2017; Xue et al. 2021a, b). In my ZSL fieldwork, I observed that the wild camels' tolerance to saline water reduces competition with other herbivores, mirroring nomadic grazing practices I learned as a child, which ensure efficient resource use (Jemmett et al. 2023). Yet, these important ecological roles in stabilizing the ecosystem and supporting pastoral communities ultimately depend on healthy rangelands.

The latter is of particular concern because the wild camel faces a plethora of human-induced threats such as climate change, habitat degradation (water scarcity, pasture degradation, illegal mining), hybridization, diseases, livestock competition, and wolf predation (Yadamsuren et al. 2012; Battsetseg et al. 2007; Dovchindorj 2022; Reading et al. 1999, 2005; Tulgat and Schaller 1992; Walzer et al. 2008; Shiirevdamba et al. 2013). In China, infrastructure, mining, hunting, and nuclear testing further isolate populations (Guodong et al. 2008; Hare 1997; Kaczensky et al. 2014; Reading et al. 2005). These severe threats disrupt the wild camel's ecological contributions and community well-being as follows:

1. *Climate change:* Climate change has made the Gobi one of Earth's fastest-growing deserts, posing a major threat to the wild camel (Han et al. 2021). This climate-driven desertification dries water points and reduces suitable habitat, with up to 44% of the wild camel's habitat in China projected to be lost by 2050 (Xue et al. 2021a, b).
2. *Water scarcity:* Of the approximately 50 recorded water points in GGASPA (Haulenbeck 2013), only around 30 provide permanent water. For example, in a 2008 survey of 35 water points, 15 had dried up, and four lacked enough water for sampling (Yadamsuren et al. 2019). Increasing water scarcity is associated with rising temperatures in the Altai Gobi, particularly the increase in hot days exceeding 30 °C (Haulenbeck 2013). Under the harsh conditions of the Gobi, this scarcity lengthens the journey for wild camels seeking water and raises predation risk due to intensified competition at water sources, with some camels succumbing to exhaustion and dehydration.
3. *Illegal mining:* While mining supports Mongolia's economy and job creation, it severely undermines biodiversity conservation, a concern herder in the Gobi deeply share, as it damages land, pollutes ecosystems, and reduces wild camel habitat. In the GGASPA's buffer zone, 33 mining licenses exist, with two operational mines in Bayankhongor (Bayarbat 2024). These activities degrade pastures, pushing livestock into protected areas. Illegal miners, known as *ninjas*, further exacerbate the issue by entering the protected area, driving away wildlife, hunting, damaging land, and polluting the environment. For instance, a 2020 gold discovery in Choigin Us triggered an influx of ninjas, which required significant efforts to curb violations, with over 150 cases detected in 6 months, leading to fines of approximately 70 million tugriks (USD19500) and criminal cases against ten individuals (Bayarbat 2024).
4. *Rangeland degradation:* The doubling of livestock to 842,000 animals in the GGASPA buffer zone over the past 20 years, including a threefold increase in Shinejinst soum and a twofold increase in Erdene soum, intensifies pressure on

pastures. This is a challenge that herders in the Gobi strive to manage (NSO 2024). This livestock growth increases fodder demand by 1.5–1.7% annually and, combined with vehicle-related road expansion from mining, degrades the steppes (NSO 2024). A 2022 rangeland survey rated pasture conditions in Bayan-Undur and Shinejinst poorly, with scores below 50 (Ariuntsetseg 2023). Overuse of saxaul, vital to the Gobi ecosystem, persists due to unresolved heat and fuel supply issues (Batkhuu and Haulenbeck 2022). Climate change and livestock pressure reduce soil moisture and structure, lowering biomass yields, decreasing species richness, and favoring unpalatable plants. All this contributes to land degradation. This pushes livestock into restricted areas, while free-roaming horses and domestic camels graze in protected zones, increasing competition with the wild camel and other grazers (Kaczensky et al. 2014; Jemmett 2023).

5. *Hybridization:* Although a distinct species, the wild camel can breed with the domestic Bactrian camel to produce hybrids. This results in fertile offspring, diluting traits essential for extreme desert survival, such as the wild camel's smaller, more agile build and flatter skull. Historically, herders castrated or slaughtered male hybrids, preventing further breeding to preserve the wild camel's genetic purity (Tulgat and Schaller 1992; Bannikov 1975; Sumikov 1937). Yet, current local policies allow herders to graze livestock within the park boundary during disasters such as drought and *dzud* (severe winter conditions causing mass livestock mortality), further exacerbating the threat of hybridization. As the GGASPA is unfenced, domestic camels can enter easily. There are therefore no reproductive barriers between wild camels and domesticated Bactrian camels, which share behavioral and physical similarities (Silbermayr et al. 2006). Additionally, the deliberate breeding of domestic females with wild camel bulls has been incentivized by the high prize money offered for camel racing, which is becoming increasingly popular. The resulting hybrids are valued for their speed and endurance, traits that are favored in these local races (Yadamsuren et al. 2019). Surveys recorded 44 hybrids from 16 households in 2006. This number rose threefold to 124 from 48 households by 2019, reflecting growing interest among Gobi herders (Dovchindorj and Nasanjargal 2022). This cross-species mating leaves a lasting genetic trace through introgression (hybridization followed by repeated back-crossing). This has been detected in microsatellite markers (Silbermayr et al. 2010a, b) and on the male-specific region of the Y chromosome (Felkel et al. 2019). Introgression has been shown to be present in approximately 10% of the Mongolian wild camel population (Jemmett et al. 2026). Illegal practices such as breeding hybrids or domesticating young wild male camels violate Mongolia's Law on Protection of Livestock Genetic Resources and Mongolia's Law on Genetic Resources (Dovchindorj and Nasanjargal 2022). Furthermore, escaping hybrids that rejoin wild herds pose a severe risk to the wild camel population through the spread of livestock diseases (Dovchindorj and Nasanjargal 2022).
6. *Disease:* Outbreaks of zoonotic diseases, such as anthrax, tuberculosis, brucellosis, or echinococcus, threaten the wild camel and local communities (Walzer

and Kaczensky 2005). These are often spread by canine species, like dogs and wolves, or are water-borne pathogens, which are transmitted via sick animals and their feces. These diseases could devastate fewer than 700 wild camels in GGASPA, as well as cause significant health and financial losses for pastoral communities. A parasite survey two decades ago found *Nematodirus* spp. in 12 camel fecal samples and *Hyalomma* spp. ticks on three camels near the Ats Ovoon River, with one deceased camel showing *Echinococcus* spp. cysts in its liver and lungs (Battsetseg et al. 2007). Zoonotic diseases have also affected Gobi herders rearing Bactrian camels, underscoring the need for strict supervision of breeding and rearing, along with urgent risk assessments and preventive measures to protect both wild camels and human health (Silbermayr et al. 2010a, b; Kaczensky et al. 2014; Walzer and Kaczensky 2005).

7. *Predation:* Predation by Gobi predators, primarily wolves but also lynx and snow leopards, may threaten wild camels, particularly the young and weak. Although a concern for the small GGASPA wild camel population, the GGASPA rangers recognize that predation is a part of the delicate balance of desert ecosystems. Between 1984 and 1989, 61% of 89 examined wild camel carcasses were attributed to wolf attacks, and only 2% were killed by snow leopards. Thirty-one percent of the 54 wolf-killed animals were young camels, with wolf stomachs confirming consumption of juveniles (Tulgat and Schaller 1992). While predation is a natural event, its impact on the small wild camel population could heighten extinction risks, especially without recent data to assess current trends (Tulgat and Schaller 1992). In the 1980s, park authorities attempted wolf population control, but its effectiveness remains unclear (Kaczensky et al. 2014). The defensive behavior of wild camels, limited to fleeing or migrating, is poorly studied, underscoring the need for updated research to inform conservation strategies (ibid).

Maintaining the genetic diversity of the wild camel is crucial for environmental health; hence, GGASPA rangers emphasize that genetic diversity is a priority for sustaining the Gobi's ecosystems. The wild camel's unique genome supports adaptations like saline tolerance, enabling its ecological roles amid climate change (Burger et al. 2019; MECC 2025). Hybridization and disease threaten this diversity, risking ecological collapse in line with global biodiversity decline (WWF 2024). To counteract these pressures, we have conducted expert workshops to train herders in monitoring hybrids, drawing on their traditional practice of castrating hybrid males with the ultimate aim of protecting the genetic purity of wild camels (Tulgat and Schaller 1992).

10.2.2 Australia: A Unique Wildlife Population

Under the conservation biology paradigm that dominates wildlife management in Australasia (Australia and New Zealand) (Palmer et al. 2023), all introduced species, i.e., animals that were moved to new environments by humans, were considered 'pests' and targeted for eradication or population control (Myers et al. 2000). However, feral camels in Australia are not a typical animal 'pest'. Usually, invasive species are highly fecund and fast-growing generalist species that attain high densities, which were usually released into disturbed habitats and which have a history of invasiveness elsewhere in the world (Clout and Russell 2008). On the contrary, the camel is poorly fecund and slow growing, exhibits low population densities, was released into an undisturbed desert, and has no history of invasiveness elsewhere (Long 2003). Their flexible physiological adaptations allow them to eat 80% of Australian plant species, and they obtain much of their water through ingesting vegetation (Crowley 2014). They are "invasive" only in the sense that they failed to go extinct once they were released from domestication in central Australia. The species is not known to have formed a feral population anywhere else in the world (Long 2003). Yet, in Australia, dromedary camels seemingly found an environment that suited them ideally. Their populations grew and grew through the twentieth century.

By the time any real resources were diverted to investigate how big this largely unseen population had become, around 2010, or roughly 80 years after their introduction, aerial surveys suggested a population close to 1 million animals (Edwards et al. 2004b). With no natural predators and an incredible ability to use the landscape to find required resources during droughts (Wells and Lethbridge 2020), Australian camels were estimated to have an annual growth rate of 10% (Edwards et al. 2004b). Camels were occupying close to 50% of the Australian continent, and impacts on environmental, economic, and cultural assets are increasingly being seen (Edwards et al. 2004b). Suddenly, they were recognized as a successful invasive species and "a nationally significant pest animal requiring a national management approach" (Woolnough et al. 2016). Hence, the Australian Feral Camel Management Project (AFCMP) was born, running for 4 years, from 2009 to 2013 (Hart and Bubb 2016). Prior to 2009, there was little coordinated management of feral camels in Australia. Approaches taken to camels were marked by ignorance, with most Australians living far from the camel range, and neglect, with wildlife managers having higher priorities such as lethal control of European rabbits (*Oryctolagus cuniculus*), a species that caused profound economic losses for agriculture (Cooke and Fenner 2002).

When active population management began under the AFCMP, several different control tools were considered and explored (Coventry et al. 2010; Lapidge et al. 2010). Eventually, the lethal tools settled upon were aerial shooting and commercial use (Fig. 10.4), with the latter achieved through capture of animals via helicopter mustering, followed by truck transport and either abattoir slaughter in Australia or live export via sea transport (Virtue et al. 2016). While AUD$19 million was spent on the program, and population densities of camels were demonstrably reduced (Lethbridge et al. 2016), eradication was not close to being achieved. In landscape-scale environments lacking natural barriers to dispersal (like inland Australia), reinvasion of areas targeted for management often quickly nullifies the effects of control efforts (Bomford and O'Brien 1995). This challenge has been the "thorn in the side"

Fig. 10.4 The management of feral dromedary camels (*Camelus dromedarius*) in contemporary central Australia: the habitat occupied by camels and the consequence of no management (**a**), mustering and trucking for slaughter in abattoirs (**b**), capture via helicopter darting for use in "Judas" animal control programs (**c**). (Photos: J. Hampton)

of sustained invasive species management (Myers et al. 2000; Cruz el al. 2009; Hampton et al. 2004) and has led to claims that culling efforts are not always easily justifiable for widespread or established introduced species (Lederman et al. 2021). Practical camel management options in Australia have largely shrunk to strategic population reduction or protection of assets at risk (e.g., protecting culturally valuable rock holes; Brim-Box et al. 2010; Hulme 2006). This approach has utility for local-scale asset management, and significant impact reduction is often achieved at small spatial and temporal scales, but this use of lethal methods has its detractors (Wallach et al. 2020).

Since 2016, very little has been heard about Australian camels in scientific circles or in the media. The exceptions to this trend include a burgeoning camel dairy industry in Australia, based on the genetic stock of recaptured free-living camels (Zou et al. 2022), and a small-scale tourism industry focused on camel rides (Hanslow-Sells et al. 2020). It would seem that now camels have returned to their long-held status of being a low-priority wildlife species, a forgotten member of Australia's ecosystems.

10.3 Current Management Strategies: Challenges and Opportunities

10.3.1 Mongolia: In Situ and Ex Situ Conservation

Conserving the wild camel requires integrated strategies to ensure its survival, maintain the ecological balance of its habitat, and support local communities. A holistic perspective, linking animal, environmental, and human well-being (One Health), guides ex situ and in situ efforts to address threats like water scarcity, hybridization, and rangeland degradation (Gibbs 2014; Zinsstag et al. 2020). As a conservation practitioner, my work blends nomadic wisdom with scientific innovation, fostering resilience for camels and Gobi herders. Ex situ breeding programs safeguard genetic diversity, while in situ initiatives such as GGASPA's water management and community collaborations sustain habitats and livelihoods, ensuring a balanced approach to conservation.

10.3.1.1 Captive Breeding

Ex situ efforts through captive breeding protect wild camels from disease and hybridization, preserve genetic diversity, and maintain a population pool in case wild populations vanish. The GGASPA Administration began captive breeding in 1987, raising the first calf in Bayantooroi by capturing or rescuing 22 young camels in Zakhuin Us (Tsogt soum, Gobi-Altai aimag) over 5 years (Dovchindorj 2017). In 2003, the Wild Camel Protection Foundation (WCPF) took over, establishing a breeding center in Zakhuin Us to act as an insurance against outright extinction. This herd is successful in longevity as well as breeding success and, as of 2025, held approximately 40 animals, having bred for eight generations.

Holding the entire ex situ population in one location limits husbandry management options, which can also compromise animal welfare and increase the risk posed by disease, genetics, and stochastic disaster. Therefore, in 2024, a second breeding center was opened in Toli Bulag (Shinejinst soum, Bayankhongor aimag) by WCPF. This second facility is supported by Prague Zoo, with the first seven camels transported from Zakhuin Us to Toli Bulag, moving them from GGSPA's western to the eastern regions. The individuals in ex situ care in Mongolia constitute approximately 5% of the total wild population in Mongolia, highlighting the importance of this population.

The Mongolian ex situ population is a long-standing, in-country project that has proven breeding success, important as insurance against extinction. While increasing the population size, this ex situ population is also important for allowing research that would otherwise be incredibly difficult. The wild camel is scarce and elusive and inhabits a vast, harsh environment, so there are limited in situ studies. Recent research conducted on the ex situ population includes genetic research, infectious disease monitoring, and animal behavior. All these research areas are important for both individual care and effective ex situ population management overall.

Genetic testing has proved that the levels of genetic diversity and inbreeding in the in situ wild population are also reflected in the ex situ stock (Jemmett et al.

2026). Understanding the genetics of wild camels helps us manage the population to maximize genetic diversity, thereby addressing issues caused by hybridization and inbreeding. For example, full genome sequencing has been undertaken for parts of the ex situ population recently. This work is ongoing but will allow for both an increased understanding of the ex situ population and for potential biobanking opportunities.

10.3.1.2 Disease and Mortality

By vaccinating captive-bred individuals against diseases, these ex situ populations are actively managed to reduce disease risk in the captive stock as well as reduce zoonotic risks for neighboring herders, e.g., from brucellosis (Battogtokh 2023). Disease monitoring, led by WCPF, includes vaccinations against anthrax and brucellosis and parasite checks for *Nematodirus* spp., reducing zoonotic risks (Battogtokh 2023). Veterinary checks at Zakhuin Us detect parasites like *Nematodirus* spp., reducing zoonotic risks to herders (Battsetseg et al. 2007). Research on health, infectious diseases, and genetic data allows for informed conservation management decisions both in the ex situ herd and, as we learn more about the species, in the in situ population.

In situ efforts within GGSPA address ecological balance and community well-being, countering threats outlined in Mongolia's draft Rare Species Conservation Plan (2025–2028) and the National Biodiversity Strategy and Action Plan (NBSAP), a flagship document implementing Mongolia's Convention on Biological Diversity commitments (MECC 2025). The plan's seven objectives, namely, improving legal frameworks, expanding protected areas, bolstering genetic diversity, adapting to climate change, advancing research, raising awareness, and enhancing resource management, guide initiatives to sustain wild camels and their habitat (MECC 2025). ZSL's "Protecting Mongolia's Gobi Desert for wild camels and herder communities" initiative has reinforced these efforts, integrating nomadic knowledge with science.

These programs stabilize camel populations and benefit herders by reducing disease transmission and preserving rangeland resources. However, there is limited veterinary awareness among herders, who self-administer medications to their livestock. The practices underpinning hybridization and managing disease outbreaks, driven by livestock pressure and herder self-medication, remain unaddressed in situ (Dovchindorj and Nasanjargal 2022; Battogtokh 2023). This poses a significant risk to One Health goals, necessitating further education (Battogtokh 2023). As such, my own work aims at safeguarding camel health and rangelands to address disease and hybridization risks. In particular, I push for stricter herder regulations during *dzud* or drought, to prevent disease transmission and to protect the wild camel's genetic purity. Additionally, ex situ disease control requires more in situ veterinary outreach, as herder self-medication ultimately risks food security (Battogtokh 2023).

Another survival-related issue is high calf mortality, which lacks dedicated research (Fig. 10.5) (Kaczensky et al. 2014). The calf survival number drops to near zero by autumn, possibly due to wolf predation. In past UNDP forums, we urged studies to identify possible causes such as predation, disease, or environmental

Fig. 10.5 A wild camel calf in the Gobi Desert. (Photo: Zoological Society Luujin)

stress. There is a lack of recent survey data, necessary for informing critical measures that are necessary for stabilizing wolf predation, e.g., through culling, the banning of which has been debated since the 1980s. Currently, PAA patrols deter wolf attacks, thereby supporting camel survival (Kaczensky et al. 2014). Unfortunately, GGSPA lacks a science-based monitoring protocol for wild camels, applying a mixed-method approach using camera traps at water points, drones in rugged terrains, transect counts, SMART patrol data, and satellite collaring. Monitoring via PAA's annual reports and SMART patrol data since 2022 tracks progress but needs public access and regular updates (ZSL 2024). SMART data enhances monitoring (Bayarbat 2024), which can be integrated with other sources, addressing monitoring gaps and ensuring data-driven conservation for camel and ecosystem health. Yet without standardization through, e.g., unified databases, PAA's erratic monitoring data hinders comprehensive wild camel population assessments. Additionally, limited transboundary cooperation with Chinese conservationists stalls population-wide data integration and migration studies across GGSPA, Lake Lop, and Taklamakan (Jemmett 2023).

10.3.1.3 Environmental Threats

Generally, good progress has been made in reducing environmental threats faced by wild camels. To combat water scarcity, the Ministry of Environment and Tourism's 2018 hydrological explorations identified new water sources, leading to the installation of eight solar-powered water pumps (Fig. 10.6) and one reservoir by ZSL (Munkhbayar et al. 2025). These pumps are used by at least ten mammal species,

Fig. 10.6 Installation of a solar-powered water pump in the Gobi Desert. (Photo: Zoological Society Luujin)

including wild camels. They have stabilized oases and support water cycle regulation (Munkhbayar 2024). Cloud-seeding trials, though unvalidated, reflect additional innovative attempts to address drying springs (Haulenbeck 2013). These measures reduce camel mortality from dehydration, ensuring ecological stability and herder access to water for livestock, which can be considered a One Health win. Yet, nationwide rangeland degradation exacerbates competition, a challenge Mongolian and donor initiatives struggle to resolve (MECC 2025). Illegal mining has been curtailed through the Protected Area Administration's (PAA) 24-h monitoring and patrols, halting artisanal *ninja* activities like those in Choigin Us in 2020 (Bayarbat 2024). These efforts protect rangelands from degradation, preserve grazing areas for camels and herders' livestock, reduce competition, and support community livelihoods.

The proposed Naran Sevstei border trade point, as outlined in Mongolia's Government Action Programme 2024–2028, further threatens GGSPA's southern border, which is adjacent to China. Vehicle roads, human travel, potential illegal hunting of wild camels, black-tailed gazelles, and wild ass, as well as pollution of fragile oases, could devastate the desert ecosystem (MECC 2025). The Mongolian Environmental Civil Council (MECC) is leading a protest with 33,000 petition signatures to reconsider this decision. My Zoological Society Luujin work supports MECC's advocacy by highlighting risks to One Health caused by disrupting camel habitats and herder water access. Luckily, the UNESCO World Heritage nomination for Gobi Desert parks, including GGASPA, led by the International Takhi Group with German funding, attracts other global investment. This will help to mitigate threats like the Naran Sevstei's proposed trade point by strengthening habitat protections for herder livelihoods (Succow Foundation 2023; MECC 2025).

Rangeland degradation is also exacerbated by an 842,000-head livestock increase, which has nevertheless stabilized since 2005 through PAA's collaboration with the Buffer Zone Council (NSO 2024; Sarnai 2024). Rangeland surveys, herder trainings, and contracts regulating livestock access to protected zones, funded by projects like UNDP's "Great Gobi Umbrella Species Conservation Project," mirror my nomadic herder lessons in seasonal grazing. The Collaborative Management Council's (CMC) landscape management model also unites Buffer Zone leaders and GGASPA administration, aligning with Mongolia's biodiversity policies and international commitments (ZSL 2024; MECC 2025). These policies focus on combatting rangeland degradation through grazing contracts. These grazing contracts reflect traditional seasonal movements, shifting livestock at least four times a year across spring, summer, fall, and winter camps with sheltered corrals. This helps to effectively utilize pastures by grazing diverse plant species, fatten livestock for harsh winters, and allow grazed pasture recovery (NSO 2024; Batbold and Suvd 2007; ZSL 2024). Community banking groups (VSLAs) and herders' conferences further bolster livelihoods, linking ecological health to economic stability (ZSL 2024). My personal project advocacy (Fig. 10.7) has strengthened this platform, and donor partnerships can sustain rangeland health, supporting camel grazing and herder prosperity. Unfortunately, funding shortages and inconsistencies for CMC meetings and outreach forums disrupt multistakeholder coordination, threatening rangeland management and community livelihoods.

10.3.1.4 Education and Outreach

Education and policy advocacy amplify in situ conservation efforts. Buffer Zone Ecoclubs, established in 2005 across five GGASPA soums, educate youth and herders. The Ecoclubs use new tools like a ZSL-funded book tailored to the GGASPA ecosystem called *My Homeland, My Great Gobi* (ZSL 2024), and a box containing fun games to learn about the Gobi biodiversity known as the *Nomadic trunk*. For example, I personally train Gobi youth to integrate nomadic knowledge with scientific research using the *Nomadic Trunk* and ZSL's educational mentorship. Fostering future conservationists is key to address monitoring gaps and study the causes of calf mortality, such as wolf predation and disease. It is also intended to enhance transboundary cooperation with Chinese conservationists for migration studies, ensuring the wild camel's role in sustaining the Gobi's ecosystems and heritage (ZSL 2024; Kaczensky et al. 2014; Jemmett 2023). These programs cultivate successors, ensuring One Health's longevity (ZSL 2024).

Overall, protecting the wild camel faces persistent challenges that threaten its survival, habitat, and the well-being of Gobi communities. Yet, opportunities to enhance conservation outcomes, driven by community engagement and innovative tools such as SMART patrolling, camera traps, and *Nomadic trunk*, offer hope for

Fig. 10.7 Tungalag Ulambayar with colleagues in the field. (Photo: T. Zoological Society Luujin)

sustaining this ecological and cultural heritage. The One Health approach, integrating camel health, environmental stability, and human prosperity (Gibbs 2014; Zinsstag et al. 2020), frames these efforts. These strategies, rooted in my nomadic heritage, ensure the wild camel's survival sustains Gobi's ecological and social fabric, a mission I carry forward as a Gobi daughter and scientist. As a conservationist rooted in nomadic traditions, I advocate for new conservationists to carry forward this mission, blending science and local wisdom to ensure the wild camel's legacy.

10.3.2 Australia: Nativeness and Belonging

To understand recent developments in Australian camel management, we need to look at critical events that have occurred in two very different eras. Firstly, we need to go back about 50,000 years to look at what other large mammalian herbivores lived on this continent in the past. This will enable some understanding of why camels seem so preadapted to Australia. Second, we need to look at very recent cultural changes that have occurred in the past 5 years in Australia, and relate to changing concepts of *nativeness*, the importance of the perspectives of Indigenous peoples, and moves towards *rewilding*.

The idea that "introduced megafauna are rewilding the Anthropocene" was proposed by Lundgren et al. (2017) 8 years ago. Wild post-domestic species were assigned to the threat status of their pre-domestic ancestor. For example, introduced feral dromedary camels originate from the domesticated form of an extinct camel species (possibly *C. thomasi*), and were therefore considered extinct in the wild in their native range (Lundgren et al. 2017). Chris Johnson (2019) argued that "the camels that roam across central Australia browse on thorny vegetation that would once have been eaten by diprotodons and other marsupial mega-herbivores, and their soft feet do less harm than the hooves of other introduced herbivores."

Australia was once populated by a diverse assemblage of large mammalian herbivores, omnivores, and carnivores. But all species weighing >80 kg abruptly disappeared from the fossil record when humans entered Australia ~50,000 years ago (Flannery 1994). Growing awareness of Australia's megafauna extinctions, and the increasing willingness of conservationists to focus on functional roles in ecosystems, rather than the origins of species (Lundgren et al. 2024), have played a crucial role here (Johnson 2002). The adaptation of Australia's arid zone to large mammalian browsers has been described as follows: "In a land now dominated by mostly grass-eating kangaroos, plants that once protected themselves against big browsers with spines or toxins find their defenses nearly redundant" (Weber 2013). The thought here is that perhaps the camel, although not a marsupial, is filling the functional ecological role of the extinct diprotodons (Johnson 2019).

The environments inhabited by camels in Australia are today mostly Indigenous-managed lands, and the perceptions of Indigenous Australians have played a crucial role in how, when, and if camels are managed (Vaarzon-Morel and Edwards 2012; Vaarzon-Morel 2017). Some camel culling has been motivated by negative impacts exerted by thirsty camels during dry seasons on cultural sites such as rock holes (water holes) in desert landscapes (Box et al. 2016). But not all Indigenous Australians have seen camels as pests. Camels are not regarded as part of the Dreaming [The Dreaming refers to Aboriginal creation stories shaping the spiritual connection with life, the land, and the laws governing them], and people do not have consubstantial relations with them. Camels have not been incorporated into traditional ceremonies by many Indigenous groups (Vaarzon-Morel 2017), but other introduced ungulates have, e.g., swamp buffalo (*Bubalus bubalis*) in Arnhem Land of Australia's Northern Territory (Riley 2013).

The concept of *nativeness*, so central to conservation biology (Soulé 1985), has been increasingly teased apart in recent years. Menzies (2023) posed the question of why camels and feral horses (*Equus caballus*) are treated so differently in Australia. "Horses also represent a particular identity associated with Britain and Europe. Compare horses with camels…Camels are culturally and historically significant for Arab peoples, yet here they are routinely culled without a second thought. Despite their long history in Australia, camels are associated with Asia and the Middle East and seen as exotic and foreign. Horses, on the other hand, have Anglo-European connotations that fit comfortably within Australia's white identity" (Menzies 2023). Some authors, concerned with the lethal methods applied to introduced animals in Australian rangelands, have referred to camels and other species such as feral donkeys (*Equus asinus*) as *illegible animals*, or *animals without a status* (Celermajer and Wallach 2019).

If Australian camels were following the scientific wildlife biology literature in Australia, their collective ears would have pricked up around the year 2020, when many Australian conservationists began referring to the dingo (*Canis familiaris* or *C. dingo*) as an Australian native animal (Banks 2021). Dingoes had traditionally been considered introduced animals, like camels, having been brought to Australia by south-east Asian traders around 5000 years ago (Jackson et al. 2021). There were many reasons for this change of heart, including ideas that dingoes might have become "functional replacements" for extinct marsupial thylacines (*Thylacinus cynocephalus*), the esteem they are held in by many Indigenous Australians, and growing popularity for the idea of rewilding, often centered around apex terrestrial predators (Johnson 2019). But why might this matter to Australian camels?

If the dingo, an animal that was categorically introduced to Australia by humans (Jackson et al. 2021) and which meets all modern criteria for being considered a *feral* animal (Lord et al. 2025), can be reimagined as a native animal deserving conservation, why not the camel? If it comes down to how long each animal has been on the continent (Crees and Turvey 2015), this amounts to a matter of degree, not of kind (Carthey and Banks 2012). And it has been argued by some that all wild animals possess biodiversity value, regardless of whether they are currently inhabiting the continent on which they first evolved, an argument that has been framed as "when all life counts in conservation" (Wallach et al. 2020):

I am not personally making the case that dromedary camels should be reimagined as native species in Australia, but I can see that it might be viewed as consistent with recent developments in how introduced species are conceived in wildlife management (Wallach et al. 2020). There is the interesting precedent of the endangered but non-native banteng (*Bos javanicus*) in northern Australia, which has been argued to demonstrate the feasibility of conserving large herbivores otherwise imperilled in their native range (Bradshaw et al. 2006). While the banteng was not domesticated, it otherwise shares many similarities with Australia's camels. I argue that, at a minimum, wildlife managers believing that dromedary camels in Australia should be eradicated or managed via lethal control should at least be prepared to have this conversation.

10.4 Future Directions

10.4.1 Mongolia: A Call to Action

Gobi communities revere the wild camel as a cultural emblem, its image in songs and stories a part of my childhood memories. My journey from herder to scientist drives my resolve to safeguard this heritage, ensuring the Gobi's ecological and cultural vitality. My journey began on the Gobi's steppes, learning seasonal grazing beside my grandparents' Bactrian camel, its steady gait my first lesson in resilience. This heritage fuels my work, ranging from supporting MECC's protests, contributing to UNESCO's nomination, to bolstering in situ conservation programs through Luujin. In UNDP forums, I've advocated for herder training and unified databases, weaving One Health into every effort to sustain camels, ecosystems, and communities. Yet my proudest contributions are the facilitation of Ecoclubs (Fig. 10.8) and

Fig. 10.8 Education of the younger generation through involvement in Ecoclubs. (Photo: Zoological Society Luujin)

the mentoring of youth via ZSL's *Nomadic Trunk*, i.e., inspiring Gobi children to blend nomadic wisdom with science.

The wild camel, a critically endangered emblem of the Gobi Desert, stands at the heart of a delicate interplay between ecological vitality and cultural heritage. With fewer than 700 individuals remaining, its survival demands a One Health approach, intertwining camel health, environmental stability, and Gobi community prosperity (Gibbs 2014; Zinsstag et al. 2020). As a conservationist shaped by nomadic roots and driven by scientific and policy advocacy, I call for a new generation of conservationists to embrace this holistic vision, ensuring the Gobi's legacy endures through collective action and innovation.

Current opportunities illuminate a path forward. The WCPF's ex situ breeding centres in Zakhuin Us and Toli Bulag safeguard genetic diversity and help curb zoonotic risks (Dovchindorj 2017; Battogtokh 2023). In situ efforts, including the eight solar-powered water pumps and PAA's mining patrols, stabilize rangelands and water access (Munkhbayar et al. 2025). The UNESCO World Heritage nomination elevates GGSPA's global value and promises funding (Succow Foundation 2023). Ecoclubs and the CMC's landscape management empower communities, fostering ecotourism and policy alignment (ZSL 2024; MECC 2025). SMART patrol data and emerging technologies like UAVs enhance monitoring, ensuring scientific data-driven conservation (Bayarbat 2024).

I urge new conservationists to view wild camel conservation holistically, supporting initiatives that benefit animals, humans, and the environment concurrently. Join programs like *Nomadic Trunk*, innovate with tools like SMART data, and amplify community voices through public engagement activities. By uniting ex situ and in situ efforts and leveraging global platforms like UNESCO, you can ensure wild camels continue to roam, sustaining the Gobi's ecological balance and cultural pride. As a Gobi daughter, I envision a future where this heritage thrives, carried forward by those who answer this call with passion and purpose.

10.4.2 Australia: A Fork in the Road

Australian feral camels currently sit at a fork in the road. Their future management (if any) seems highly uncertain. Land managers could commit strongly to the principle of removing all introduced wildlife populations, and attempt eradication, as per the *status quo* for wildlife management in Australia (Edwards et al. 2004a). Alternatively, the voices behind *Compassionate Conservation* and the emerging reality of assisted migration in the face of rapid anthropogenic climate change may necessitate a rethink of the moral necessity to remove wild animals when humans have had a hand in their populations reaching a new environment (McLachlan et al. 2007). Pivotal crossroads have been noted for other feral ungulates about which communities feel conflicted, such as feral horses in the USA (Garrott and Oli 2013). The ramifications of which path is chosen at these crossroads can have profound long-term consequences for wild animal populations, and the humans charged with managing them (Hennig et al. 2023).

The functional replacement rewilding ideas proposed by Johnson (2019) and his colleagues currently sit well outside of mainstream conservation thinking in Australia. But change has become rapid in this space, and which animals are considered *native* is no longer so fixed (van Eeden et al. 2025). Most Australian wildlife managers now accept the idea that the dingo should be considered an Australian native animal and conserved as such (Banks 2021). Perhaps the dromedary camel, too, may soon find itself reimagined as an Australian native, and conserved for its capacity to coarsely fill the vacant ecological niche of Australia's extinct megafauna browsers.

I am not sure what the future holds for Australia's camels. A part of me would love to see a return to Gondwanan ecosystems for the landmasses of Australasia, as envisioned by ambitious programs such as *Predator Free New Zealand*. Yet another part of me is affected by my family's history with camels on this continent and wonders whether Chris Johnson and his colleagues might be right about the dromedary camel being as close as we are likely to get to a functional replacement for the extinct megafauna. I do feel confident that, even if the worst projections come to pass, and Australia's inland ecosystems continue to become hotter and drier, the incredible physiology of the dromedary is likely to see it persist so far from its evolutionary home for many decades to come—as long as humans allow it to. Then again, perhaps Australia's camels will increasingly return to domestic husbandry, with commercialization through camel dairies and increasing tourism operations. We shall see.

10.5 Conclusion

Based on these two examples, what is the outlook for camels in terms of conservation? As of today, many anthropogenic pressures are driving detrimental changes to wild animals, specific ecosystems and biodiversity. Extinction rates today are dramatically higher than natural background rates, disproportionately affecting certain taxonomic groups, which may represent what is increasingly being considered as the Earth's Sixth Mass Extinction (Cowie et al. 2022). In the late Pleistocene, megafauna extinctions in the past 50,000 years were mainly caused by direct human activities, namely, hunting, which were exacerbated by climatic changes associated with the advance and retreat of major ice caps or ice sheets, causing the reduction of favorable habitats. More recently, threats to biodiversity and species extinction, largely affecting smaller species, are linked to modern human activities such as habitat disruption and destruction (land use change), the introduction of non-native species, and anthropogenic climate change.

Recent colonization of geographic regions such as Australia, the Americas and many islands seem to have led to severe extinction events compared to gradual extinction events. This is likely because existing megafauna was not adapted to human predation (Svenning et al. 2024). Megafauna such as camels play a disproportionally large role in maintaining ecosystem structure and processes. The loss of large mammals thus results in the disruption of key roles in the ecosystem that cannot be replaced. For example, the loss of top-down control of ecosystem structure and function, e.g., through intact predator-prey relationships, results in food web disruption (Thakur and Eisenhauer 2015). Likewise, the loss of ecosystem engineers alters the habitat or landscape (Hyvarinen et al. 2021; Trepel et al. 2024). And, most importantly in relation to camels, the loss of top herbivores like camels causes a drastic change to vegetation communities, build-up of biomass, often increased fire frequency, and reduction of nutrient and seeds dispersal (Sinclair 2003; Ripple et al. 2015). Overall, such major ecological changes impact species composition and structure in each ecosystem, leading to the long-term disruption of ecological processes and a shift towards a new state.

While anthropogenic activities have a global impact, some ecosystems are more susceptible to change that can have larger and more lasting consequences. Drylands, the ecosystem type that camels naturally inhabit, are one of the most vulnerable ecosystems affected by anthropogenic activities, such as livestock grazing, and climate change (Burrell et al. 2020). Drylands are very fragile ecosystems due to water scarcity and are simultaneously considered crucial ecosystems for global food security. Over 35 years, 6% of the world's drylands underwent desertification, mainly driven by unsustainable land use practices such as livestock overgrazing and anthropogenic climate change altering the temperature and rain patterns (Burrell et al. 2020).

The Gobi Desert is no exception when it comes to experiencing the impact of anthropogenic activities on its ecosystem stability and resilience (Liang et al. 2021), with a gradual increase of desertification. The remaining wild camels are considered an umbrella species for these ecosystems, occupying a unique and fundamental role in shaping the landscape inter alia through seed dispersal. Nevertheless, many human activities have had a significantly negative impact on the wild camel, resulting in a dramatic population decrease by ~70% since the last century (Kaczensky et al. 2014). In general, educational and conservation programs have shown a positive impact on the wild camel population and combatting local biodiversity loss. The species' functional role is being safeguarded by the ex situ population, which currently acts as an insurance against extinction in the wild. This could, in the future, serve a restorative purpose, either through reinforcing the in situ population by introducing individuals from a captive breeding program or rewilding purposes through the reintroduction of individuals in an area where they have gone extinct.

In case these conservation efforts are unsuccessful, the wild two-humped camel's extinction would leave a gaping hole in the Gobi Desert ecosystem. As a result, the Gobi would not remain as we know it today. Its loss would likely result in further desertification, leading to harsher environmental conditions and impacting human livelihoods. Conversely, the Australian dromedary might have filled a vacant ecological role, even if just for a short period of time in evolutionary terms (Johnson 2019). Correspondingly, free-ranging camels in the Middle East, Africa, and Asia likely fill a similar role. In this context, domesticated camels may serve as functional analogues to restore some of the ecological functions, such as shaping vegetation structure, redistributing nutrients, and maintaining landscape heterogeneity which were once performed by their extinct wild ancestors and other extinct megafaunal herbivores (Trepel et al. 2024; Hyvarinen et al. 2021).

As the preservation of desert ecosystems is gaining interest in many Middle Eastern countries like Saudi Arabia, e.g., through their Vision 2030, Saudi's National Environmental Strategy and the Saudi Green Initiative, the protection of unique desert wildlife is emphasized. The dromedary, currently a domesticated species, is a unique desert icon, playing a crucial ecosystem role that cannot be replaced by any other ungulate. The protection of the dromedary as wildlife may seem like a wild idea, given its long history of domestication and close relationship with humankind, which will be hard to unwind. Yet, it could be an interesting concept to consider the reintroduction of camels as large megafauna in areas where the native herbivore megafauna have gone extinct.

Extending this thought experiment, recent rewilding and reintroduction efforts for other charismatic large herbivores like the Arabian Oryx can give a glimpse at the type of role rewilded dromedaries could play. In the Middle East, the Arabian Oryx (*Oryx leucoryx*) has been the most successful example of rewilding. The species went extinct in the wild in 1972, was successfully bred in captivity, and was finally reintroduced on the Arabian Peninsula (Oman and Saudi Arabia). It is currently estimated that there are around 1000 wild individuals and 6000–7000 captive individuals. Due to the genetic founder effect resulting from the small initial breeding population, the genetic diversity of the Arabian Oryx is still under debate.

Therefore, the current recommendation is to improve this species' management through building a reference gene and mapping the genetic diversity (Qais Al Rawahi et al. 2022).

Given that there are three extant camel species for which reference genomes are available, with further pan-genomic work being conducted, species management for conservation can already benefit from this foundation. Additional research on their immunology, behavior, and health is needed to inform management strategies for both domesticated and wild animals. Coupling this data with in situ environmental data will further enable gaining a better understanding of the camel's resilience to environmental changes, which is critical given the accelerated warming, particularly in desert ecosystems (Zhou 2016). As we learn more about the biology of the camel through studies in genomics and immunology, we gain important insights into the camel's resilience, its adaptation to the extreme environment it inhabits, and the future relationship to humans it can hold.

It is important to note that by supporting research and conservation efforts, we invest not only into the future of camels but also into improving the health and livelihoods of their surrounding environment and those within. The camels' metabolic capacities, enabling adaptation to deserts, make them an ideal candidate for the sustainable development of nomadic populations and the sustainable production of milk, meat, and wool (Faye 2016). Yet, scaling camel-based industries should be approached with caution, leveraging the learnings from mistakes made through the mass production of animals for human consumption and the extraction of animal products and resources that have significantly contributed to major One Health challenges, such as antimicrobial resistance. After all, the well-being of camels is linked to that of their environment and to humans, requiring a careful balance between extraction from camels for human well-being and the conservation of camels to help combat global biodiversity loss and protect crucial ecosystem functioning. Ultimately, safeguarding the future of camels is inseparable from safeguarding the resilience of the ecosystems and communities that depend on them. Today, we face a pivotal moment to shape that future by investing in research, conservation, and science-based stewardship to guide a symbiotic and sustainable coexistence.

References

Al Rawahi Q, Mijangos JL, Khatkar MS, Al MA, Abri MH, AlJahdhami JK, Senn H, Brittain K, Gongora J (2022) Rescued back from extinction in the wild: past, present and future of the genetics of the Arabian oryx in Oman. R Soc Open Sci 9(3):210558. https://doi.org/10.1098/rsos.210558

Ariuntsetseg L (2023) ГИДЦГ-ын А хэсгийн бэлчээрийн төлөв байдал болон даацыг тодорхойлсон судалгааны тайлан [GIDTsG-yn A khesgiin belcheeriin tulov baidal bolon daatsyg todorkhoilson sudalgaany tailan; Report of the rangeland study to determine current condition and carrying capacity in the GGASPA] [Research report]. School of Science, National University of Mongolia

Banks PB (2021) An eco-evolutionary rationale to distinguish alien and native status: why the dingo is a native species on mainland Australia. Aust Zool 41:358–366

Bannikov AG (1975) Дикий верблюд—Хавтагай [Dikiy verblyud—Khavtagay; Wild Camel—Havtagaj]. Priroda 2:63–69

Batbold J, Suvd B (2007) Final evaluation report on UNDP/GEF Conservation of the Great Gobi ecosystem and its umbrella species project. UNDP

Batkhuu N, Haulenbek A (2022) Говь-Алтай аймгийн Цогт, Эрдэнэ сумын заган ойн хашсан талбайн харьцуулсан судалгаа [Govi-Altai aimgiin Tsogt, Erdene sumyn zagan oin khashsan talbain khar'tsuulsan sudalgaa; Report on comparative study of the fenced sites in saxaul forest in Tsogt and Erdene soums of Gobi-Altai aimag] [Research report]

Battogtokh N (2023, November) Малчид "Ивомек"-ийг дур мэдэн малдаа тарьж, доозыг нь хэтрүүлж байна [Malchid "Ivomek"-iig dur meden maldaa tar'j, doozyg ni khetrüülj baina; Herders are indiscriminately injecting "Ivomek" into their livestock, overdosing]. Daily News. https://dnn.mn/news/129931

Battsetseg G, Batsukh Z, Naranbaatar K, Khatanbaatar I (2007, October 12) The results of study on ecto and endoparasites on wild camels [Conference session]. In: International Conference on Conservation and Management of the Wild Bactrian Camels, Ulaanbaatar, Mongolia

Bayarbat CH (2024, October 22) Говийн их дархан цаазат газрын А бүсийн дарга Ч.Баярбат: Анх "Та нарыг харна аа" гэдэг байсан. Ямартай ч бид хамгаалж чадаж байна [Goviin ikh darkhan tsaazat gazryn A büsiin darga Ch.Bayarbat: Ankh "Ta naryg kharna aa" gedeg baisan. Yamartai ch bid khamgaalj chadaj baina; Head of Zone A of the Gobi Desert Ch. Bayarbat: At first, they said, "We'll see how you do." However, we were able to protect] (E. Enkhbold, Interviewer). Eguur News. https://eguur.mn/541177/

Boardman WS, Lethbridge MR, Hampton JO, Smith I, Woolnough AP, McEwen M-M, Miller GW, Caraguel CG (2014) Evaluation of medetomidine-ketamine and medetomidine-ketamine-butorphanol for the field anesthesia of free-ranging dromedary camels (*Camelus dromedarius*) in Australia. J Wildl Dis 50:873–882

Bomford M, O'Brien P (1995) Eradication or control for vertebrate pests? Wildlife Soc Bull (1973–2006) 23:249–255

Box JB, McBurnie G, Strehlow K, Guest T, Campbell M, Bubb A, McConnell K, Willy S, Uluru R, Kulitja R (2016) The impact of feral camels (*Camelus dromedarius*) on remote waterholes in Central Australia. Rangel J 38:191–200

Bradshaw CJ, Isagi Y, Kaneko S, Bowman DM, Brook BW (2006) Conservation value of non-native banteng in northern Australia. Conserv Biol 20:1306–1311

Brim-Box J, Guest T, Barker P, Jambrecina M, Moran S, Kulitja R (2010) Camel usage and impacts at a permanent spring in Central Australia: a case study. Rangel J 32:55–62

Burger PA, Ciani E, Faye B (2019) Old World camels in a modern world—a balancing act between conservation and genetic improvement. Anim Genet 50(6):598–612. https://doi.org/10.1111/age.12858

Burrell AL, Evans JP, De Kauwe MG (2020) Anthropogenic climate change has driven over 5 million km^2 of drylands towards desertification. Nat Commun 11:3853. https://doi.org/10.1038/s41467-020-17710-7

Carthey AJ, Banks PB (2012) When does an alien become a native species? A vulnerable native mammal recognizes and responds to its long-term alien predator. PLoS One 7:e31804

Celermajer D, Wallach A (2019) The fate of the illegible animal: the case of the Australian wild donkey. Anim Stud J 8:229–258

Chuluunbat B, Charruau P, Silbermayr K, Khorloojav T, Burger PA (2014) Genetic diversity and population structure of Mongolian domestic Bactrian camels (*Camelus bactrianus*). Anim Genet 45(4):550–558. https://doi.org/10.1111/age.12158

Clout MN, Russell JC (2008) The invasion ecology of mammals: a global perspective. Wildl Res 35:180–184

Cooke BD, Fenner F (2002) Rabbit haemorrhagic disease and the biological control of wild rabbits, *Oryctolagus cuniculus*, in Australia and New Zealand. Wildl Res 29:689–706

Coventry J, Edwards G, Zeng B (2010) The odyssey of managing feral camels and their impacts—is there an Achilles' heel? Aust Zool 35:251–264

Cowie RH, Bouchet P, Fontaine B (2022) The Sixth Mass Extinction: fact, fiction or speculation? Biol Rev 97:640–663. https://doi.org/10.1111/brv.12816
Crees JJ, Turvey ST (2015) What constitutes a 'native' species? Insights from the Quaternary faunal record. Biol Conserv 186:143–148
Crowley SL (2014) Camels out of place and time: the dromedary (*Camelus dromedarius*) in Australia. Anthrozoös 27:191–203
Cruz F, Carrion V, Campbell KJ, Lavoie C, Donlan JC (2009) Bio-economics of large-scale eradication of feral goats from Santiago Island, Galapagos. J Wildl Manag 73:191–200
Dementyev GP, Tsevegmid D (1963) Wild camel of Mongolia [Дикий верблюд в Монголии; Dikiy verblyud v Mongolii]. Priroda 1:71–74
Dovchindorj G (2022) Ecological aspects of wild camels in Mongolia. Admon Print, Ulaanbaatar, Mongolia
Dovchindorj G (2017) Монгол оронд тэмээн гөрөөс буюу хавтгайг гаршуулсан нь [Mongol orond temeen goroos buyuu khavtgaig garshuulsan ni; Domestication of wild camels in Mongolia]. Soyombo Printing
Dovchindorj B, Nasanjargal T (2022, November 18) Монгол орны бэсрэг тэмээний өнөөгийн байдал [Mongol orny besreg temeenii onoo-giin baidal; Current status of hybrid camels in Mongolia] [Conference presentation]. Scientific conference of the Inner Altai Gobi 2022, Tsengeg Usnii Tuv, Ulaanbaatar, Mongolia
Edwards G, Pople A, Saalfeld K, Caley P (2004a) Introduced mammals in Australian rangelands: future threats and the role of monitoring programmes in management strategies. Austral Ecol 29:40–50
Edwards GP, Saalfeld K, Clifford B (2004b) Population trends of feral camels in the Northern Territory, Australia. Wildl Res 31:509–517
Enkhjargal T (2023) Water quality analysis of saline springs in the Great Gobi A Strictly Protected Area. Environ Monit Assess 195(6):Article 712. https://doi.org/10.1007/s10661-023-09876-4
Eregdendagva D (1954) Rare animals of Western Mongolia [Редкие животные Западной Монголии]. Priroda 9:105. Available at https://priroda.ras.ru/
Faye B (2016) The camel, new challenges for a sustainable development. Trop Anim Health Prod 48:689–692. https://doi.org/10.1007/s11250-016-0995-8
Felkel S, Wallner B, Chuluunbat B, Yadamsuren A, Faye B, Brem G, Walzer C, Burger PA (2019) A first Y-chromosomal haplotype network to investigate male-driven population dynamics in domestic and wild Bactrian camels. Front Genet 10:Article 423
Flannery T (1994) The future eaters: an ecological history of the Australasian lands and people. Reed New Holland, London
Garrott RA, Oli MK (2013) A critical crossroad for BLM's wild horse program. Science 341:847–848
Gibbs EPJ (2014) The evolution of One Health: a decade of progress and challenges for the future. Vet Rec 174(4):85–91. https://doi.org/10.1136/vr.g143
Guodong L, Naiang W, Qiusheng S, Junhua Z (2008) Research on distribution, living environment and protection of wild Bactrian camel. J Camel Pract Res 15(2):211–217
Hampton JO, Spencer PBS, Alpers DL, Twigg LE, Woolnough AP, Doust J, Higgs T, Pluske JR (2004) Molecular techniques, wildlife management and the importance of genetic population structure and dispersal: a case study with feral pigs. J Appl Ecol 41:735–743
Hampton JO, Cowled BD, Perry AL, Miller CJ, Jones B, Hart Q (2014) Quantitative analysis of animal-welfare outcomes in helicopter shooting: a case study with feral dromedary camels (*Camelus dromedarius*). Wildl Res 41:127–135
Hampton JO, Jones B, Perry AL, Miller CJ, Hart Q (2016) Integrating animal welfare into wild herbivore management: lessons from the Australian Feral Camel Management Project. Rangel J 38:163–171
Han J, Dai H. Gu Z (2021) Sandstorms and desertification in Mongolia, an example of future climate events: a review. Environ Chem Lett 19:4063–4073. https://doi.org/10.1007/s10311-021-01285-w

Hanslow-Sells E, Perry DN, Carey L, Krikheli L, Drakopoulos T, Heath A, Vargas C (2020) Camel commercialisation in the goldfields region of Western Australia: an exploratory scoping review. La Trobe University, Melbourne. https://doi.org/10.26181/10271300

Hare J (1997) The wild Bactrian camel *Camelus bactrianus ferus* in China: the need for urgent action. Oryx 31(1):45–48. https://doi.org/10.1046/j.1365-3008.1997.d01-4.x

Hare J (2008) *Camelus ferus*. The IUCN Red List of Threatened Species 2008: e.T63543A12689285. https://doi.org/10.2305/IUCN.UK.2008.RLTS.T63543A12689285.en

Hart Q, Bubb A (2016) Effective collaboration and governance processes to underpin large-scale natural resource management projects: the Australian Feral Camel Management Project experience. Rangel J 38:117–123

Haulenbeck A (2013) Говийн баянбүрдүүдийн байршил, өнөөгийн төлөв байдал, нөхөн сэргээх боломж: Алтайн Өвөр Говийн жишээн дээр [Goviin bayanbürdüüdiin bairshil, onoogiin tolov baidal, nokhon sergeekh bolomj: Altain Övör Goviin jisheen deer; Location, current status, and potential for restoration of Gobi oases: the case of the Inner Altai Gobi] [Research report]. Desertification Research Center, Institute of Geoecology, Mongolian Academy of Sciences. https://eic.mn/dldbase/upload/2014/tadesertbook/3/20141205_3511_3.pdf

Hennig JD, Duchardt CJ, Esmaeili S, Fuhlendorf SD, Beck JL, Francisco TI, Scasta JD (2023) A crossroads in the rearview mirror: the state of United States feral equid management in 2023. Bioscience 2023:biad033

Hulme PE (2006) Beyond control: wider implications for the management of biological invasions. J Appl Ecol 43:835–847

Hyvarinen O, Te Beest M, le Roux E, Kerley G, de Groot E, Vinita R, Cromsigt JPGM (2021) Megaherbivore impacts on ecosystem and Earth system functioning: the current state of the science. Ecography 44:1579–1594. https://doi.org/10.1111/ecog.05703

IUCN (2025) The IUCN Red List of Threatened Species (Version 2025-1) [Dataset]. https://www.iucnredlist.org

Jackson SM, Fleming PJ, Eldridge MD, Archer M, Ingleby S, Johnson RN, Helgen KM (2021) Taxonomy of the dingo: it's an ancient dog. Aust Zool 41:347–357

Jemmett AM (2023) Conservation genetics and population surveys of the Critically Endangered Wild Camel (*Camelus ferus*) in Mongolia. Doctoral dissertation, University of Kent Repository

Jemmett AM, Groombridge JJ, Hare J, Yadamsuren A, Burger PA, Ewen JG (2023) What's in a name? Common name misuse potentially confounds the conservation of the wild camel *Camelus ferus*. Oryx 57(2):175–179. https://doi.org/10.1017/S0030605322000642

Jemmett AM, Ewen JG, Yadamsuren A et al (2026) Challenges of introgression in conservation: genetic diversity of the endangered wild camel (*Camelus ferus*) in Mongolia. Ecol Evol 16(4):e73293. https://doi.org/10.1002/ece3.73293

Ji R, Cui P, Ding F, Geng J, Gao H, Yu J, Hu S, Meng H (2009) Monophyletic origin of domestic Bactrian camel (*Camelus bactrianus*) and its evolutionary relationship with the extant wild camel (*Camelus bactrianus ferus*). Anim Genet 40(4):377–382. https://doi.org/10.1111/j.1365-2052.2008.01848.x

Jirimutu, Wang Z, Ding G, Chen G, Sun Y, Sun Z, Zhang H, Wang L, Hasi S, Zhang Y, Li J, Shi Y, Xu Z, He C, Yu S, Li S, Zhang W, Batmunkh M, Ts B et al (2012) Genome sequences of wild and domestic bactrian camels. Nat Commun 3:Article 1202. https://doi.org/10.1038/ncomms2192

Johnson CN (2002) Determinants of loss of mammal species during the Late Quaternary 'megafauna' extinctions: life history and ecology, but not body size. Proc R Soc Lond Ser B Biol Sci 269:2221–2227

Johnson C (2019) Rewilding Australia. Australas Sci 40:25–27

Joly K, Gurarie E, Sorum MS, Kaczensky P, Cameron MD, Jakes AF, Borg BL, Nandintsetseg D, Hopcraft JGC, Buuveibaatar B (2019) Longest terrestrial migrations and movements around the world. Sci Rep 9(1), Article 15333. https://doi.org/10.1038/s41598-019-51884-5

Jones P, Kenny A (2010) Australia's Muslim cameleers: pioneers of the inland, 1860s–1930s. Wakefield Press, Adelaide

Kabir NA (2005) The economic plight of the Afghans in Australia, 1860–2000. Islam Stud 44:229–250

Kaczensky P, Adiya Y, von Wehrden H, Mijiddorj B, Walzer C, Güthlin D, Enkhbileg D, Reading RP (2014) Space and habitat use by wild Bactrian camels in the Transaltai Gobi of southern Mongolia. Biol Conserv 169:311–318. https://doi.org/10.1016/j.biocon.2013.11.033

Ladygin VF (1900) О пресечении Гоби от Далын туру в Су-чжоу [On crossing the Gobi from Dalyn Turu to Suzhou]. Известия Императорского Русского Географического Общества (Изв. РГО), 36. (Proceedings of the Imperial Russian Geographical Society).

Lapidge SJ, Eason CT, Humphrys ST (2010) A review of chemical, biological and fertility control options for the camel in Australia. Rangel J 32:95–115

Lederman Z, Magalhães-Sant'Ana M, Voo TC (2021) Stamping out animal culling: from anthropocentrism to one health ethics. J Agric Environ Ethics 34:27

Lethbridge M, Saalfeld KW, Edwards GP (2016) Measured reductions in the density of camels under the Australian Feral Camel Management Project. Rangel J 38:173–179

Liang X, Li P, Wang J, Shun Chan FK, Togtokh C, Ochir A, Davaasuren D (2021) Research progress of desertification and its prevention in Mongolia. Sustainability 13:6861. https://doi.org/10.3390/su13126861

Long J (2003) Introduced mammals of the world. CSIRO Publishing, Melbourne

Lord KA, Larson G, Allaby RG, Karlsson EK (2025) A universally applicable definition for domestication. Proc Natl Acad Sci 122:e2413207122

Lundgren EJ, Ramp D, Ripple WJ, Wallach AD (2017) Introduced megafauna are rewilding the Anthropocene. Ecography 41:857–866

Lundgren EJ, Bergman J, Trepel J, Le Roux E, Monsarrat S, Kristensen JA, Pedersen RØ, Pereyra P, Tietje M, Svenning J-C (2024) Functional traits—not nativeness—shape the effects of large mammalian herbivores on plant communities. Science 383:531–537

McCarthy C, Hoshino B, Dashnyam B (2022) Preserving the Gobi: Identifying potential UNESCO world heritage in Mongolia's Gobi Desert. J Asia-Pac Biodivers 11(4):312–325

McKnight TL (1969) The camel in Australia. Melbourne University Press, Parkville

McLachlan JS, Hellmann JJ, Schwartz MW (2007) A framework for debate of assisted migration in an era of climate change. Conserv Biol 21:297–302

MECC (2025) Нэн ховор, ховор хөхтөн амьтан хамгаалах төлөвлөгөө (төсөл) [Nen khovor, khovor khökhtön am'tan khamgaalakh tolovlogoo (tosol); Draft action plan for the protection of very rare and endangered mammals]. Ministry of Environment and Climate Change, Government of Mongolia

Menzies I (2023) The brumby debate will never be settled until we face the role horses played in colonisation. The Conversation. The Conversation Media Group, Carlton, Australia

Mohandesan E, Fitak RR, Corander J, Yadamsuren A, Chuluunbat B, Abdelhadi O, Raziq A, Nagy P, Stalder G, Walzer C, Faye B, Burger PA (2017) Mitogenome sequencing in the genus *Camelus* reveals evidence for purifying selection and long-term divergence between wild and domestic Bactrian camels. Sci Rep 7:Article 9970. https://doi.org/10.1038/s41598-017-08995-8

Mönkhbayar M, Bilgüün B, Tungalag U, Bolor R, Lkhagvasüren D, Bayaraa M (2025) Говийн Их Дархан Цаазат Газрын "А" хэсгийн том хөхтөн амьтдын ус ашиглалт [Goviin Ikh Darkhan Tsaazat Gazryn "A" khesgiin tom khökhtön am'tdyn us ashiglat; Water use by large mammals in the Great Gobi SPA "A" section]. Unpublished manuscript

Munkhbayar, T. (2024, November). Goviin Ikh Darkhan Tsaazat Gazryn "A" khesgiin tom khokhton am´tdyn us ashiglalt [Water usage of large mammals in the "A" section of the Gobi Great Strictly Protected Area] [Conference presentation]. Khureltogoot 2024: International Conference of Young Scientists (Biology and Agriculture Section), Ulaanbaatar, Mongolia

Munkhbayar S, Bilguun B, Tungalag U, Bolor R, Lkhagvasuren D, Bayaraa, M (2025) Water use of large mammals in Part A of the Great Gobi Strictly Protected Area [Говийн Их Дархан Цаазат Газрын "А" хэсгийнтом хөхтөн амьтдын ус ашиглалт]

Myers JH, Simberloff D, Kuris AM, Carey JR (2000) Eradication revisited: dealing with exotic species. Trends Ecol Evol 15:316–320

NSO (2024) Малын тоо толгой 2024 [Malyn too tolgoi 2024; Livestock census 2024] [Dataset]. National Statistics Office of Mongolia

Palmer C, Fischer B, Gamborg C, Hampton JO, Sandøe P (2023) Wildlife ethics: ethics in wildlife management and conservation. Wiley-Blackwell, Oxford

Przewalski N (1879) From Kulja, across the Tian Shan, to Lob-Nor. Nature 20(496):4–4. https://doi.org/10.1038/020004a0

Przhevalsky, N. M. (1878). Ot Kul'dzhi za Tyan'-Shan' i na Lob-Nor: Puteshestvie N. M. Przheval'skogo v 1876 i 1877 gg. [From Kulja across the Tian Shan to Lob-Nor: N. M. Przhevalsky's journey in 1876 and 1877]. Imperial Russian Geographical Society

Reading RP, Mix H, Lhagvasuren B, Blumer ES (1999) Status of wild Bactrian camels and other large ungulates in south-western Mongolia. Oryx 33(3):247–255. https://doi.org/10.1046/j.1365-3008.1999.00067.x

Reading RP, Blumer ES, Mix H, Adiya J (2005) Wild Bactrian camel conservation. Proceedings of the Symposium "Ecosystem Research in the Arid Environments of Central Asia: Results, Challenges, and Perspectives" 91–100

Riley S (2013) 'Buffalo belong here, as long as he doesn't do too much damage': indigenous perspectives on the place of alien species in Australia. Australas J Nat Resour Law Policy 16:157–196

Ripple WJ, Newsome TM, Wolf C, Dirzo R et al (2015) Collapse of the world's largest herbivores. Sci Adv 1:e1400103

Sarnai B (2024) Говийн бүсийн бэлчээрийн даацыг хиймэл дагуулын мэдээ ашиглан тооцох нь: Говийн их дархан цаазат газрын А хэсгийн орчны бүсийн жишээн дээр [Goviin büsiin belcheeriin daatsyg khiimel daguulyn medee ashiglan tootsoh ni: Goviin ikh darkhan tsaazat gazryn A khesgiin orchiny büsiin jisheen deer; Estimating the pasture carrying capacity of the Gobi region using satellite data: a case of the buffer zone of GGASPA]. Master's thesis, National University of Mongolia

Scriver P (2004) Mosques, ghantowns and cameleers in the settlement history of colonial Australia. Fabrications 13:19–41

Shiirevdamba C, Shagdarsuren O, Erdenejav G, Amgalan C, Cecegmaa C (eds) (2013) МонголынУлаан ном (3rd ed.) [Mongolian Red Book]. Ministry of Environment and Green Development.

Silbermayr K, Burger P, Schlötterer C, Vogl C, Schwarzenberger F, Walzer C (2006) Genetic diversity and introgression in the Mongolian wild Bactrian camel (*Camelus bactrianus ferus*): conceptual design [Conference session]. International Workshop on the Wild Bactrian Camel, Ulaanbaatar, Mongolia

Silbermayr K, Orozco-terWengel P, Charruau P, Enkhbileg D, Walzer C, Vogl C, Schwarzenberger F, Kaczensky P, Burger PA (2010a) High mitochondrial differentiation levels between wild and domestic Bactrian camels: a basis for rapid detection of maternal hybridization. Anim Genet 41(3):315–318. https://doi.org/10.1111/j.1365-2052.2009.01993.x

Silbermayr K, Tero N, Charruau P, Enkhbileg D, Walzer C, Burger PA (2010b) Isolation and characterization of nine new microsatellite loci in the domestic Bactrian camel (*Camelus bactrianus*) and amplification in the wild Bactrian camel (*C. ferus*). Mol Ecol Resour 10(6):1106–1108

Sinclair ARE (2003) The role of mammals as ecosystem landscapers. Alces 39:161–176

Soulé ME (1985) What is conservation biology? Bioscience 35:727–734

Spencer P, Hampton JO, Pacioni C, Kennedy MS, Saalfeld K, Rose K, Woolnough AP (2015) Genetic relationships within social groups influence the application of the Judas Technique: a case study with wild dromedary camels. J Wildl Manag 79:102–111

State Library South Australia (n.d.) Afghans with resting camels

Stevens C (2002) Tin mosques & ghantowns: a history of Afghan cameldrivers in Australia. Paul Fitzsimons, Alice Springs

Succow Foundation (2023) UNESCO World Heritage nomination dossier for Gobi Desert parks. Michael Succow Foundation

Sumikov AD (1937) По пустыням Западной Гоби [Po pustynyam Zapadnoy Gobi; Across the deserts of the Western Gobi]. Sovremennaya Mongolia 6:44–57

Svenning JC, Lemoine RT, Bergman J, Buitenwerf R, Le Roux E, Lundgren E et al (2024) The late-Quaternary megafauna extinctions: patterns, causes, ecological consequences and implications for ecosystem management in the Anthropocene. Camb Prism Extinct 2:e5

Tan ET, Al Jassim R, D'Arcy BR, Fletcher MT (2016) Level of natural hepatotoxin (Indospicine) contamination in Australian camel meat. Food Addit Contam: Part A 33:1587–1595

Thakur MP, Eisenhauer N (2015) Plant community composition determines the strength of top-down control in a soil food web motif. Sci Rep 5:9134

Trepel J, le Roux E, Abraham AJ, Buitenwerf R, Kamp J, Kristensen JA, Tietje M, Lundgren EJ, Svenning J-C (2024) Meta-analysis shows that wild large herbivores shape ecosystem properties and promote spatial heterogeneity. Nat Ecol Evol 8:705–716. https://doi.org/10.1038/s41559-024-02327-6

Tulgat R, Schaller GB (1992) Status and distribution of wild Bactrian camels *Camelus bactrianus ferus*. Biol Conserv 62(1):11–19. https://doi.org/10.1016/0006-3207(92)91147-K

Vaarzon-Morel P (2017) Alien relations: ecological and ontological dilemmas posed for indigenous Australians in the management of "feral" camels on their lands. In: Dussart F, Poirier S (eds) Entangled territorialities: negotiating indigenous lands in Australia and Canada. University of Toronto Press, Toronto, pp 186–211

Vaarzon-Morel P, Edwards G (2012) Incorporating Aboriginal people's perceptions of introduced animals in resource management: insights from the feral camel project. Ecol Manag Restor 13:65–71

van Eeden LM, Martin JV, Fisk J, Lehnen L, Ellis EC, Gavin MC, Landon AC, Larson LR, Leong KM, Linklater W (2025) Species nativeness as a cultural paradigm in conservation. Biol Conserv 311:111415

Virtue J, Gee P, Secomb N, O'Leary P, Grear B (2016) Facilitating feral camel removal in Australia through commercial use. Rangel J 38:143–151

Wallach AD, Lundgren E, Batavia C, Nelson MP, Yanco E, Linklater WL, Carroll SP, Celermajer D, Brandis KJ, Steer J (2020) When all life counts in conservation. Conserv Biol 34:997–1007

Walzer C, Kaczensky P (2005) Wild camel training and collaring mission for the Great Gobi a strictly protected area in Mongolia [Report]. Wild Camel Collaring and Data Collection Report, pp 1–32

Walzer C, Kaczensky P, Dulmantseren E, Ganbold D, Yadamsuren A (2008, October) Through the eye of the desert to capture and collar wild Bactrian camels [Conference session]. In: Proceedings AAZV ARAV Joint Conference, Los Angeles, CA

Weber L (2013) Plants that miss the megafauna. Wildlife Aust 50:22–25

Wells CR, Lethbridge M (2020) Intensive and extensive movements of feral camels in Central Australia. Rangel J 42:195–210

Woolnough AP, Edwards G, Hart Q (2016) Feral camels: a nationally significant pest animal requiring a national management approach. Rangel J 38:109–115

Wu H, Guang X, Al-Fageeh MB, Cao J, Pan S, Zhou H, Zhang L, Abutarboush MH, Xing Y, Xie Z et al (2014) Camelid genomes reveal evolution and adaptation to desert environments. Nat Commun 5:Article 5188. https://doi.org/10.1038/ncomms6188

WWF (2024) Living planet report 2024—a system in peril. World Wide Fund for Nature. https://www.worldwildlife.org/publications/living-planet-report-2024

Xue Y, Li J, Li D (2021a) The wild camel (*Camelus ferus*) in China: current status and conservation implications. J Nat Conserv 60:Article 125979. https://doi.org/10.1016/j.jnc.2021.125979

Xue Y, Li J, Zhang Y, Li D, Yuan L, Cheng Y, Liu S, Hacker CE (2021b) Assessing the vulnerability and adaptation strategies of wild camel to climate change in the Kumtag Desert of China. Glob Ecol Conserv 29:Article e01725. https://doi.org/10.1016/j.gecco.2021.e01725

Yadamsuren A, Enkhbileg D, Reading RP (2012) The conservation status and management of wild camels in Mongolia. Camels in Asia and North Africa, 43–50

Yadamsuren, Ya. (2017). Mongol orny khavtgai temee (Camelus ferus., Przewalski, 1878)-nii ekologiin zarim ontslog, khamgaallyn asuudal [Some ecological characteristics and conservation issues of the wild Bactrian camel (Camelus ferus., Przewalski, 1878) in Mongolia] [Doctoral dissertation, National University of Mongolia]

Yadamsuren A, Daria O, Liu S (2019) The seasonal distribution of wild camels (*Camelus ferus*) in relation to changes of the environmental conditions in Mongolia. Open J Ecol 9(8):293–314. https://doi.org/10.4236/oje.2019.98021

Zhang Y, Zhong Y, Hong Y, Xue Y, Li D, Zhou C, Liu S (2019) RAD-Seq data advance captive-based conservation of wild bactrian camels (*Camelus ferus*). Conserv Genet 20:817–824. https://doi.org/10.1007/s10592-019-01173-5

Zhou L (2016) Desert amplification in a warming climate. Sci Rep 6:31065. https://doi.org/10.1038/srep31065

Zinsstag J, Utzinger J, Probst-Hensch N, Shan L, Zhou X-N (2020) Towards integrated surveillanceresponse systems for the prevention of future pandemics. Infect Dis Poverty 9(5):87–92. https://doi.org/10.1186/s40249-020-00676-y

Zou Z, Duley JA, Cowley DM, Reed S, Arachchige BJ, Shaw PN, Bansal N (2022) Comprehensive biochemical and proteomic characterization of seasonal Australian camel milk. Food Chem 381:132297

ZSL (2024) Protecting wild camels with Great Gobi nomadic herders: project report 2024. Zoological Society of London

Glossary[1]

16S rRNA gene A gene encoding a ribosomal component, widely used as a *barcode gene* in *metabarcoding* to identify bacteria and archaea.

Ablynx A biopharmaceutical company (now part of Sanofi) that developed the first *nanobody*-based drug, *Caplacizumab*.

Abortion storm A sudden wave of spontaneous abortions across a large proportion of pregnant animals in a herd, characteristic of diseases such as *Rift Valley Fever*.

Acheb An Arabic term for *ephemeral* desert plants that appear only after rainfall and complete their life cycle before the soil dries out.

Adaptive heterothermy The ability of an animal to allow its body temperature to fluctuate rather than maintaining a fixed set-point, reducing the need for evaporative cooling.

Aedes A type of mosquitoes that act as vectors for several viral diseases, including *Rift Valley Fever*.

Affinity maturation The process by which the immune system generates antibodies with progressively higher binding strength for a target *antigen* through iterative mutation and selection.

Agroecosystem (AES) A system integrating plants, animals, and their environment to produce food and other agricultural products, with environmental, socio-economic, and public health aspects.

Aimag A province-level administrative division in Mongolia.

Allelic diversity The number and frequency of different variants of a gene within a population. Low allelic diversity at immune loci may limit pathogen recognition.

Alternative splicing A process by which a single gene produces multiple mRNA variants by including or excluding specific exons during RNA processing.

Ambulatory grazing A feeding mode in which the animal moves continuously between plants, taking small bites from each. This dispersed pattern reduces localized grazing pressure.

Amplicon A segment of DNA or RNA amplified by *PCR*. In *metabarcoding*, amplicons of marker genes are sequenced to profile microbial communities.

Amplification host A species in which a pathogen replicates to high levels, increasing its prevalence and facilitating transmission to other hosts.

Anthrax An acute infectious disease caused by *Bacillus anthracis*, transmissible to humans through contact with infected animals or contaminated products.

---

[1] Words in italics within definitions have their own glossary entry.

S. A. Gordine et al. (eds.), *Camels and One Health*,
https://doi.org/10.1007/978-981-92-2246-9

Anthropocene A proposed geological epoch defined by the dominant influence of human activities on the Earth's climate and ecosystems.

Antigen-binding fragment (Fab) The portion of a conventional antibody containing the antigen-binding site. In *HCAbs*, the Fab is replaced by a single *VHH* domain.

Arabian oryx A large antelope native to the Arabian Peninsula. The species went extinct in the wild in 1972 but was successfully reintroduced through captive breeding programs.

Arginine vasopressin A peptide hormone that promotes water reabsorption in the kidneys, producing concentrated urine.

Artiodactyla The mammalian order of even-toed ungulates, including camels, cattle, pigs, and deer.

Autochory A mode of seed dispersal in which seeds are ejected by the parent plant over a short distance, limiting geographic spread compared with *endozoochory*.

Backcrossing The mating of a hybrid with a member of one of its parental populations.

Bactrian camel The domesticated two-humped camel (*Camelus bactrianus*), found primarily in Central Asia.

Banteng A wild bovid native to South-East Asia.

Barcode gene A standardized genetic marker used to identify organisms in *metabarcoding* studies.

Bedouin Traditionally nomadic Arab peoples of the deserts of the Arabian Peninsula, North Africa, and the Levant.

Berlin Principles of One Health Ten principles (2019) updating the *Manhattan Principles* to address emerging pathogens, climate change, *antimicrobial resistance*, and biodiversity loss.

Best Linear Unbiased Prediction (BLUP) A statistical method used in animal breeding to estimate genetic merit (*EBVs*) by accounting for genetic and environmental factors.

Betacoronavirus A genus of coronaviruses that includes *MERS-CoV*, *SARS-CoV-1*, and *SARS-CoV-2*, causing respiratory illness in humans.

Bioactive component A compound in food with biological effects beyond basic nutrition, such as antimicrobial or immunomodulatory properties.

Biobanking The collection and long-term storage of biological samples for future research or conservation use.

Biopanning The iterative selection process in *phage display* that enriches for phage particles displaying antibodies or peptides that bind a target antigen.

Bisulfite sequencing A technique for mapping *DNA methylation* at single-nucleotide resolution by chemically converting unmethylated cytosines to uracil.

Brevilinear A compact, robust body type.

Bridge host A host species that carries a pathogen between distant geographic areas or between otherwise separate host populations.

Brucellosis A bacterial *zoonosis* caused by *Brucella* species. In livestock, it causes reproductive losses; in humans, undulating fever, joint pain, and fatigue.

Calotropis procera An invasive flowering shrub (Apple of Sodom) expanding rapidly under climate change in African and Asian rangelands.

Camel Belt The geographic band from North and East Africa through the Middle East to Central and South Asia where camel populations are concentrated.

Camel chymosin A milk-clotting enzyme produced by genetic engineering, which allows cheese production from camel milk, which has low *κ-casein* content.

Camel prion disease (CPD) A *transmissible spongiform encephalopathy* identified in dromedaries. The implications of *PRNP* gene differences in camelids remain unclear.

Camel wrestling A traditional sport practiced in Türkiye in which male camels, engage in wrestling bouts during organized festivals.

Cameleer A person who drives, tends, or works with camels.

Camelidae The taxonomic family encompassing all camelids, divided into *Camelini* (Old World camels) and *Lamini* (New World camelids).

Camelini The tribe within *Camelidae* comprising the three Old World camel species: dromedary, *Bactrian camel*, and *wild Bactrian camel*.

Camelpox A contagious viral disease caused by an *orthopoxvirus*, producing skin lesions. One of the few diseases with a camel-specific vaccine.

Camelus bactrianus The Latin binomial for the domesticated Bactrian (two-humped) camel.

Camelus dromedarius The Latin binomial for the dromedary (one-humped) camel and primary animal *reservoir* of *MERS-CoV*.

Camelus ferus The Latin binomial for the *wild camel*, a critically endangered species in Central Asia.

Camelus knoblochi An extinct camel species whose paleogenomic data reveals historical *hybridization* among Old World camel lineages.

Caplacizumab The first *nanobody*-based drug approved for human use, targeting *vWF* for thrombotic thrombocytopenic purpura.

CAR-T cell therapy Immunotherapy in which a patient's T cells are engineered to express a chimeric antigen receptor directing them to kill cancer cells. *VHH* domains can serve as the recognition component.

Cellulolytic bacteria Microorganisms that break down cellulose, the main structural carbohydrate in plant cell walls, so that herbivores can extract nutrients from fibrous vegetation.

CH1 domain The first constant domain of the antibody heavy chain. In camelid *HCAbs*, a mutation deletes this domain, connecting the *VHH* directly to the hinge region.

Chromatin The complex of DNA and *histone* proteins that packages the genome. Its compaction state determines gene accessibility.

Cisternal milk The milk fraction stored in the gland cistern, immediately available at milking without requiring a *milk let-down* reflex.

Clade A group of organisms sharing a common ancestor. In virology, clades describe evolutionary branches of a virus.

Cloud seeding A weather modification technique in which substances (typically silver iodide) are dispersed into clouds to encourage precipitation.

Codex Alimentarius Internationally recognized food safety standards established by *FAO* and *WHO*, governing the quality and trade of food products.

Colostrum The first milk produced after birth, rich in *immunoglobulins*. It provides passive immunity to the newborn.

Comorbidity The simultaneous presence of two or more chronic diseases in a patient, increasing the risk of severe outcomes from infections such as *MERS-CoV*.

Complementarity-determining region (CDR) The hypervariable loops within an *antibody* that directly contacts an antigen. In *VHH* domains, CDR3 is typically elongated, giving them access to cryptic *epitopes*.

Conformational epitope An antigenic site formed by amino acids that are distant in primary sequence but spatially close in the folded protein. *VHHs* preferentially recognize conformational epitopes.

Copy number variation (CNV) A structural genetic variant in which a DNA segment is present in a different number of copies compared with a reference genome.

CpG island A genomic region rich in cytosine-guanine dinucleotides, often near gene promoters. CpG methylation status regulates gene expression.

Crimean-Congo hemorrhagic fever (CCHF) A tick-borne viral disease causing severe hemorrhagic illness in humans, with camels linked to its ecology.

CRISPR-Cas9 A gene-editing tool that uses a guide RNA to direct the Cas9 enzyme to a specific DNA location, to make precise genomic modifications.

Cryptic epitope A molecular region on a target protein that is sterically inaccessible to conventional antibodies but reachable by smaller molecules such as *nanobodies*.

Cuticle (plant) A waxy layer on plant surfaces that limits water loss through transpiration. Desert plants develop thick cuticles as a drought adaptation.

CX3CR1 A chemokine receptor on macrophages that promotes immune surveillance at mucosal sites by facilitating macrophage recruitment to infected tissues.

Dermatophytosis A fungal skin infection caused by dermatophyte fungi such as *Trichophyton* and *Microsporum* species.

Desertification The degradation of land in arid and semiarid regions, resulting in desert-like conditions.

Dingo A canid brought to Australia by South-East Asian traders.

Dipeptidyl peptidase 4 (DPP4) A cell-surface receptor used by *MERS-CoV* to enter host cells. Its expression varies across species and tissues, influencing susceptibility.

Diprotodon An extinct genus of large marsupial herbivores (*megafauna*) that inhabited Australia.

Disseminated intravascular coagulation (DIC) A life-threatening condition of widespread abnormal blood clotting, depleting clotting factors and causing uncontrolled bleeding.

Diyah Blood compensation in pre-Islamic and Islamic Arabian legal tradition.

DNA methylation An *epigenetic* modification in which a methyl group is added to a cytosine base, generally associated with gene silencing.

Dreaming (the) The foundational spiritual and cultural framework of Indigenous Australian peoples, encompassing creation narratives, law, and relationships between people, land, and species.

Drylands Regions characterized by water scarcity, including arid, semi-arid, and dry sub-humid zones.

Dzud A Mongolian term for severe winter conditions typically involving heavy snow, extreme cold, or ice cover on pastures that cause mass livestock mortality.

Ecopathology A holistic approach to animal health studying how biological, management, and environmental factors interact to influence disease patterns within production systems.

Effective population size (N_e) A genetic measure of the number of breeding individuals contributing to the next generation's genetic diversity. Low N_e indicates vulnerability to drift and inbreeding.

Ehrlichiosis A tick-borne bacterial disease caused by *Ehrlichia* species, recognized as an emerging *zoonotic* threat.

El Niño A climate phenomenon of warm Pacific sea-surface temperatures that disrupts global rainfall patterns and can trigger disease outbreaks.

Endozoochory Seed dispersal through ingestion and fecal excretion by animals, which may also improve germination through digestive scarification.

Environmental Orientalism A concept describing how Western colonial narratives portrayed Middle Eastern and North African landscapes as degraded and in need of Western intervention.

Ephemeral plant A plant that completes its entire life cycle in a short period following rainfall. They provide seasonal forage in deserts.

Epigenome The complete set of *epigenetic* modifications on an organism's genome at a given time, including *DNA methylation* and *histone modifications*.

Epigenomics The study of heritable changes in gene function that occur without altering the DNA sequence, including *DNA methylation* and *histone modifications*.

Epigraphic Relating to epigraphy, the study and interpretation of ancient inscriptions carved or engraved on durable materials such as stone, metal, or clay tablets.

Epitope The specific part of an *antigen* recognized and bound by an antibody or immune cell receptor.

Erythrocyte A red blood cell. Camelid erythrocytes are distinctively oval-shaped, long-lived, and resistant to osmotic stress.

Estimated Breeding Value (EBV) A numerical estimate of an animal's genetic merit for a specific trait, calculated using methods such as *BLUP*.

Ethnoveterinary knowledge Traditional, community-based knowledge of animal health and disease, accumulated through generations of observational learning.

Eutrophication Enrichment of water with excess nutrients, causing algal blooms and oxygen depletion.

Ex situ (Latin: "out of place") Conservation or management of a species outside its natural habitat, such as in zoos or breeding centers.

Exome The protein-coding portion of the genome. *Whole-exome sequencing* targets this fraction for efficient detection of disease-related mutations.

Exosome A small membrane-bound vesicle secreted by cells, carrying proteins, lipids and nucleic acids with potential bioactive and immune-modulating properties.

Exposome The totality of environmental exposures experienced by an organism over its lifetime, linking environmental data with health outcomes.

Five Freedoms (animal welfare) A *WOAH*-adopted framework: freedom from hunger/thirst, fear/distress, physical discomfort, pain/disease, and freedom to express natural behaviours.

Follicular wave synchronization (FWsynch) A reproductive technique using *GnRH* and *PGF2α* to synchronize follicular development for timed breeding in induced-ovulation species.

Foot-and-mouth disease A highly contagious viral disease of cloven-hoofed animals caused by an aphthovirus, characterized by fever and oral/pedal blisters.

Founder effect The loss of genetic variation that occurs when a new population is established from a very small number of individuals.

Frankincense An aromatic resin obtained from trees of the genus *Boswellia.*

Functional genomics The branch of genomics that characterizes gene function and regulation through expression studies and gene-editing experiments.

General Systems Theory A framework (von Bertalanffy, 1968) recognizing shared principles of organization, self-regulation, and differentiation across biological and social systems.

Genome-wide association study (GWAS) A method scanning the genomes of many individuals to find *SNP* variants statistically associated with a trait or disease.

Genomic selection A breeding strategy using genome-wide marker data to predict genetic merit, which allows faster and more accurate selection than traditional *progeny testing*.

Genotype-phenotype coupling The physical linkage of a gene with its encoded protein on the same particle, so that functional selection is possible in phage display.

Genotyping-by-sequencing (GBS) A cost-effective method for discovering and genotyping *SNPs* across many individuals using restriction enzymes and *NGS*.

Gift economy A system of exchange in which goods are given without explicit reciprocal payment. Camel milk was traditionally part of such systems before commercialization.

Gondwana An ancient supercontinent comprising what are now South America, Africa, Antarctica, Australia, and the Indian subcontinent.

Granzyme A serine protease in cytotoxic immune cell granules that triggers programmed cell death in infected or abnormal target cells.

Hadith Recorded sayings and actions attributed to Prophet Mohammed.

Hemolysis The rupture of red blood cells, releasing their contents into plasma. Rapid rehydration after dehydration can cause hemolysis in most mammals.

Halophyte A plant adapted to saline soils, forming part of the diverse vegetation of arid rangelands.

Hapten A small molecule that binds antibodies but cannot alone elicit an immune response unless conjugated to a larger carrier protein.

Havtgai The Mongolian name for the wild camel, meaning "flat," referring to the species' distinctively flatter skull compared to the domestic *Bactrian camel.*

Heartwater disease A tick-borne disease of ruminants caused by *Ehrlichia ruminantium*, transmitted by *Amblyomma* ticks, with high mortality.

Heat shock protein A molecular chaperone produced under heat stress that protects cellular proteins from denaturation, maintaining cellular homeostasis.

Heavy-chain-only antibody (HCAb) An antibody naturally produced by camelids that lacks light chains. Its antigen-binding domain (*VHH*) is the basis of *nanobodies*.

Hema system A traditional Arabian system of rotational grazing and rangeland conservation, cited in the Qur'an and supported by Islamic law.

Hepatic lipidosis Excessive fat accumulation in liver cells, impairing function. It can result from overfeeding on high-energy diets in intensive systems.

Heritage diplomacy The strategic use of cultural heritage and traditions as instruments of international diplomacy and economic engagement.

Heterozygosity The proportion of genetic loci at which an individual carries two different alleles. Higher heterozygosity is generally associated with greater immune responsiveness.

Hi-C A method that captures chromosome-wide physical interactions, revealing three-dimensional *chromatin* folding and long-range regulatory contacts.

Histone modification A chemical change to *histone* proteins that influences gene expression without altering the DNA sequence.

Homology modelling A computational technique predicting protein structure based on the known structure of a related protein.

Humanization (antibody) Genetic engineering of non-human antibody sequences to resemble human antibodies, reducing therapeutic *immunogenicity*.

Hyalomma A genus of hard ticks that vector several pathogens including *CCHF* virus and *Rickettsial* bacteria.

Hybridization (interspecific) Interbreeding between individuals of different species or genetically distinct populations, which can affect genetic integrity and immune diversity.

Hybridoma A cell line produced by fusing a B lymphocyte with a myeloma cell, for continuous production of a single *monoclonal antibody*.

Immune cell engager A multi-specific antibody construct that bridges immune effector cells to tumor cells, redirecting the immune response against cancer.

Immunogenicity The capacity of a substance to provoke an immune response. In antibody therapeutics, low immunogenicity is desirable to avoid drug neutralization.

Immunoglobulin A Y-shaped protein (antibody) produced by the immune system to recognize and neutralize foreign substances (*antigens*).

In silico Analyses conducted entirely through computational methods, without laboratory or living-organism experimentation.

Incense Roads Ancient caravan routes dating from approximately the 7th–3rd century BCE, connecting southern Arabia to the Mediterranean and Mesopotamia.

Introgression The incorporation of genetic material from one species into another through repeated *backcrossing* following *hybridization*.

Judas technique A wildlife management method in which a captured animal is fitted with a tracking device and released so that it leads managers to other members of its species for removal.

Lactoferrin An iron-binding glycoprotein in milk whey with antimicrobial, antioxidant, and immunomodulatory properties.

Lactophorin A *whey protein* (PP3) that inhibits pathogen growth, supporting immune protection in lactating animals and their young.

Lamini The tribe within *Camelidae* comprising New World camelids: llama, alpaca, guanaco, and vicuña.

Laminitis A painful inflammatory condition of the laminar tissues within foot pads, often triggered by dietary carbohydrate excess.

Lignification The reinforcement of plant cell walls with lignin, increasing structural rigidity and resistance to decomposition.

Lignocellulose The complex structural material of plant cell walls, composed of cellulose, hemicellulose, and lignin.

Longilinear A tall, slender body type.

Lysozyme C An antimicrobial enzyme in milk and other secretions that target Gram-positive bacteria by breaking down cell walls.

Mahr A marriage dowry or bridal gift in Arabian and Islamic tradition.

Major histocompatibility complex (MHC) Cell-surface proteins that present pathogen fragments to immune cells, so the immune system can distinguish self from nonself.

Mange A skin disease caused by parasitic mites, causing itching, hair loss, and crusting. High stocking density facilitates spread.

Manhattan Principles Twelve recommendations (2004) calling for recognition of the link between human, animal and wildlife health. Later updated by the *Berlin Principles*.

Mastitis Inflammation of the mammary gland, usually bacterial, causing milk loss and quality decline. An important production disease in dairy systems.

Maximum Residue Limit (MRL) The maximum concentration of a substance legally permitted in a food product, set by bodies such as *Codex Alimentarius*.

Mazayna Traditional camel beauty pageants in Gulf countries, epidemiologically significant due to the concentration of animals and people.

Megafauna Large-bodied animals.

Meharist A soldier serving in a camel-mounted military unit. Méhariste companies were established by French and British colonial administrations for desert patrol and border surveillance.

MERS-CoV Middle East respiratory syndrome coronavirus, first identified in 2012.

Metabarcoding A *metagenomics* method using *PCR* to amplify specific marker genes from environmental samples, followed by sequencing to identify organisms present.

Metabolomics The study of small molecules in biological samples, capturing active biochemical processes and metabolic shifts.

Metagenomics The study of all genetic material from an environmental or biological sample, which allows detection of unknown or unculturable micro-organisms.

Methanogenic (emissions) Producing methane as a by-product of digestive fermentation. Methane output varies by species and feeding efficiency.

Microsatellite A short, repetitive DNA sequence used as a genetic marker for assessing population diversity and detecting *introgression*.

Milk let-down The neuroendocrine reflex by which milk is ejected from the alveoli into the ducts and cistern, making it available for milking.

Monoclonal antibody (mAb) An antibody from a single B-cell clone, recognizing one *epitope*. Widely used in diagnostics and therapeutics.

Multi-omics An integrative approach combining data from two or more *omics* disciplines for a comprehensive view of biological systems.

Myrrh An aromatic resin obtained from trees of the genus *Commiphora*, harvested in the Dhofar region and Hadramaut and transported along the *Incense Roads*.

Nabataean Kingdom An ancient Arab kingdom centered on Petra, which flourished from the fourth century BCE to the second century CE.

Naïve library (antibody) A *phage display* library from nonimmunized animals, providing a universal source of binders without prior immunization.

Nanobody The smallest functional *antigen-binding fragment*, derived from camelid *HCAbs*, consisting of a single *VHH* domain. Characterized by high stability, solubility, and access to *cryptic epitopes*.

Natural killer (NK) cell An innate immune cell that destroys virus-infected or abnormal cells without prior sensitization.

Neolithic The later part of the Stone Age, characterized by the development of agriculture, animal domestication, and permanent settlements.

Next-generation sequencing (NGS) High-throughput technologies that read millions of DNA or RNA sequences simultaneously, for rapid genome and transcriptome analysis.

Normalized Difference Vegetation Index (NDVI) A satellite-derived index measuring vegetation greenness and density from reflectance data.

Omics A collective term for large-scale analyses of biological molecules: genomics, *transcriptomics*, *proteomics*, *metabolomics*, *epigenomics*, and *metagenomics*.

One Health An integrated approach recognizing the interdependence of human, animal, and ecosystem health, promoting cross-sectoral collaboration.

Open reading frame (ORF) A stretch of DNA or RNA with the potential to encode a protein. Accessory ORFs in viruses can influence immune evasion and virulence.

Orthopoxvirus A genus of large DNA viruses including the agents of smallpox and *camelpox*.

Oxidative stress An imbalance between reactive oxygen species production and antioxidant defenses, causing cellular and tissue damage.

Pan-genome The complete set of genes present across all individuals or strains of a species, rather than a single reference genome.

Panmictic population A population in which individuals interbreed freely across regions, resulting in low geographic genetic differentiation.

Pan-PCR A broad-spectrum *PCR* technique using degenerate primers to detect a wide range of related pathogens within a taxonomic group.

Parietal compounds Structural components of the plant cell wall (cellulose, hemicellulose, lignin) that reduce digestibility for most herbivores.

Pasteurellosis A bacterial disease caused by *Pasteurella* species, commonly affecting the respiratory tract of livestock.

Pasteurization A heat treatment that destroys pathogenic micro-organisms in milk or other foods while preserving nutritional and sensory quality.

Pastoralism Livestock management based on herding across natural rangelands, often with seasonal movement in search of grazing and water.

Peptide-binding promiscuity The ability of a single *MHC* molecule to bind and present a wide range of pathogen-derived peptides, compensating for limited allelic diversity.

Peptidoglycan recognition protein (PGRP) An innate immune protein that binds bacterial peptidoglycan, regulating immune responses and participating in bacterial killing.

Perennial plant (desert context) A plant persisting for multiple years through adaptations such as deep roots, reduced leaf surface, and thick *cuticles*.

Perforin A pore-forming protein released by cytotoxic immune cells, facilitating the entry of *granzymes* into target cells to trigger cell death.

Peri-urbanization The settlement of pastoral herders near towns, driven by drought and market access, leading to partial intensification of production.

Petroglyphs Images carved or incised into rock surfaces, as distinct from painted rock art.

Phage display A technique in which genes encoding proteins of interest are fused to phage coat-protein genes, displaying the protein on the phage surface for screening.

Pharyngitis Inflammation of the throat, causing sore throat and difficulty swallowing. A distinctive sign of plague acquired from raw camel meat.

Phylogenetics The study of evolutionary relationships among organisms, inferred from DNA, RNA, or protein sequence comparisons and represented as branching trees.

Phylogeography A discipline using genetic data to study the geographic distribution of evolutionary lineages and infer migration and population history.

Pleistocene A geological epoch spanning between 2.58 million and 11,700 years BP, characterized by repeated glacial and interglacial cycles.

Population genomics The study of genome-wide genetic variation within and among populations to understand selection, drift, and migration.

Posttranslational modification (PTM) A chemical change to a protein after synthesis that regulates its function, stability, and interactions.

Progeny testing Evaluation of an animal's genetic merit by measuring offspring performance. Slow and costly in species with long generation intervals.

Prosopis juliflora An invasive tree native to the Americas, spreading aggressively in degraded rangeland across Africa and Asia.

Proteomics The large-scale study of proteins, including their abundance, structure, modifications, and interactions.

Pseudoscabies A transient skin infestation in humans caused by animal-specific *Sarcoptes scabiei* mites that cannot complete their life cycle on human hosts.

Q fever A *zoonosis* caused by *Coxiella burnetii*, transmissible from livestock (including camels) to humans.

Quasispecies A population of closely related but genetically variable viral genomes within a single host, generated by high mutation rates.

Recombination (viral) Exchange of genetic material between viral variants coinfecting the same cell, generating new combinations that may alter virulence or transmissibility.

Red Data Book A catalogue of threatened species within a particular country or region.

Reflexive modernity A sociological concept developed by Anthony Giddens in which modern societies continually reexamine and reform their traditions in light of new knowledge and changing circumstances.

Reservoir (disease) A species or population in which a pathogen persists long-term, serving as a source of infection for other species.

Restocking model (domestication) A domestication model in which wild individuals are repeatedly added to existing herds, rather than a single founding event.

Rhinorrhea Excessive nasal mucus discharge, a nonspecific clinical sign of upper respiratory infection.

Ribosome display A cell-free in vitro selection technique linking mRNA, ribosome, and encoded protein for screening large libraries without bacterial transformation.

Rickettsial Relating to *Rickettsiales*, obligate intracellular bacteria transmitted by arthropod vectors (ticks, fleas, lice) with significant *zoonotic* potential.

Rift Valley fever (RVF) A mosquito-borne viral *zoonosis* affecting livestock and humans, causing *abortion storms* in pregnant animals and febrile to hemorrhagic disease in humans.

Rub al-Khali The "Empty Quarter," a vast sand desert covering much of the southern Arabian Peninsula.

Rumen (C1 compartment) The largest forestomach compartment, housing *cellulolytic bacteria* that ferment fibrous plant material. In camelids, the C1 functions analogously to the true ruminant rumen.

Runs of homozygosity (ROH) Long stretches of homozygous genotypes resulting from inbreeding or population bottlenecks, potentially harboring both beneficial and deleterious alleles.

Sanger sequencing A DNA sequencing method, developed in 1977, slower than *NGS* but remains the gold standard for sequence validation.

Sarcoptic mange A contagious skin disease caused by the burrowing mite *Sarcoptes scabiei*. When transmitted to humans, it is termed *pseudoscabies*.

Saxaul A drought-resistant shrub or small tree native to Central Asian deserts and a keystone species of the Gobi ecosystem.

Sclerophyllous Having hard, leathery leaves adapted to limit water loss, common in arid-environment plants.

Sedentarization The process by which nomadic or seminomadic groups settle permanently, often involving a shift from extensive to intensive livestock management.

Seed bank (soil) The reserve of viable, ungerminated seeds in and on the soil. In deserts, seeds may remain dormant for years until conditions trigger germination.

Seed dormancy A condition preventing germination despite favorable conditions, due to physical or physiological inhibitors. Digestive scarification by herbivores can break dormancy.

Selection signature (selective sweep) A genomic pattern indicating that selection has increased the frequency of a beneficial variant, reducing variation in the surrounding region.

Sentinel species A species used as an indicator of environmental health or early warning of ecological change.

Seroprevalence The proportion of a population testing positive for antibodies against a specific pathogen, indicating past or current exposure.

Ship of the desert A long-standing epithet for the dromedary, attributed to Arab peoples, reflecting the camel's indispensable role in traversing vast desert expanses, much as a ship crosses the sea.

Shotgun metagenomics An untargeted sequencing approach in which all nucleic acids from a sample are sequenced, providing a comprehensive microbial community profile.

Single nucleotide polymorphism (SNP) A single-position DNA variation among individuals, the most common type of genetic marker used in *GWAS* and population studies.

Single-chain variable fragment (scFv) An engineered antibody format linking VH and VL domains with a peptide linker, widely used before *VHH* technology.

Single-domain antibody (VHH / nanobody) The variable domain of camelid *heavy-chain-only antibodies*, small, stable, and able to access *cryptic epitopes*.

Sixth Mass Extinction The ongoing, human-driven loss of biodiversity considered comparable in scale to the five previous mass extinction events in Earth's geological history.

Somatic hypermutation A process introducing point mutations into *immunoglobulin* genes during immune cell development, diversifying the antibody repertoire.

Soum A district-level administrative division in Mongolia, roughly equivalent to a municipality.

Spike protein A surface glycoprotein on coronaviruses mediating host-cell entry by binding to receptors. Mutations can alter transmissibility.

Spillover Transmission of a pathogen from its usual host species to a new host, often at the human-animal interface.

Sternal pad A thickened callous on the sternum on which camels rest when couched, insulating the body from hot ground and allowing air circulation underneath.

Structural genomics The branch of genomics concerned with genome sequencing, assembly, and annotation.

Subclinical infection An infection in which the host harbors the pathogen without showing clinical signs, complicating surveillance and control.

Sunnah The body of traditional social and legal custom based on the recorded practices and way of life of Prophet Mohammed.

Surra A parasitic disease caused by *Trypanosoma evansi*, transmitted by biting flies. Characterized by anemia, wasting, and immunosuppression.

Synthetic/semisynthetic library An antibody library with artificially introduced diversity, constructed without animal immunization for rapid access to binders during outbreaks.

Systems dynamics A modelling method that characterizes system components and their interactions over time, applicable to disease and health system modelling.

Systems science A transdisciplinary approach to understanding complex systems, providing frameworks for integrating human, animal, and environmental health.

Systems thinking An approach recognizing that systems are adaptive with interdependent components, shifting from reductionist to holistic analysis.

Thylacine An extinct marsupial carnivore, also known as the Tasmanian tiger.

Toll-like receptor (TLR) An innate immune receptor recognizing conserved microbial molecular patterns and triggering immune responses. TLR polymorphisms may influence disease susceptibility.

Topologically associating domain (TAD) A self-interacting genomic region where DNA sequences contact each other more frequently than sequences outside the domain, influencing gene regulation.

Toxoplasmosis A *zoonotic* infection caused by *Toxoplasma gondii*. Many warm-blooded animals can serve as intermediate hosts.

Transcriptomics The study of all RNA molecules produced by the genome under specific conditions, revealing which genes are actively expressed.

Transgenic mouse (for VHH production) A genetically altered mouse bearing camelid VHH-encoding genes, so that HCAbs can be produced upon immunization without a large camelid species.

Transhumance Seasonal movement of livestock between fixed grazing areas, typically between highland and lowland pastures.

Transmissible spongiform encephalopathy (TSE) A group of fatal neurodegenerative diseases caused by misfolded prion proteins, including *camel prion disease*.

Transovarial transmission Passage of a pathogen from an infected female arthropod to her offspring via the eggs, and the virus can persist between epidemics.

Trypanosomiasis A parasitic disease caused by *Trypanosoma* species, transmitted by blood-sucking insects. *T. evansi* causes *surra* in livestock.

Tuberculosis A chronic bacterial infection caused by the *Mycobacterium tuberculosis* complex, *zoonotic* in its bovine form (*M. bovis*).

Tylopoda The suborder of *Artiodactyla* containing the *Camelidae*, named for the padded feet characteristic of camelids.

Ultra-high-temperature (UHT) processing Milk sterilization at ~135–150 °C for a few seconds, giving long shelf life. Not all milks tolerate standard UHT parameters.

Umbrella species A species whose conservation indirectly protects many co-occurring species by preserving shared habitat requirements.

Ungulate A hoofed mammal.

VHH (variable heavy-chain domain) The antigen-binding domain of camelid *HCAbs*. Also known as *nanobodies*.

Virome The complete collection of viruses in or on an organism or in an environmental sample.

Vision 2030 A set of national strategic plans adopted by several Gulf states (notably Saudi Arabia) aimed at diversifying their economies beyond oil dependency.

von Willebrand factor (vWF) A blood glycoprotein required for platelet adhesion and clot formation. Target of the first approved *nanobody* drug, *Caplacizumab*.

Wadi A valley or dry riverbed in arid regions, containing water only during and shortly after rainfall, supporting richer vegetation than the surrounding desert.

Wasum The traditional Arabian practice of branding camels with a tribal mark to signify ownership. Each tribe has a distinctive wasum, and the practice serves as a public declaration of lineage and property rights.

Wesselsbron virus A mosquito-borne flavivirus causing disease in livestock. Newly recognized in dromedaries in Ethiopia (2024).

Whey protein Milk proteins remaining soluble after casein coagulation, including *bioactive components* such as *lactoferrin* and *lysozyme C*.

Whole exome sequencing (WES) Sequencing targeting only protein-coding regions, faster and cheaper than *WGS* while capturing most disease-related mutations.

Whole-genome sequencing (WGS) Determination of the complete DNA sequence of an organism's genome, identifying all types of genetic variation.

Wild camel A critically endangered species (*Camelus ferus*) distinct from the domestic *Bactrian camel*, living in Mongolia and China.

Xerophyte A plant adapted to dry conditions through features such as reduced leaf surface, thick *cuticles*, and deep root systems.

Zoonosis An infectious disease transmissible from animals to humans. Key livestock-associated zoonoses include *MERS-CoV*, *brucellosis*, plague, *RVF*, and *Q fever*.

γδ T cell A T-lymphocyte subset bearing gamma and delta receptors, recognising non-classical antigens and expanding immune surveillance beyond the conventional αβ T cell repertoire.

κ-Casein (kappa-casein) A casein protein that stabilizes the casein micelle in milk and is the target of chymosin during cheese coagulation.

MIX
Papier aus verantwortungsvollen Quellen
Paper from responsible sources
FSC® C105338

If you have any concerns about our products,
you can contact us on
ProductSafety@springernature.com

In case Publisher is established outside the EU,
the EU authorized representative is:
Springer Nature Customer Service Center GmbH
Europaplatz 3, 69115 Heidelberg, Germany

Printed by Libri Plureos GmbH
in Hamburg, Germany